Ingo Mül

Extended Thermodynamics

With 43 Illustrations

Springer-Verlag
New York Berlin Heidelberg London Paris
Tokyo Hong Kong Barcelona Budapest

Ingo Müller
Fachbereich 10
Verfahrenstechnik und Energietechnik
Technische Universität Berlin
Straβe des 17. Juni 135
1000 Berlin 12
Germany

Tommaso Ruggeri
Università degli Studi di Bologna
C.I.R.A.M.
Centro Interdipartimentale di Ricerca
per le Applicazioni della Matematica
Via Saragozza, 8
40123 Bologna
Italy

Mathematics Subject Classification (1991): 35L65, 82C35, 82C40

Library of Congress Cataloging-in-Publication Data
Müller, Ingo, 1936–
Extended thermodynamics/Ingo Müller & Tommaso Ruggeri.
p. cm. – (Springer tracts in natural philosophy; v. 37)
Includes bibliographical references.
ISBN-13: 978-1-4684-0449-4
1. Thermodynamics. I. Ruggeri, Tommaso. II. Title.
III. Series.
QC311.M783 1993
536′.7–dc20 92-26754

Printed on acid-free paper.

Production managed by Francine Sikorski; manufacturing supervised by Vincent Scelta.
Photocomposed copy prepared using LaTeX.

ISBN-13: 978-1-4684-0449-4 e-ISBN-13: 978-1-4684-0447-0
DOI: 10.1007/978-1-4684-0447-0

Springer Tracts in Natural Philosophy

Volume 37

Edited by C. Truesdell

Springer Tracts in Natural Philosophy

Vol. 1 Gundersen: Linearized Analysis of One-Dimensional Magnetohydrodynamic Flows
With 10 figures. X, 119 pages. 1964.

Vol. 2 Walter: Differential- und Integral-Ungleichungen und ihre Anwendung bei Abschatzungs- und Eindeutigkeitsproblemen
With 18 figures. XIV, 269 pages. 1964.

Vol. 3 Gaier: Konstruktive Methoden der Konformen Abbildung
With 20 figures and 28 tables. XIV, 294 pages. 1964.

Vol. 4 Meinardus: Approximation von Funktionen und Ihre Numerische Behandlung
With 21 figures. VIII, 180 pages. 1964.

Vol. 5 Coleman/Markovitz/Noll: Viscometric Flows of Non-Newtonian Fluids: Theory and Experiment
With 37 figures. XII, 130 pages. 1966.

Vol. 6 Eckhaus: Studies in Non-Linear Stability Theory
With 12 figures. VIII, 117 pages. 1965.

Vol. 7 Leimanis: The General Problem of the Motion of Coupled Rigid Bodies About a Fixed Point
With 66 figures. XVI, 337 pages. 1965.

Vol. 8 Roseau: Vibrations Non Linéaires et Théorie de la Stabilité
With 7 figures. XII, 254 pages. 1966.

Vol. 9 Brown: Magnetoelastic Interactions
With 18 figures. VIII, 155 pages. 1966.

Vol. 10 Bunge: Foundations of Physics
With 5 figures. XII, 312 pages. 1967.

Vol. 11 Lavrentiev: Some Improperly Posed Problems of Mathematical Physics
With 1 figure. VIII, 72 pages. 1967.

Vol. 12 Kronmuller: Nachwirkung in Ferromagnetika
With 92 figures. XIV, 329 pages. 1968.

Vol. 13 Meinardus: Approximation of Functions: Theory and Numerical Methods
With 21 figures. VIII, 198 pages. 1967.

Vol. 14 Bell: The Physics of Large Deformation of Crystalline Solids
With 166 figures. X, 253 pages. 1968.

Vol. 15 Buchholz: The Confluent Hypergeometric Function with Special Emphasis on Its Applications
XVIII, 238 pages. 1969.

Vol. 16 Slepian: Mathematical Foundations of Network Analysis
XI, 195 pages. 1968.

Vol. 17 Gavalas: Nonlinear Differential Equations of Chemically Reacting Systems
With 10 figures. IX, 107 pages. 1968.

Preface

Physicists firmly believe that the differential equations of nature should be hyperbolic so as to exclude action at a distance; yet, the equations of irreversible thermodynamics—those of Navier–Stokes and Fourier—are parabolic. This incompatibility between the expectation of the physicist and the classical laws of thermodynamics has prompted the formulation of extended thermodynamics.

The main ingredients of extended thermodynamics are

— field equations of balance type,
— constitutive quantities depending on the present local state, and
— entropy as a concave function of the state variables.

This set of assumptions leads to first-order quasi-linear symmetric hyperbolic field equations, guarantees the well-posedness of initial value problems, and permits the treatment of shock waves.

Several tenets of irreversible thermodynamics had to be changed in subtle ways to make extended thermodynamics work. Thus, the entropy is allowed to depend on nonequilibrium variables, the entropy flux is a general constitutive quantity, and the equations for stress and heat flux contain inertial terms. New insight is therefore provided into the principle of material frame indifference.

With these modifications an elegant formal structure can be set up in which, just as in classical thermostatics, all restrictive conditions—derived from the entropy principle—take the form of integrability conditions.

Also the modifications made by extended thermodynamics render the theory fully consistent with the kinetic theory of gases, in particular, Grad's 13-moment version of the kinetic theory of gases. In fact, extended thermodynamics is most restrictive for gases or, more generally, for bodies whose constituent particles have large mean free paths. Most of this book, therefore, deals with gases: classical ideal gases, degenerate gases, relativistic gases, and mixtures of gases. It puts into perspective the various phenomena called *second sound*, viz. heat propagation, propagation of shear stress, and the second sound in superfluid helium.

Phonons and photons may have large mean free paths as well, and therefore, they are amenable to a treatment by extended thermodynamics. Two brief chapters describe the present status of the systematic theory in this field, which is still progressing.

A certain disappointment with extended thermodynamics of 13 or 14 fields is created by the observation that it describes resonance experiments and light scattering data only slightly better than the conventional theory. These data require further extensions to many fields, as is demonstrated in the last chapters of the book.

We may now say that extended thermodynamics provides a hierarchy of symmetric hyperbolic equations that are relevant for the solutions of initial and boundary value problems provided that the frequencies and wave numbers of their Fourier spectra do not exceed the appropriate limits.

Berlin, Germany
Bologna, Italy

Ingo Müller
Tommaso Ruggeri

Contents

Chapter 1

Early Version of Extended Thermodynamics and Kinetic Theory of Gases

The laws of Fourier, Fick, and Navier–Stokes were proposed by ad hoc arguments to describe the phenomena of heat conduction, diffusion, and shear stresses in fluids. This was before there was any non-equilibrium thermodynamics. It was much later—essentially when Eckart's papers [1], [2], [3] appeared—that these laws became incorporated into thermodynamics of irreversible processes. That theory, while not entirely systematic, is simple and has great suggestive value. We start out with a brief description of how thermodynamics of irreversible processes arrives at the laws of Fourier and Navier–Stokes.

Once these laws are applied to the equations of balance of mechanics and thermodynamics, a system of parabolic equations that predicts infinite speeds for the propagation of heat and shear stress appears. This phenomenon has been called a paradox, and upon reflection it is clear that the laws of Fourier and Navier–Stokes are to blame.

Cattaneo [4] was the first to propose a remedy. He was concerned with heat conduction and, basing his arguments on molecular kinetics, he formulated a modification of Fourier's law, which is now called the Cattaneo equation. Thus Cattaneo arrived at a hyperbolic differential equation for the propagation of temperature. Much later Müller [5], [6] made an attempt to incorporate the Cattaneo equation *and* a corresponding modification of the Navier–Stokes equation into thermodynamics of irreversible processes. He was thus able to predict finite speeds for heat waves and shear waves. The theory that emerged has been called *extended* thermodynamics of irreversible processes.

That extended theory is characterized by non-equilibrium terms in the entropy and by a modified entropy flux, both ad hoc assumptions. Those assumptions can be shown, however, to have specific counterparts in the kinetic theory of gases at the level of Grad's 13-moment approximation [7]. While the extended theory is not properly invariant under a change of frame, the kinetic theory of gases provides the arguments by which that flaw may be remedied.

1 Paradoxa of Heat Conduction and Shear Diffusion

1.1 Heuristic Derivation of the Laws of Fourier and Navier–Stokes

One objective of irreversible thermodyamics of fluids is the determination of the 5 fields of

$$\begin{array}{ll} \text{mass density} & \varrho(\mathbf{x},t), \\ \text{velocity} & v_i(\mathbf{x},t), \\ \text{temperature} & T(\mathbf{x},t) \end{array} \tag{1.1}$$

in all points of the fluid and at all times.

For this purpose we need field equations, and these are based upon the equations of balance of mechanics and thermodynamics, viz. the conservation laws of mass and momentum and the equation of balance of internal energy

$$\begin{aligned} \dot{\varrho} + \varrho\frac{\partial v_j}{\partial x_j} &= 0, \\ \varrho\dot{v}_i - \frac{\partial t_{ij}}{\partial x_j} &= 0, \\ \varrho\dot{\epsilon} + \frac{\partial q_j}{\partial x_j} &= t_{ij}\frac{\partial v_i}{\partial x_j}. \end{aligned} \tag{1.2}$$

While these are 5 equations, they are not field equations for the fields (1.1). Indeed, T does not even appear in (1.2), and instead they contain new quantities, viz.

$$\begin{array}{ll} \text{(symmetric) stress tensor} & t_{ij}, \\ \text{heat flux} & q_i, \\ \text{specific internal energy} & \epsilon. \end{array} \tag{1.3}$$

In order to close the system we must find relations between t_{ij}, q_i, ϵ and the fields (1.1), so-called constitutive relations.

In thermodynamics of irreversible processes, a theory universally known as TIP, such relations are derived in a heuristic manner from an entropy inequality that is based upon the Gibbs equation of equilibrium thermodynamics

$$\dot{s} = \frac{1}{T}\left(\dot{\epsilon} - \frac{p}{\varrho^2}\dot{\varrho}\right). \tag{1.4}$$

Here s is the specific entropy, and ϵ and the pressure p are considered to be functions of ϱ and T as prescribed by the caloric and thermal equations of state. Elimination of ϵ and ϱ between (1.4) and $(1.2)_{1,3}$ and some rearrangement leads to the equation

$$\varrho\dot{s} + \frac{\partial}{\partial x_i}\left(\frac{q_i}{T}\right) = q_i\frac{\partial(1/T)}{\partial x_i} + \frac{1}{T}t_{\langle ij\rangle}\frac{\partial v_{\langle i}}{\partial x_{j\rangle}} + \frac{1}{T}\left(\frac{1}{3}t_{ii} + p\right)\frac{\partial v_i}{\partial x_i}, \tag{1.5}$$

which may be interpreted as an equation of balance of entropy. Such an interpretation implies that

$$\begin{aligned} \varphi_i &= \frac{q_i}{T} && \text{is the entropy flux,} \\ \Sigma &= -\frac{q_i}{T^2}\frac{\partial T}{\partial x_i} + t_{ij}\frac{\partial v_{<i}}{\partial x_{j>}} + \left(\frac{1}{3}t_{ii} + p\right)\frac{\partial v_n}{\partial x_n} && \text{is the entropy production.} \end{aligned} \tag{1.6}$$

Inspection shows that the entropy production is a sum of products of

thermodynamic fluxes	and	thermodynamic forces	
heat flux q_i		temperature gradient $\partial T/\partial x_i$	(1.7)
stress deviator $t_{<ij>}$		deviatoric velocity gradient $\partial v_{<i}/\partial x_{j>}$	
dynamic pressure $\pi = \frac{1}{3}t_{ii} + p$		divergence of velocity $\partial v_n/\partial x_n$.	

The entropy production must be non-negative. Assuming only linear relations between forces and fluxes TIP ensures that inequality by constitutive equations—phenomenological equations in the jargon of TIP—of the type

$$\begin{aligned} q_i &= -\kappa \frac{\partial T}{\partial x_i}, & \kappa &\geq 0, \\ t_{<ij>} &= 2\mu \frac{\partial v_{<i}}{\partial x_{j>}}, & \mu &\geq 0, \\ \pi &= \lambda \frac{\partial v_n}{\partial x_n}, & \lambda &\geq 0. \end{aligned} \tag{1.8}$$

Along with the thermal and caloric equations of state $p = p(\varrho, T)$, $\epsilon = \epsilon(\varrho, T)$ the equations (1.8) are the constitutive equations of TIP. They are known as the laws of Fourier and Navier–Stokes with κ being the heat conductivity and μ and λ the shear and bulk viscosity respectively. All of these coefficients may be functions of ϱ and T.

1.2 Parabolic Laws of Heat Conduction and Shear Diffusion

Insertion of (1.8) and of the equations of state into the balance equations (1.2) leads to field equations for the determination of the fields (1.1). The field equations are parabolic in character. In order to emphasize this fact we investigate two special cases:

1. the case of a fluid at rest with constant density; and
2. the case of a fluid with constant density and temperature and with a velocity field of the form $\mathbf{v} = (0, v(x^1, t), 0)$.

In these cases the system of equations (1.2) reduces to a single partial differential equation for T and v respectively. Neglecting non-linear terms in gradients and time-derivatives we obtain

$$\frac{\partial T}{\partial t} = \frac{\kappa}{\varrho \epsilon_T} \Delta T \quad \text{and} \quad \frac{\partial v}{\partial t} = \frac{\mu}{\varrho} \frac{\partial^2 v}{\partial x_1^2}, \tag{1.9}$$

where $\epsilon_T = \left(\frac{\partial \epsilon}{\partial T}\right)_\varrho$ is the specific heat at constant volume, a positive quantity. Both equations are of the same type; they are known as diffusion equations.

We may write the solution of $(1.9)_1$ for an initial value problem in infinite space in the form

$$T(\mathbf{x}, t) = \frac{1}{(4\pi D t)^{3/2}} \int_{-\infty}^{\infty} T(\mathbf{y}, 0) \exp\left(-\frac{(\mathbf{y} - \mathbf{x})^2}{4Dt}\right) d\mathbf{y}, \tag{1.10}$$

where D stands for $\kappa/\varrho\epsilon_T$. Equation $(1.9)_2$ has the same general solution, if we make allowance for its one-dimensional character. This solution implies that $T(\mathbf{x}, t)$ is unequal to zero for all $\mathbf{x}$ and $t > 0$ even though $T(\mathbf{x}, 0)$ may have support on a finite interval only. Thus the temperature and the shear velocity spread throughout the whole space infinitely fast, a phenomenon that has been called a paradox.

It is fair to say that few people cared, because the equations (1.9) were perfectly well-suited for the problems of engineers and physicists. Only a few mathematicians and theoretical physicists

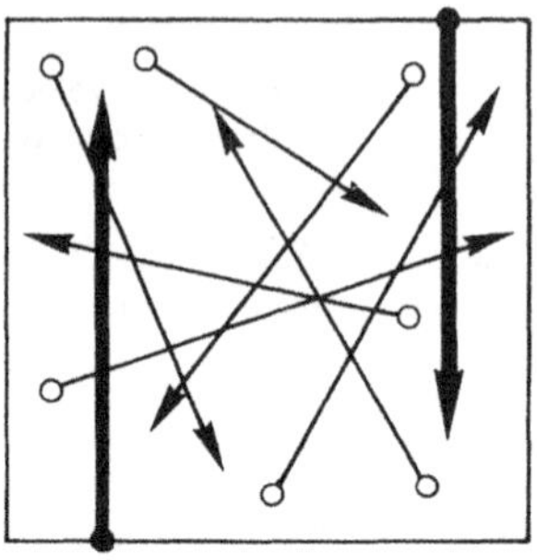

Figure 1. On Fourier's law.

were concerned. However, the desire to have hyperbolic equations in thermodynamics and, hence, finite speeds was the main motivation for the development of extended thermodynamics. It began with Cattaneo.

2 Paradox Removed

2.1 The Cattaneo Equation

In 1948 Cattaneo wrote a paper entitled "Sulla Conduzione del Calore" [4] in which he addressed the question of the paradox of heat conduction. He modified Fourier's law and based his argument on the elementary kinetic theory of gases. The reasoning proceeds as follows:

Cattaneo first reviewed an old argument of Maxwell's that gives a molecular interpretation to heat conduction. To understand the gist of that argument we focus attention upon a small volume element of linear dimensions of the mean free path of the molecules in which a temperature gradient prevails, see Figure 1.

A particle moving upward will, in the mean, carry more energy than a particle moving downward. Therefore, the pair of particles carry a net energy upward, i.e., opposite to the temperature gradient. This amounts to a heat flux across the middle layer, and we obtain the

$$\textit{stationary}\text{ Fourier law:}\qquad q_i = -\kappa\frac{\partial T}{\partial x_i}. \tag{2.1}$$

Cattaneo changed this argument slightly. He argued that there is a time-lag between the start of the particles at their point of departure and the time of passage through the middle layer. If the temperature changes in time, it is thus clear that the heat flux at a certain time depends on the temperature gradient at an earlier time. It therefore seemed reasonable to formulate the

$$\textit{instationary}\text{ Fourier law:}\qquad q_i = -\kappa\left(\frac{\partial T}{\partial x_i} - \tau\left(\frac{\partial T}{\partial x_i}\right)^{\bullet}\right). \tag{2.2}$$

If the equations (2.1) and (2.2) are inserted into the energy balance $(1.2)_3$ for a body at rest and with constant density, viz.

$$\varrho\epsilon_T\dot{T} + \frac{\partial q_i}{\partial x_i} = 0, \tag{2.3}$$

we obtain differential equations for the heat condition, namely,

$$\dot{T} = \frac{\kappa}{\varrho\epsilon_T}\Delta T \qquad \text{[from (2.1)]}, \tag{2.4}$$

$$\dot{T} = \frac{\kappa}{\varrho\epsilon_T}(\Delta T - \tau(\Delta T)^\bullet) \qquad \text{[from (2.2)]}. \tag{2.5}$$

Neither of these two equations are hyperbolic, and therefore both predict infinite speeds for the propagation of thermal disturbances. Cattaneo must have noticed this, because he proceeds to modify equations (2.2) assuming that the operator $\tau(d/dt)$ is small such that

$$\left(1 - \tau\frac{d}{dt}\right)^{-1} \approx 1 + \tau\frac{d}{dt}.$$

In this approximation the equation assumes the form

$$q_i + \tau\dot{q}_i = -\kappa\frac{\partial T}{\partial x_i}. \tag{2.6}$$

The last equation is often called the *Cattaneo equation*; it is again a modification of Fourier's law, and if we combine it with the energy equation (2.3), we obtain

$$\tau\ddot{T} + \dot{T} = \frac{\kappa}{\varrho\epsilon_T}\Delta T. \tag{2.7}$$

This is the telegraph equation. It is hyperbolic, if $\tau > 0$ holds, and it predicts the propagation of heat pulses at the finite speed

$$V = \pm\sqrt{\frac{\kappa}{\varrho\epsilon_T\tau}}.$$

We conclude that the transition from (2.2) to (2.6) is all-important for finite speeds and that it changes a parabolic equation into a hyperbolic one. Yet that transition is impossible to justify rigorously. Cattaneo had this to say about it:

> *Nel risultato ottenuto approfitteremo della piccolezza del parametro τ per trascurare il termine che contiene a fattore il suo quadrato, conservando per altro il termine in cui τ compare a primo grado. Naturalmente, per delimitare la portata delle conseguenze che stiamo per trarre converrà precisare un pò meglio le condizioni in cui tale approssimazione è lecita. Allo scopo ammetteremo esplicitamente che il fenomeno di conduzione calorifica avvenga nell'intorno di uno stato stazionario o, in altri termini, che durante il suo svolgersi si mantengano abbastanza piccole le derivate temporali delle varie grandezze in gioco.*

Thus he argues that his derivation is correct, if both τ and d/dt are small. Lack of rigour in the derivation is not the only criticism to be leveled at Cattaneo's equation, because it is also not invariant under a change of frame. We shall come back to this point later; see Paragraph 2.4 and Section 3.

2.2 Extended TIP

Nevertheless, Cattaneo pointed the direction and his work motivated Müller [5], [6] to try to derive modifications of the laws of Fourier and Navier–Stokes from the principles of TIP. Müller noticed that the arguments of TIP (as presented above in Paragraph 1.1) could be used to derive Cattaneo's equation. All one would need is the assumption that the specific entropy s, which heretofore was considered to be a function of ϵ and ϱ only, might depend on q_i as well. Actually, however, if q_i is a possible variable in s, there is no reason why s should not also depend on $t_{<ij>}$ and π. Of course, these dependencies must be such that s is a scalar and that s has a maximum in equilibrium, where q_i, $t_{<ij>}$, and π all vanish. Therefore, restricting attention to quadratic terms Müller allowed s to depend on the scalars $q_i q_i$, $t_{<ij>} t_{<ij>}$, and π^2 with negative coefficients a, b, c, and thus he generalized the Gibbs equation (1.4) to read

$$\dot{s} = \frac{1}{T}\left(\dot{\epsilon} - \frac{p}{\varrho^2}\dot{\varrho} + 2aTq_i\dot{q}_i + 2bTt_{<ij>}\dot{t}_{<ij>} + 2cT\pi\dot{\pi}\right). \tag{2.8}$$

Now again, $(1.2)_{1,3}$ are used to eliminate $\dot{\varrho}$ and $\dot{\epsilon}$ and, after some rearrangement, one obtains an entropy balance of the form [1]

$$\varrho\dot{s} + \frac{\partial}{\partial x_i}\left(\frac{q_i}{T}\right) = q_i\left(\frac{\partial(1/T)}{\partial x_i} + 2\varrho a\dot{q}_i\right) + t_{<ij>}\left(\frac{1}{T}\frac{\partial v_{<i}}{\partial x_{j>}} + 2\varrho b\dot{t}_{<ij>}\right) + \pi\left(\frac{1}{T}\frac{\partial v_i}{\partial x_i} + 2\varrho c\dot{\pi}\right). \tag{2.9}$$

We may again interpret q_i/T as the entropy flux and the right side of (2.9) as the entropy production, which is a sum of products of the thermodynamic fluxes $q_i, t_{<ij>}, \pi$ and modified thermodynamic forces, viz.

$$\frac{\partial T}{\partial x_i} - 2\varrho T^2 a\dot{q}_i, \qquad \frac{\partial v_{<i}}{\partial x_{j>}} + 2\varrho Tb\dot{t}_{<ij>}, \qquad \frac{\partial v_i}{\partial x_i} + 2\varrho Tc\dot{\pi}. \tag{2.10}$$

This is not all: There is an inherently weak point in TIP as it proceeds from the Gibbs equation to an entropy balance, because the interpretation of the entropy flux is ambiguous. Indeed, it is possible to add the term $Kt_{<ij>}q_j + L\pi q_i$, with arbitrary coefficients K and L, to the usual entropy flux q_i/T without upsetting the structure of the entropy production as a sum of products of forces and fluxes. If this is done, the entropy flux assumes the form

$$\varphi_i = \frac{q_i}{T} + Kt_{<ij>}q_j + L\pi q_i, \tag{2.11}$$

and the entropy balance reads

$$\begin{aligned}\varrho\dot{s} + \frac{\partial\varphi_i}{\partial x_i} = {} & q_i\left(\frac{\partial(1/T)}{\partial x_i} + 2\varrho a\dot{q}_i + K\frac{\partial t_{ij}}{\partial x_j} + L\frac{\partial\pi}{\partial x_i}\right) \\ & + t_{<ij>}\left(\frac{1}{T}\frac{\partial v_{<i}}{\partial x_{j>}} + 2\varrho b\dot{t}_{<ij>} + K\frac{\partial q_{<j}}{\partial x_{i>}}\right) + \pi\left(\frac{1}{T}\frac{\partial v_i}{\partial x_i} + 2\varrho c\dot{\pi} + L\frac{\partial q_i}{\partial x_i}\right).\end{aligned} \tag{2.12}$$

The fluxes are the same as always, namely, $q_i, t_{<ij>}, \pi$, but the thermodynamic forces have changed from (2.10) again, because we have to compensate for the changed entropy flux by also changing the entropy production. We still assume, however, linear phenomenological relations between the fluxes and the forces and obtain

$$\begin{aligned} q_i &= -\kappa\left\{\frac{\partial T}{\partial x_i} - 2\varrho T^2 a\dot{q}_i \underline{- KT^2\frac{\partial t_{<ij>}}{\partial x_j} - LT^2\frac{\partial\pi}{\partial x_i}}\right\}, \\ t_{<ij>} &= 2\mu\left\{\frac{\partial v_{<i}}{\partial x_{j>}} + 2\varrho Tb\dot{t}_{<ij>} \underline{+ KT\frac{\partial q_{<i}}{\partial x_{j>}}}\right\}, \\ \pi &= \lambda\left\{\frac{\partial v_i}{\partial x_i} + 2\varrho Tc\dot{\pi} \underline{+ LT\frac{\partial q_i}{\partial x_i}}\right\}. \end{aligned} \tag{2.13}$$

Thus the Fourier law $(1.8)_1$ is now replaced by $(2.13)_1$, which—without the underlined terms that were introduced by the modification of the entropy flux—is equivalent to the Cattaneo equation (2.6). Also, a corresponding change has appeared in the Navier–Stokes equations, because we now

[1]Terms of third order in the entropy balance are ignored as they should be in TIP.

have $(2.13)_{2,3}$ instead of $(1.8)_{2,3}$. Thus, as we shall see, the theory provides finite speeds not only for heat conduction but also for shear-diffusion.

This theory is called extended TIP, because it has extended the list of variables on which the specific entropy and the entropy flux depend. After this alteration of ordinary TIP a large number of papers have appeared, all using Müller's methodology and devoted in particular to gases. A recent review article of this type of *Irreversible Thermodynamics* is due to Jou, Casas-Vázques, and Lebon [8]. For rigid heat conductors a different procedure was outlined. In fact since the Cattaneo equation can be expressed as an integral over the history of the temperature gradient, viz.

$$\mathbf{q} = -\frac{\kappa}{\tau}\int_{-\infty}^{t} \exp\left(-\frac{t-t'}{\tau}\right)\frac{\partial T(\mathbf{x}t')}{\partial x_i}\,dt',$$

some authors developed the idea that the Cattaneo equation is a particular case of materials with memory. See in particular the papers of Gurtin and Pipkin [9], Green and Lindsay [10], and Grioli [11]. A review paper of this approach was recently given by Joseph and Preziosi [12]. A limitation of this *functional approach* lies in the fact that, when the material is not rigid, we have interaction between the thermal and mechanical fields. In such cases it is difficult to interpret the Cattaneo equation and its counterparts in (2.13) as integral equations.

2.3 Finite Pulse Speeds in Extended TIP

Comparing the phenomenological equations (1.8) and (2.13) we realize that extended TIP has replaced the simple algebraic equation between q_i, t_{ij} and $\frac{\partial T}{\partial x_i}, \frac{\partial v_i}{\partial x_j}$ by a complex set of partial differential equations. The complete set of equations is represented by the balance laws (1.2), by the phenomenological equations (2.13), and by the thermal and caloric equations of state $p = p(\varrho, T)$ and $\epsilon = \epsilon(\varrho, T)$. It is impossible now to characterize a reasonably realistic special case in which these equations decouple to give simple equations like (1.9) for T or for the shear velocity v. Therefore it is not immediately obvious whether the speeds of propagation are finite.

In order to find this out we consider plane harmonic waves of small amplitude assuming two types of solutions, namely,

$$\text{I: } \varrho = \tilde{\varrho} + \varrho E,\ v_i = \begin{bmatrix}\bar{v}\\0\\0\end{bmatrix} E,\ T = \tilde{T} + \bar{T}E,\ t_{<ij>} = \begin{bmatrix}\bar{P} & 0 & 0\\ 0 & -\frac{1}{2}\bar{P} & 0\\ 0 & 0 & -\frac{1}{2}\bar{P}\end{bmatrix} E,\ \pi = \bar{\pi}E,\ q_i = \begin{bmatrix}\bar{q}\\0\\0\end{bmatrix} E; \tag{2.14}$$

$$\text{II: } \varrho = \tilde{\varrho}, \quad v_i = \begin{bmatrix}0\\\bar{v}\\0\end{bmatrix} E, \quad T = \tilde{T}, \quad t_{<ij>} = \begin{bmatrix}0 & \bar{P} & 0\\ \bar{P} & 0 & 0\\ 0 & 0 & 0\end{bmatrix} E, \quad \pi = 0, \quad q_i = \begin{bmatrix}0\\\bar{q}\\0\end{bmatrix} E. \tag{2.15}$$

For a generic field g we have $\tilde{g}$ as a constant field, while $\bar{g}$ denotes small amplitudes of which products are neglected. E in (2.14), (2.15) stands for

$$E = e^{i(\omega t - kx)}, \tag{2.16}$$

where ω is the frequency and k is the wave number. The latter is generally complex.

The wave (2.14) is longitudinal in velocity and heat flux and it propagates disturbances in density and temperature, hence, also in pressure. The wave (2.15), on the other hand, is transversal in velocity and heat flux, and we may say that it is a shear wave.

If (2.14) is introduced into the system of equations (1.2) and (2.13), all but 6 of these equations are identically satisfied. The remaining 6 form a linear homogeneous algebraic system for the determination of $\bar{\varrho}$ through $\bar{q}$, viz.

$$\begin{bmatrix} 1 & -\varrho\frac{k}{\omega} & 0 & 0 & 0 & 0 \\ -p_\varrho\frac{k}{\omega} & \varrho & -p_{\mathrm{T}}\frac{k}{\omega} & \frac{k}{\omega} & \frac{k}{\omega} & 0 \\ -\frac{T}{\varrho}p_{\mathrm{T}} & 0 & \varrho\epsilon_{\mathrm{T}} & 0 & 0 & -\frac{k}{\omega} \\ 0 & 0 & \kappa\frac{k}{\omega} & -\kappa KT^2\frac{k}{\omega} & -\kappa LT^2\frac{k}{\omega} & \left[2\kappa\varrho T^2 a + i\frac{1}{\omega}\right] \\ 0 & -\frac{4}{3}\mu\frac{k}{\omega} & 0 & \left[4\mu\varrho Tb + i\frac{1}{\omega}\right] & 0 & -\frac{4}{3}\mu KT\frac{k}{\omega} \\ 0 & -\lambda\frac{k}{\omega} & 0 & 0 & \left[2\lambda\varrho Tc + i\frac{1}{\omega}\right] & -\lambda LT\frac{k}{\omega} \end{bmatrix} \begin{bmatrix} \bar{\varrho} \\ \bar{v} \\ \bar{T} \\ \bar{P} \\ \bar{\pi} \\ \bar{q} \end{bmatrix} = 0. \quad (2.17)$$

Here p_{T}, p_ϱ, and ϵ_{T} stand for the partial derivatives of p and ϵ with respect to T and ϱ. The determinant of this matrix must vanish. This condition gives rise to the dispersion relation from which we may calculate the permissible wave numbers for a given frequency ω. The phase speed v_{ph} and the attenuation α may be calculated from

$$v_{\mathrm{ph}} = \frac{\omega}{\mathrm{Re}(k)} \quad \text{and} \quad \alpha = -\mathrm{Im}(k). \quad (2.18)$$

Later in Chapter 11 when we treat specific fluids we shall calculate the phase speed and the attenuation as functions of ω. Here, however, we concentrate on the maximum speed of propagation allowed by the system of equations (1.2), (2.13). We ask whether that speed of propagation, the pulse speed V, is finite or infinite. It can be proved that V is the limiting value of the phase speed for $\omega \to \infty$, see Chapter 7.

For $\omega \to \infty$ the dispersion relation simplifies considerably and k is real. We introduce the ordinary sound speed

$$v_{\mathrm{ph}}(0) = \sqrt{p_\varrho + \frac{T}{\varrho^2\epsilon_{\mathrm{T}}}{p_{\mathrm{T}}}^2} \quad (2.19)$$

and obtain for the pulse speed

$$\overset{\mathrm{I}}{V} = v_{\mathrm{ph}}(0)\sqrt{1 - \frac{1}{12\varrho^2 T v_{\mathrm{ph}}^2(0)}\frac{6ab + 4ac + 6(T\epsilon_{\mathrm{T}})^{-1}bc - 2TK^2c - 3TL^2b}{abc}}. \quad (2.20)$$

We recall that a, b, c, are negative so that $\overset{\mathrm{I}}{V}$ is always greater than $v_{\mathrm{ph}}(0)$, but it is *finite*.

Introducing (2.15) into the system of equations (1.2) and (2.13) we note that all but 3 equations are satisfied and that the remaining 3 form a linear homogeneous system of equations for the determination of $\bar{v}$, $\bar{P}$, and $\bar{q}$, viz.

$$\begin{bmatrix} \varrho & \frac{k}{\omega} & 0 \\ 0 & -\kappa KT^2\frac{k}{\omega} & \left[2\kappa\varrho T^2 a + i\frac{1}{\omega}\right] \\ -2\mu\frac{k}{\omega} & \left[4\mu\varrho Tb + i\frac{1}{\omega}\right] & -2\mu KT\frac{k}{\omega} \end{bmatrix} \begin{bmatrix} \bar{v} \\ \bar{P} \\ \bar{q} \end{bmatrix} = 0. \quad (2.21)$$

We obtain the dispersion relation by setting the determinant of the matrix equal to zero. Again we determine the pulse speed V as ω/k for $\omega \to \infty$. We obtain

$$\overset{\text{II}}{V} = \sqrt{\frac{1}{\varrho^2 T}\frac{-4a + 2K^2T}{ab}}, \tag{2.22}$$

which is a finite speed as long as a, b, and K are non-zero.

We finish this consideration of pulse speeds by treating the special case in which $\lambda = 0$ holds, so that by $(2.13)_3$ we also have $\pi = 0$. This case, though singular, is of considerable importance, because it is realized in mon-atomic gases. We assume a solution of the form

$$\text{III:} \qquad \varrho = \tilde{\varrho} + \bar{\varrho}E, \qquad v_i = \begin{bmatrix} \bar{v} \\ 0 \\ 0 \end{bmatrix} E, \qquad T = \tilde{T} + \bar{T}E,$$

$$t_{<ij>} = \begin{bmatrix} \bar{P} & 0 & 0 \\ 0 & -\frac{1}{2}\bar{P} & 0 \\ 0 & 0 & -\frac{1}{2}\bar{P} \end{bmatrix} E, \qquad q_i = \begin{bmatrix} \bar{q} \\ 0 \\ 0 \end{bmatrix} E \tag{2.23}$$

and obtain the homogeneous system

$$\begin{bmatrix} 1 & -\varrho\frac{k}{\omega} & 0 & 0 & 0 \\ -p_\varrho\frac{k}{\omega} & \varrho & -p_T\frac{k}{\omega} & \frac{k}{\omega} & 0 \\ -\frac{T}{\varrho}p_T & 0 & \varrho\epsilon_T & 0 & -\frac{k}{\omega} \\ 0 & 0 & \kappa\frac{k}{\omega} & -\kappa K T^2\frac{k}{\omega} & \left[2\kappa\varrho T^2 a + i\frac{1}{\omega}\right] \\ 0 & -\frac{4}{3}\mu\frac{k}{\omega} & 0 & \left[4\mu\varrho T b + i\frac{1}{\omega}\right] & -\frac{4}{3}\mu K T\frac{k}{\omega} \end{bmatrix} \begin{bmatrix} \bar{\varrho} \\ \bar{v} \\ \bar{T} \\ \bar{P} \\ \bar{q} \end{bmatrix} = 0. \tag{2.24}$$

The pulse speed is calculated as before and we obtain

$$\overset{\text{III}}{V} = v_{\text{ph}}(0)\sqrt{1 - \frac{1}{12\varrho^2 T v_{\text{ph}}^2(0)}\frac{4a + 6\frac{1}{T\epsilon_T}b - 2TK^2}{ab}}, \tag{2.25}$$

which is finite as long as a and b are non-zero.

2.4 Conclusion and Criticism

Extended TIP has provided finite speeds, but in so doing it had to make the unconventional assumption that the entropy depends on t_{ij} and q_i. Also the entropy flux was assumed to be given by the complex expression (2.11) rather then simply by q_i/T. Clearly it is desirable to have some support for these assumptions; indeed, such support can be derived from the kinetic theory of mon-atomic gases as we shall now see.

There is a severe criticism to be leveled both at the Cattaneo equation (2.6) and at the phenomenological equations (2.13) of extended TIP. Indeed, neither of these can be true physical laws, because they are not properly invariant. The time derivatives $\dot{q}_i$ and $\dot{t}_{<ij>}$ in (2.6) and (2.13) are not objective tensors, while all other terms in those equations are objective tensors; therefore, even if the equations hold in an inertial frame, they do not hold in a non-inertial frame. The kinetic theory will furnish an interpretation of this phenomenon and, in doing so, will provide the basis for a new systematic version of extended thermodynamics.

Also the system of equations (1.2), (2.13) is awkward in a mathematical sense. Although we have seen that small disturbances of a constant equilibrium state propagate with a finite speed, it is not sure at this time that the non-linear system is hyperbolic so that the speeds of non-linear waves are finite as well.[2]

It seems desirable, both from the mathematical and the physical points of view, to deal with field equations that guarantee existence, uniqueness, and continuous dependence for the solutions of Cauchy problems. That is, we want equations for which Cauchy problems with reasonable data are well-posed. The present system (1.2), (2.13), however, is complicated to a degree that there exists no theorem guaranteeing well-posedness of Cauchy problems.

Finally we know that in quasi-linear systems even smooth data may in time lead to the formation of shocks, thus calling for jump conditions across the shock front. Such conditions can only be had if the field equations are of balance type; and while the equations (1.2) are indeed balance equations, this seems not to be the case for the phenomenological equations (2.13). Here again, the kinetic theory will help us to find the correct interpretation.

3 Kinetic Theory of Mon-Atomic Gases

3.1 Boltzmann Equation and Moments

The kinetic theory describes the state of a gas by the phase density $f(\mathbf{x}, \mathbf{c}, t)$ such that

$$f(\mathbf{x}, \mathbf{c}, t)d\mathbf{c} \tag{3.1}$$

is the number density of atoms at the point $\mathbf{x}$ and at time t that have velocities between $\mathbf{c}$ and $\mathbf{c} + d\mathbf{c}$. The phase density obeys the Boltzmann equation

$$\frac{\partial f}{\partial t} + c_i \frac{\partial f}{\partial x_i} + (f_i + \underset{ci}{i}) \frac{\partial f}{\partial c_i} = \int (f' f^{1'} - f f^1) \sigma \, g \sin\theta \, d\theta \, d\varepsilon d\mathbf{c}^1. \tag{3.2}$$

Here f_i is the specific external body force, assumed independent of c_i, and $\underset{ci}{i}$ is the inertial acceleration of an atom, viz.

$$\underset{ci}{i} = 2W_{ik}(c_k - \dot{b}_k) - W_{ik}^2(x_k - b_k) + \dot{W}_{ik}(x_k - b_k) + \ddot{b}_i, \tag{3.3}$$

where W_{ik} is the matrix of the angular velocity of the frame with respect to an inertial frame and b_i is the distance vector between the origins of the two frames. The right-hand side of (3.2) is due to collisions between the atoms and for its derivation we refer the reader to books on the kinetic theory of gases, for example, [13], [14], or [15]. f, f^1, f', and $f^{1'}$ are the values of the phase density for the velocities $\mathbf{c}, \mathbf{c}^1, c'$, and $c^{1'}$ for two atoms before and after the collision.

Most macroscopic thermodynamic quantities are formed by taking moments of the phase density. The generic moment has the form

$$F_{i_1 i_2 \cdots i_N} = \int m c_{i_1} c_{i_2} \cdots c_{i_N} f \, d\mathbf{c}. \tag{3.4}$$

Thus, in particular, we have

$$\begin{aligned} &\text{mass density} && F = \varrho, \\ &\text{momentum density} && F_i = \varrho v_i. \end{aligned} \tag{3.5}$$

With the velocity v_i of the gas we form the *peculiar velocity* $C_i = c_i - v_i$ of an atom and use this to define *the central moments*

$$\varrho_{i_1 i_2 \cdots i_N} = \int m C_{i_1} C_{i_2} \cdots C_{i_N} f d\mathbf{x}. \tag{3.6}$$

[2]See Chapter 7 for the properties of hyperbolic systems.

The first few moments and central moments have a suggestive name and interpretation, viz.

$$\begin{array}{llll} \frac{1}{2}F_{ii} = \varrho e & \text{energy density,} & \frac{1}{2}\varrho_{ii} = \varrho\epsilon & \text{internal energy density,} \\ F_{ij} & \text{momentum flux,} & -\varrho_{ij} = t_{ij} & \text{stress tensor,} \\ \frac{1}{2}F_{ijj} = J_i & \text{energy flux,} & \frac{1}{2}\varrho_{ijj} = q_i & \text{heat flux.} \end{array} \tag{3.7}$$

Since $c_i = C_i + v_i$ holds, there are simple relations between the moments and the central moments, namely,

$$\begin{aligned} F &= \varrho \\ F_i &= \varrho v_i \\ F_{ij} &= \varrho_{ij} + \varrho v_i v_j \\ F_{ijk} &= \varrho_{ijk} + 3\varrho_{(ij}v_{k)} + \varrho v_i v_j v_k \\ F_{ijk\ell} &= \varrho_{ijk\ell} + 4\varrho_{(ijk}v_{l)} + 6\varrho_{(ij}v_k v_{\ell)} + \varrho v_i v_j v_k v_\ell. \end{aligned} \tag{3.8}$$

Round brackets indicate symmetrization, that is, the sum of the permutations of all N indices divided by $N!$. Prototypes of the relations (3.8) are the well-known decompositions of momentum flux, energy density, and energy flux into convective and non-convective parts which are more commonly called kinetic and internal parts. We have

$$\begin{aligned} F_{ij} &= -t_{ij} + \varrho v_i v_j, \\ \varrho e &= \varrho\epsilon + \frac{\varrho}{2}v^2, \\ J_i &= q_i + \varrho\left(\epsilon + \frac{v^2}{2}\right)v_i - t_{ij}v_j. \end{aligned} \tag{3.9}$$

In a mon-atomic ideal gas there is no dynamic pressure; therefore, $t_{ii} = -3p$ holds and from (3.7) we conclude

$$p = \frac{2}{3}\varrho\epsilon. \tag{3.10}$$

This is a well-known relation between the thermal and caloric equations of state in a mon-atomic gas. The kinetic theory confirms this relation, which shows that we have chosen the definitions (3.7) properly.

We know that in mon-atomic ideal gases we have $\epsilon = \frac{3}{2}\frac{k}{m}T$, and this relation may be used to define temperature in the kinetic theory by

$$\frac{3}{2}kT = \frac{1}{\varrho}\int \frac{m}{2}C^2 f d\mathbf{c}. \tag{3.11}$$

3.2 Equations of Balance for Moments

Multiplication of the Boltzmann equation by a generic function $\psi(\mathbf{x}, \mathbf{c}, t)$ and integration over all $\mathbf{c}$ gives an equation of balance for the mean value

$$\overline{\psi} = \frac{\int \psi f d\mathbf{c}}{\int f d\mathbf{c}} = \frac{m}{\varrho}\int \psi f\, d\mathbf{c}. \tag{3.12}$$

The equation is derived under the assumption that the phase density falls off rapidly for large values of $\mathbf{c}$ and it reads

$$\begin{aligned}&\frac{\partial \varrho\bar{\psi}}{\partial t}+\frac{\partial \varrho\overline{\psi c_i}}{\partial x_i}-\overline{\varrho(f_i+\underset{ci}{i})\frac{\partial \psi}{\partial c_i}}\\&=\overline{\varrho\left(\frac{\partial \psi}{\partial t}+c_i\frac{\partial \psi}{\partial x_i}\right)}+\frac{1}{4}m\int(\psi+\psi^1-\psi'-\psi^{1'})(f'f^{1'}-ff^1)\sigma g\sin\theta\, d\theta\, d\mathbf{c}_1\, d\mathbf{c}.\end{aligned} \tag{3.13}$$

This equation is an equation of balance in which the third term represents the supply due to external body forces and inertial forces, while the right-hand side represents the production.

Of particular interest in thermodynamics are the equations of balance for moments. We set

$$\psi = c_{i_1}c_{i_2}\cdots c_{i_N} \tag{3.14}$$

and obtain

$$\frac{\partial F_{i_1\cdots i_N}}{\partial t}+\frac{\partial F_{i_1\cdots i_N j}}{\partial x_j}-NF_{(i_1\cdots i_{N-1}}\left(f_{i_N)}+i^0_{i_N)}\right)-NF_{k(i_1\cdots i_{N-1}}2W_{i_N)k}=S_{i_1\cdots i_N}, \tag{3.15}$$

where i^0_j is the part of the inertial acceleration that is independent of the velocity. The right-hand side represents the collision production. Comparison with (3.13) shows that S, S_i, and S_{ii} must vanish, because mass, momentum, and energy of a particle are conserved in a collision. More explicitly we write the balance equations for the first thirteen moments $F=\varrho$, $F_i=\varrho v_i$, F_{ij}, and F_{ijj}. They read

$$\begin{aligned}&\frac{\partial \varrho}{\partial t}+\frac{\partial \varrho v_i}{\partial x_i} &&=0,\\&\frac{\partial \varrho v_i}{\partial t}+\frac{\partial F_{ij}}{\partial x_j}-\varrho(f_i+i^0_i)-2\varrho v_k W_{ik} &&=0,\\&\frac{\partial F_{ij}}{\partial t}+\frac{\partial F_{ijk}}{\partial x_k}-2\varrho v_{(i}(f_{j)}+i^0_{j)})-4F_{k(i}W_{j)k} &&=S_{<ij>},\\&\frac{\partial F_{ijj}}{\partial t}+\frac{\partial F_{ijjk}}{\partial x_k}-3F_{(ij}(f_{j)}+i^0_{j)})-6F_{k(ij}W_{j)k} &&=S_{ijj}.\end{aligned} \tag{3.16}$$

In this explicit form it is immediately obvious that the first two of the equations (3.15) represent the conservation laws of mass and momentum, while the trace of the third equation is the conservation law of energy. We note that the contributions from body and inertial forces have the familiar form, well known in mechanics. The remaining equations in (3.16) are balance laws for the deviatoric momentum flux and for the energy flux. These balance equations also contain contributions due to external forces and inertia, but above all they are not conservation laws, because they contain productions due to atomic collisions. For some atomic interaction forces the collision terms on the right-hand sides of $(3.16)_{3,4}$ may be calculated easily. Thus, for instance, for Maxwellian atoms we have

$$\begin{aligned}S_{<ij>}&=-\frac{3}{2}\varrho B t_{<ij>}=-\frac{3}{2}\varrho B(F_{<ij>}-\varrho v_{<i}v_{j>}),\\S_{ijj}&=-2\varrho B q_i\quad=-\varrho B(F_{ijj}-3F_{(ij}v_{j)}-\varrho v_i v^2),\end{aligned} \tag{3.17}$$

where B is a positive constant whose value depends on the strength of the atomic interaction.

Note that upon first sight the system (3.16) with (3.17) appears to be a system of equations for the thirteen moments $\varrho, v_i, F_{ij}, F_{ijj}$, but the system is not closed, because of the appearance of the fluxes $F_{<ijk>}$ and F_{ijjk} in $(3.16)_{3,4}$. Nor does it help to proceed to an equation for the higher moments, because the divergence term in (3.15) will always bring in a moment of the next order. This constitutes the closure problem of the kinetic theory and we shall come back to it later.

3.3 Balance of Entropy and Possible Equilibria

If we insert

$$\psi = -\frac{k}{m} \ln f \tag{3.18}$$

into the general equation of balance (3.13), we obtain the equation

$$\frac{\partial(-k\int \ln f f dc)}{\partial t} + \frac{\partial(-k\int c_i \ln f f dc)}{\partial x_i} = \frac{m}{4}\int \ln \frac{f' f^{1\prime}}{f f^1}(f' f^{1\prime} - f f^1)\sigma \sin\theta \, d\theta \, d\varepsilon \, dc^1 \, dc. \tag{3.19}$$

The collision production on the right-hand side is obviously non-negative, which suggests that (3.19) represents the balance of entropy with

$$\begin{aligned} s &= -\frac{k}{\varrho}\int \ln f f dc && \text{as the specific entropy,} \\ h_i &= -k\int c_i \ln f f dc && \text{as the entropy flux,} \\ \Sigma &= \frac{m}{4}\int \ln \frac{f' f^{1\prime}}{f f^1}(f' f^{1\prime} - f f^1)\sigma g \sin\theta \, d\theta \, d\varepsilon \, dc_1 \, dc && \text{as the entropy production.} \end{aligned} \tag{3.20}$$

We expect the entropy production to vanish in equilibrium, and by $(3.20)_3$ this requires that the equilibrium phase density f_E satisfy the relation

$$\ln f_E + \ln f_E^1 = \ln f_E' + \ln f_E^{1\prime}, \tag{3.21}$$

that is, $\ln f_E$ is an invariant in the collision. There are, however, only 5 independent collisional invariants, namely, m, mc_i, and $m/2c^2$, the atomic masses, momenta, and energies. Therefore, $\ln f_E$ must be the linear combination

$$\ln f_E = \alpha + \beta_i c_i + \gamma c^2, \tag{3.22}$$

where α, β_i, and γ are coefficients that may be determined from (see (3.5) and (3.11))

$$\varrho = \int m f \, dc, \qquad \varrho v_i = \int m c_i f \, dc, \qquad \varrho \frac{3}{2} kT = \int \frac{m}{2} C^2 f \, dc.$$

A simplo calculation shows that f_E has the form

$$f_E = \frac{\varrho}{m}\sqrt{\frac{m}{2\pi kT}}^{3} \exp\left(-\frac{m(\mathbf{c}-\mathbf{v})^2}{2kT}\right), \tag{3.23}$$

which is called the Maxwell distribution.

The Maxwell distribution makes the entropy production $(3.20)_3$ vanish irrespective of the form of the fields $\varrho(\mathbf{x},t)$, $v_i(\mathbf{x},t)$, and $T(\mathbf{x},t)$, but not all such fields satisfy the Boltzmann equation. In fact, those that satisfy the Boltzmann equation must be solutions of the equations

$$\begin{aligned} \frac{\partial \varrho}{\partial t} + \frac{\partial \varrho v_i}{\partial x_i} &= 0, \\ \varrho \dot{v}_i + \frac{\partial \varrho \frac{k}{m} T}{\partial x_i} &= \varrho(f_i + i_i), \quad \text{where} \quad i_i = i_i^0 + 2W_{ik} v_k, \\ \frac{\partial T}{\partial t}\delta_{ij} + 2T\frac{\partial v_{(i}}{\partial x_{j)}} &= 0, \\ \frac{\partial T}{\partial x_i} &= 0. \end{aligned} \tag{3.24}$$

Thus equilibria in a mon-atomic gas need not be characterized by constant and uniform fields of ϱ, v_i, and T. Rather the equilibrium fields contain rigid rotations, isotropic expansions, a wide variety of density gradients, and even a time-dependent temperature. The temperature gradient must vanish.

3.4 The Grad Distribution

Without essential loss of generality the phase density f may be written as

$$f = \left(a + a_i \frac{\partial}{\partial c_i} + a_{ij} \frac{\partial^2}{\partial c_i \partial c_j} + a_{ijk} \frac{\partial^3}{\partial c_i \partial c_j \partial c_k} + \cdots \right) f_{\mathrm{E}}. \tag{3.25}$$

Indeed, by (3.23) this equation is equivalent to

$$\begin{aligned} f = f_{\mathrm{E}} \Bigg\{ & a - a_i \frac{m}{kT} C_i + a_{ij} \left(\frac{m}{kT}\right)^2 \left(C_i C_j - \frac{kT}{m} \delta_{ij} \right) \\ & - a_{ijk} \left(\frac{m}{kT}\right)^3 \left(C_i C_j C_k - \frac{kT}{m} (\delta_{ij} C_k + \delta_{jk} C_i + \delta_{ki} C_j) \right) + \cdots \Bigg\}, \end{aligned} \tag{3.26}$$

which is an expansion in terms of Hermite polynomials that form a complete set of functions. The coefficients $a, a_i, \ldots$ in this expansion must be determined by inserting (3.26) into (3.4). The most well-known application of this expansion is Grad's 13-moment distribution f_{G} in which only the first 13 polynomials are considered. After a little calculation one obtains

$$f_{\mathrm{G}} = f_{\mathrm{E}} \left\{ 1 - \frac{1}{2p} t_{<ij>} \left(\frac{m}{kT} C_i C_j - \delta_{ij} \right) - \frac{1}{p} \frac{m}{kT} q_i C_i \left(1 - 5 \frac{m}{kT} C^2 \right) \right\}. \tag{3.27}$$

This is called the Grad distribution.

3.5 Entropy and Entropy Flux in Grad's 13-Moment Theory

We recall from the considerations of Paragraph 3.3 that the entropy balance in the kinetic theory has the form

$$\varrho \dot{s} + \frac{\partial \varphi_i}{\partial x_i} \geq 0 \quad \text{with} \quad \varrho s = -k \int \ln f f d\mathbf{c}, \; \varphi_i = -k \int C_i \ln f f d\mathbf{c}. \tag{3.28}$$

If we replace f in $\ln f$ by the Grad distribution f_{G} given in (3.27), we obtain for processes close to equilibrium, where f_{G} differs little from f_{E},

$$s = \frac{k}{m} \ln \frac{T^{3/2}}{\varrho} + \varsigma - \frac{t_{<ij>} t_{<ij>}}{4 p \varrho T} - \frac{q_i q_i}{5 p^2 T}, \tag{3.29}$$

$$\varphi_i = \frac{q_i}{T} + \frac{2}{5pT} t_{<ij>} q_j. \tag{3.30}$$

Thus indeed, the specific entropy of the kinetic theory in the Grad approximation depends on $t_{<ij>}$ and q_i. Also the entropy flux is not simply equal to q_i/T, but it contains a term proportional to $t_{<ij>} q_j$. These facts confirm the conjectures of extended TIP. Moreover, the kinetic theory identifies the coefficients a, b, c and K, L of extended TIP appropriate for a mon-atomic ideal gas. We have

$$a = -\frac{1}{5p^2 T}, \qquad b = -\frac{1}{4p\varrho T}, \qquad c = 0, \qquad K = \frac{2}{5pT}, \qquad L = 0. \tag{3.31}$$

3.6 Phenomenological Equations Derived from the Kinetic Theory

We now ask whether the phenomenological equations (2.13) of extended TIP have a counterpart in the kinetic theory. Upon reflection it is clear that such equations in the kinetic theory can only come from the equations of balance $(3.16)_{3,4}$ that, by (3.17), already have $t_{<ij>}$ and q_i on their right-hand sides. Those equations, however, contain the higher moments $F_{<ij>n}$ and F_{ijjk}, so the system of equations (3.16) is not closed. We propose to close the system by calculating $F_{<ij>n}$ and F_{ijjk} or the corresponding non-convective quantities

$$\varrho_{<ij>k} = \int mC_{<i}C_{j>}C_k f d\mathbf{c} \quad \text{and} \quad \varrho_{ijjk} = \int mC_iC^2C_k f d\mathbf{c} \tag{3.32}$$

from Grad's 13-moment distribution (3.27). Thus we obtain

$$\varrho_{<ij>k} = \frac{2}{5}\left(q_i\delta_{jk} + q_j\delta_{ik} - \frac{2}{3}q_k\delta_{ij}\right), \quad \varrho_{ijjk} = 5p\frac{kT}{m}\delta_{ij} - \frac{7}{2}t_{<pq>}(\delta_{ip}\delta_{jq} + \delta_{iq}\delta_{jp}). \tag{3.33}$$

When these expressions are introduced into $(3.16)_{3,4}$ and when all the other moments $\mathbf{F}$ occurring in those equations are replaced by the central moment ϱ as dictated by (3.8) we obtain after some trivial calculation

$$t_{<ij>} = \frac{4}{3}\frac{p}{\varrho}\frac{1}{B}\left\{\frac{\partial v_{<i}}{\partial x_{j>}} - \frac{1}{2p}\left[\dot{t}_{<ij>} + 2t_{<n<i>}\frac{\partial v_{j>}}{\partial x_n} - 4t_{<n<i>}W_{j>n}\right] + \frac{2}{5p}\frac{\partial q_{<i}}{\partial x_{j>}}\right\} \boxed{-\frac{2}{3\varrho B}t_{<ij>}\frac{\partial v_n}{\partial x_n},}$$

$$q_i = -\frac{5}{2}\frac{k}{m}\frac{p}{\varrho}\frac{1}{B}\left\{\frac{\partial T}{\partial x_i} + \frac{2}{5}\frac{m}{k}\frac{1}{p}\left[\dot{q}_i + q_n\frac{\partial v_i}{\partial x_n} - 2q_nW_{in}\right] + \frac{2}{5}\frac{T}{p}\frac{\partial t_{<in>}}{\partial x_n}\right\}$$
$$\boxed{-\frac{1}{\varrho B}\left(\frac{7}{5}q_i\frac{\partial v_n}{\partial x_n} + \frac{4}{5}q_n\frac{\partial v_{(i}}{\partial x_{n)}} - \frac{7}{2}\frac{k}{m}t_{<iq>}\frac{\partial T}{\partial x_q} - \frac{1}{\varrho}t_{<ik>}\frac{\partial t_{<kn>}}{\partial x_n} + \frac{1}{\varrho}t_{<ik>}\frac{\partial p}{\partial x_k}\right).} \tag{3.34}$$

Comparison of these equations with the equations $(2.13)_{1,2}$ reveals enough similarity to justify the suggestion that the balance laws $(3.16)_{3,4}$, from which we have derived (3.34), correspond to the phenomenological relations of extended TIP. Of course, there is no counterpart in (2.13) to the framed terms in (3.34); TIP would have neglected such terms as non-linear anyhow, so let us ignore them in the comparison. This leaves only the curly brackets on the right-hand sides of (3.34).

We introduce

$$\mu = \frac{2}{3}\left(\frac{k}{m}T\right)\frac{1}{B} \quad \text{and} \quad \kappa = \frac{5}{2}\frac{k}{m}\left(\frac{k}{m}T\right)\frac{1}{B} \tag{3.35}$$

into (3.34) and the special values (3.31) into (2.13) to obtain a significant similarity between the two sets of equations. Indeed the only difference is that $\dot{q}_i$ and $\dot{t}_{<ij>}$ in (2.13) are now replaced by

$$\dot{q}_i + \underbrace{q_n\frac{\partial v_i}{\partial x_n} - 2q_nW_{in}} \quad \text{and} \quad \dot{t}_{<ij>} + \underbrace{2t_{<n<i>}\frac{\partial v_{j>}}{\partial x_n} - 4t_{<n<i>}W_{j>n}}. \tag{3.36}$$

Thus the kinetic theory suggests that the non-invariance of the equations (2.13) that was discussed in Paragraph 2.4 must be mended by joining the braced terms in (3.36) to $\dot{q}_i$ and $\dot{t}_{<ij>}$. It is indeed true that the sum of the three non-objective terms in $(3.36)_{1,2}$ form an objective vector and an objective tensor, respectively.

There is no way in extended TIP that would lead us to conjecture the presence of the expressions (3.36) in the phenomenological relations. Indeed, had the non-invariance been noticed in extended TIP, it would probably have been mended by replacing $\dot{q}_i$ and $\dot{t}_{<ij>}$ by the objective derivatives

$$\dot{q}_i - q_n \frac{\partial v_i}{\partial x_n} \quad \text{and} \quad \dot{t}_{<ij>} - 2t_{<n<i>} \frac{\partial v_{j>}}{\partial x_n}, \tag{3.37}$$

the so-called Jaumann derivatives, but the kinetic theory insists on inertial terms in (3.34), namely, the terms with $\mathbf{W}$. Later, in Chapter 2, we shall discuss those terms and find them eminently sensible.

3.7 Pulse Speeds

If we introduce the specific values (3.31) for a, b, and K into the general expressions for the pulse speeds $\overset{\text{II}}{V}$ and $\overset{\text{III}}{V}$ in (2.22) and (2.25), we obtain

$$\begin{array}{ll} \text{for the transversal shear wave} & \overset{\text{II}}{V} = \sqrt{\frac{7}{5}\frac{k}{m}T}, \\ \text{for the longitudinal sound wave} & \overset{\text{III}}{V} = \sqrt{5.18\frac{k}{m}T}. \end{array} \tag{3.38}$$

3.8 Conclusions

We have learned one thing above all from the kinetic theory of gases, namely, that the phenomenological relations of extended TIP are, in fact, equations of balance. Indeed, the system (1.2), (2.13) of extended TIP assumes the elegant form (3.16) in the kinetic theory in which all equations are equations of balance with

$$\begin{array}{ccc} \text{densities} & \text{fluxes} & \text{productions} \\ \varrho, \varrho v_i, F_{ij}, F_{ijj} & \varrho v_i, F_{ij}, F_{ijk}, F_{ijjk} & S_{<ij>}, S_{ijj}. \end{array} \tag{3.39}$$

Also in the kinetic theory the densities, fluxes, and productions depend algebraically on the variables

$$\varrho, \ v_i, \ \epsilon, \ t_{<ij>}, \ q_i, \tag{3.40}$$

which we may consider as the basic fields to be determined by thermodynamics.

Moreover, there is an entropy inequality of balance form, viz. by (3.19), (3.20)

$$\frac{\partial \varrho s}{\partial t} + \frac{\partial h_i}{\partial x_i} = \Sigma \geq 0 \tag{3.41}$$

in which the density ϱs and the flux h_i also depend algebraically on the variables (3.40).

These observations will guide us, in Chapter 2, in the formulation of the systematic theory of extended thermodynamics, which will be the basis of the mathematical theory of extended thermodynamics that is explained in Chapter 6.

Chapter 2

Extended Thermodynamics of Mon-Atomic Gases

The new version of extended thermodynamics to be described in this chapter is based on Rational Thermodynamics whereas the old version is based on TIP. There are several ingredients that make rational thermodynamics more powerful than TIP, viz.

i. the explicit recognition that the entropy inequality must hold for all solutions of the field equations and that the constitutive relations must ensure this validity; this was clearly expressed by Coleman and Mizel [16].

ii. the realization that the entropy flux is a constitutive quantity in its own right unrelated, a priori, to the heat flux divided by temperature; this was proposed by Müller [17].

iii. the use of Lagrange multipliers in the exploitation of the entropy inequality as proved by Liu [18].

iv. the recognition that first-order field equations of balance type form a symmetric hyperbolic system, if the Lagrange multipliers are chosen as variables; this was shown by Ruggeri [19].

After all this was known, it was possible to revisit the problem of the formulation of extended thermodynamics in a systematic way. Liu and Müller [20] formulated such a theory for mon-atomic gases. The theory considers the thirteen variables ϱ, v_i, ϵ, $t_{<ij>}$, and q_i as basic and derives field equations from the conservation laws of mass, momentum energy, and from balance equations for the fluxes of momentum and energy. It is capable of calculating specific forms for all constitutive coefficients except two, and those can be related to the measurable values of viscosity and heat conductivity.

Full agreement is achieved with the results of the kinetic theory of gases in the Grad 13-moment approximation even though extended thermodynamics is a macroscopic phenomenological theory with no explicit reference to the kinetic theory in its tenets. In particular, the entropy and its flux are proved to contain the non-linear terms in $t_{<ij>}$ and q_i postulated by extended TIP and the corresponding coefficients are specific.

Extended thermodynamics also derives the inertial terms in the equations for stress and heat flux, which we noted in the kinetic theory.

The theory applies to all monatomic gases with a thermal equation of state of the form

$$p = T^{5/2} F\left(\frac{\varrho}{T^{3/2}}\right),$$

where F is some function. This includes the classical ideal gas as well as degenerate gases of Bose and Fermi particles.

1 The Equations of Extended Thermodynamics of Mon-Atomic Gases

1.1 Thermodynamic Processes

Extended thermodynamics is a field theory whose principal objective is the determination of the thirteen fields of densities

$$\begin{aligned} \text{mass density} \quad & F = \varrho, \\ \text{momentum density} \quad & F_i = \varrho v_i, \\ \text{momentum flux density} \quad & F_{ij}, \\ \text{energy flux density} \quad & \tfrac{1}{2}F_{ppi}.^1 \end{aligned} \tag{1.1}$$

The necessary field equations are based on the equations of balance of mass, momentum, momentum flux, and energy flux, viz.

$$\begin{aligned} \frac{\partial F}{\partial t} + \frac{\partial F_k}{\partial x_k} &= 0, \\ \frac{\partial F_i}{\partial t} + \frac{\partial F_{ik}}{\partial x_k} &= 0, \\ \frac{\partial F_{ij}}{\partial t} + \frac{\partial F_{ijk}}{\partial x_k} &= S_{<ij>}, \\ \frac{\partial F_{ppi}}{\partial t} + \frac{\partial F_{ppik}}{\partial x_k} &= S_{ppi}. \end{aligned} \tag{1.2}$$

All tensors are symmetric and $S_{<ij>}$ is traceless, because the trace of $(1.2)_3$ represents the conservation law of energy.

In order to close this system we need constitutive relations for the quantities

$$F_{<ijk>}, \;\; F_{ppik}, \;\; S_{<ij>}, \;\; S_{ppi}. \tag{1.3}$$

$F_{<ijk>}$ is the trace-less part of F_{ijk}. In extended thermodynamics the constitutive quantities (1.3) at one point and time depend on the values of the fields (1.1) at that point and time, so we have

$$\begin{aligned} F_{<ijk>} &= \hat{F}_{<ijk>} \, (F, F_i, F_{ij}, F_{ppi}), \\ F_{ppik} &= \hat{F}_{ppik} \, (\underline{\qquad\qquad}), \\ S_{<ij>} &= \hat{S}_{<ij>} \, (\underline{\qquad\qquad}), \\ S_{ppi} &= \hat{S}_{ppi} \, (\underline{\qquad\qquad}). \end{aligned} \tag{1.4}$$

If we know the constitutive functions $\hat{F}_{<ijk>}$ through $\hat{S}_{ppi}$, we can eliminate $F_{<ijk>}$, F_{ppik}, $S_{<ij>}$, and S_{ppi} between (1.3) and (1.4) and obtain a system of field equations. This is a system of quasi-linear first-order differential equations. Each solution is called a *thermodynamic process.*

1.2 Discussion

The balance equations (1.2) are rather special in several ways. First, the flux in one equation appears as the density in the subsequent one. In particular, this means that the energy density e is related to the trace F_{ii} of the momentum flux by $e = \frac{1}{2}F_{ii}$. This is a relation which we know to be true for mon-atomic gases but false otherwise. Second, the symmetry of all tensors is a restrictive

[1] We use this synthetic notation in order to illustrate the simple structure of the theory. The notation recalls that in the kinetic theory the fields (1.1) are moments, although that fact is not used explicitly in this chapter.

assumption. It may not hold true for real gases and liquids, in particular, in regard to F_{ijk}. Third, body forces and radiation have been ignored in the formulation of the balance equations (1.2), and the equations are written in an inertial frame so that there are no inertial contributions. The form of such inertial contributions can be determined after the velocity dependence of the fluxes, as dictated by Galilean invariance, has been established. We proceed to do this.

1.3 Galilean Invariance: Convective and Non-Convective Fluxes

We require that the balance laws (1.2) hold in every inertial frame so that they must be invariant under a Galilean transformation, that is, a time-independent rotation O_{ij} and a translation with the constant velocity V_i, viz.

$$x_i^* = O_{ij}x_j + V_i t. \tag{1.5}$$

This requirement will permit us to determine the velocity dependence of the **F**'s and **S**'s explicitly. We shall prove that the following decompositions are valid:

$$\begin{aligned}
F &= \varrho \\
F_i &= \varrho v_i \\
F_{ij} &= \varrho_{ij} + \varrho v_i v_j \\
F_{ijk} &= \varrho_{ijk} + 3\varrho_{(ij}v_{k)} + \varrho v_i v_j v_k \\
F_{ppij} &= \varrho_{ppij} + 4\varrho_{(ijp}v_{p)} + 6\varrho_{(ip}v_p v_{j)} + \varrho v^2 v_i v_j \\
S_{\langle ij\rangle} &= s_{\langle ij\rangle} \\
S_{ppi} &= s_{ppi} + 2s_{\langle ip\rangle}v_p.
\end{aligned} \tag{1.6}$$

The quantities ϱ and **s** in (1.6) are Galilean tensors so that their components in the two inertial frames are related by the equations

$$\overset{*}{\varrho}_{i_1 i_2\cdots i_n} = O_{i_1 j_1}\cdots O_{i_n j_n}\varrho_{j_1\cdots j_n} \quad \text{and} \quad \overset{*}{s}_{i_1\cdots i_n} = O_{i_1 j_1}\cdots O_{i_n j_n}s_{j_1\cdots j_n}. \tag{1.7}$$

The general proof of the decomposition (1.6) is given in Chapter 6—for systems of an arbitrary number of equations—following Ruggeri [21]. Here we give a short proof appropriate for the simple system of equations (1.2).

The relations $(1.6)_{1,2}$ require no proof; after all, F and F_i are the densities of mass and momentum. For the proof of the other relations (1.6) it is convenient to decompose the **F**'s into convective and non-convective parts according to

$$F_{i_1\cdots i_n k} = F_{i_1\cdots i_n}v_k + G_{i_1\cdots i_n k}. \tag{1.8}$$

For $n = 0$ we have

$$G_k = 0, \tag{1.9$_1$}$$

because there is no non-convective flux of mass nor is there a momentum density in the rest frame. Therefore the mass balance $(1.2)_1$ may be written in the form

$$\dot{\varrho} + \varrho\frac{\partial v_k}{\partial x_k} = 0. \tag{1.10$_1$}$$

All terms are Galilean scalars so that this equation holds in all inertial frames. By use of $(1.9)_1$, $(1.10)_1$ the momentum balance $(1.2)_2$ assumes the form

$$\varrho\dot{v}_i + \frac{\partial G_{ik}}{\partial x_k} = 0. \tag{1.10$_2$}$$

The first term is a Galilean vector, so G_{ik} must be a Galilean tensor, if the momentum balance is to be valid in all inertial frames. We emphasize this conclusion in the notation by writing

$$G_{ik} = \varrho_{ik}. \tag{1.9$_2$}$$

By use of $(1.9)_{1,2}$ and $(1.10)_{1,2}$ we write the momentum flux balance $(1.2)_3$ in the form

$$\varrho\left[\left(\frac{\varrho_{ij}}{\varrho}\right)^{\bullet}+2\frac{\varrho_{k(i}}{\varrho}\frac{\partial v_{j)}}{\partial x_k}\right]+\frac{\partial(G_{ijk}-2v_{(i}\varrho_{j)k})}{\partial x_k}=S_{<ij>}. \qquad (1.10)_3$$

The square bracket is a Galilean tensor, so $G_{ijk}-2v_{(i}\varrho_{j)k}$ and $S_{<ij>}$ must also be Galilean tensors, if Galilean invariance of the balance law is required. We therefore write

$$G_{ijk}=\varrho_{ijk}+2v_{(i}\varrho_{j)k} \quad \text{and} \quad S_{<ij>}=s_{<ij>}. \qquad (1.9)_3$$

By use of $(1.9)_{1,2,3}$ and $(1.10)_{1,2,3}$ the energy flux balance $(1.2)_4$ may be rewritten as

$$\varrho\left[\left(\frac{\varrho_{ppi}}{\varrho}\right)^{\bullet}+\frac{\varrho_{ppk}}{\varrho}\frac{\partial v_i}{\partial x_k}\right]+2\varrho_{ijk}\frac{\partial v_j}{\partial x_k}-2\frac{\varrho_{ij}}{\varrho}\frac{\partial\varrho_{jk}}{\partial x_k}-\frac{\varrho_{pp}}{\varrho}\frac{\partial\varrho_{ik}}{\partial x_k}$$

$$+\frac{\partial\left(G_{ppik}-2\varrho_{kij}v_j-\varrho_{ppk}v_i-\varrho_{ki}v^2-\varrho_{kl}v_lv_i\right)}{\partial x_k}=S_{ppi}-2S_{<ij>}v_j. \qquad (1.10)_4$$

Once again, the first line is a Galilean vector. Therefore, Galilean invariance of the balance requires that $G_{ppik}-2\varrho_{kij}v_j-\varrho_{ppk}v_i-\varrho_{ki}v^2-\varrho_{kl}v_lv_i$ and $S_{ppi}-2S_{<ij>}v_j$ be Galilean tensors. We denote them by ϱ_{ppik} and s_{ppi} and have

$$G_{ppik}=\varrho_{ppik}+2\varrho_{kij}v_j+\varrho_{ppk}v_i+\varrho_{ki}v^2+\varrho_{kl}v_lv_i \quad \text{and} \quad S_{ppi}=s_{ppi}+2S_{<ij>}v_j. \qquad (1.9)_4$$

The combination of (1.8) with the relations $(1.9)_{1-4}$ provides the decomposition (1.6) after an easy rearrangement.

Some of the ϱ's are well-known quantities in continuum mechanics and thermodynamics. For those we shall replace the synthetic notation in much of the sequel and use the canonical letters. Thus

$$t_{ij}=-\varrho_{ij}, \qquad \varrho\epsilon=\frac{1}{2}\varrho_{ii}, \qquad q_i=\frac{1}{2}\varrho_{ppi} \qquad (1.11)$$

are the stress, the internal energy density, and the heat flux. The pressure p is defined as $-\frac{1}{3}t_{ii}$, and therefore from $(1.11)_{1,2}$ we conclude

$$p=\frac{2}{3}\varrho\epsilon, \qquad (1.12)$$

a relation that is typical for mon-atomic gases.

1.4 Euclidean Invariance: Inertial Effects

We require that the fluxes ϱ and productions $\mathbf{s}$ introduced in (1.6) are objective tensors rather than merely Galilean tensors; that is, in an Euclidean transformation

$$x_i^*=O_{ij}(t)x_j+b_i(t) \qquad (1.13)$$

they transform according to the rule

$$\varrho^*_{i_1\cdots i_n}=O_{i_1j_1}\cdots O_{i_nj_n}\varrho_{j_1\cdots j_n} \quad \text{and} \quad s^*_{i_1\cdots i_n}=O_{i_1j_1}\cdots O_{i_nj_n}s_{j_1\cdots j_n}. \qquad (1.14)$$

The Euclidean transformation represents a change of frame from an inertial, or Galilean frame to a non-inertial, or Euclidean frame. $\mathbf{O}(t)$ is a time-dependent rotation and $\mathbf{b}(t)$ describes the time-dependent relative translation of the frames.

We shall prove that in a non-inertial frame the equations of balance (1.2) assume the following forms:

$$\begin{aligned}
&\frac{\partial F^*}{\partial t^*} + \frac{\partial F_k^*}{\partial x_k^*} &&= 0,\\
&\frac{\partial F_i^*}{\partial t^*} + \frac{\partial F_{ik}^*}{\partial x_k^*} - F^* i_i^0 - 2F_k^* W_{ik} &&= 0,\\
&\frac{\partial F_{ij}^*}{\partial t^*} + \frac{\partial F_{ijk}^*}{\partial x_k^*} - 2F_{(i}^* i_{j)}^0 - 4F_{k(i}^* W_{j)k} &&= S_{\langle ij\rangle}^*,\\
&\frac{\partial F_{ppi}^*}{\partial t^*} + \frac{\partial F_{ppik}^*}{\partial x_k^*} - 3F_{(ij}^* i_{j)}^0 - 6F_{k(ij}^* W_{j)k} &&= S_{ppi}^*.
\end{aligned} \tag{1.15}$$

$\mathbf{W} = \dot{\mathbf{O}}\mathbf{O}^T$ is the matrix of angular velocity of the non-inertial frame with respect to an inertial one, and $\mathbf{i}^0$ is the velocity-independent part of the specific inertial force, namely,

$$i_i^0 = -2W_{ik}\dot{b}_k - W_{ik}^2(x_k^* - b_k) + \dot{W}_{ik}(x_k^* - b_k) + \ddot{b}_i. \tag{1.16}$$

Note that, except for the mass balance, all balance equations have inertial contributions. The inertial force in $(1.15)_2$ is well known; it consists of the Coriolis, centrifugal, and Euler forces and of the force of relative translation.

For the proof of the relations (1.15) we need to realize that by (1.13) the velocity transforms according to the rule

$$v_i = O_{ji}(v_j^* - R_j) \quad \text{where } R_j = W_{jk}(x_k^* - b_k) + \dot{b}_j. \tag{1.17}$$

It is then easy to show from (1.6) that the fluxes $\mathbf{F}$ and productions $\mathbf{S}$ transform as follows.

$$\begin{aligned}
F &= F^*,\\
F_i &= O_{ri}(F_r^* - F^* R_r),\\
F_{ij} &= O_{ri}O_{sj}(F_{rs}^* - 2F_{(r}^* R_{s)} + F^* R_r R_s),\\
F_{ijk} &= O_{ri}O_{sj}O_{tk}(F_{rst}^* - 3F_{(rs}^* R_{t)} + 3F_{(r}^* R_s R_{t)} - F^* R_r R_s R_t),\\
F_{ppij} &= O_{ri}O_{sj}(F_{pprs}^* - 4F_{(rsp}^* R_{p)} + 6F_{(rp}^* R_p R_{s)} + 6F_{(r}^* R_s R_p R_{p)} + F^* R_r R_s R_p R_p),\\
S_{\langle ij\rangle} &= O_{ri}O_{sj} S_{\langle rs\rangle}^*,\\
S_{ppi} &= O_{ri}(S_{ppr}^* - 2S_{(rp)}^* R_p).
\end{aligned} \tag{1.18}$$

Space- and time-derivatives in the two frames are related by the equations

$$\frac{\partial}{\partial x_k} = O_{nk}\frac{\partial}{\partial x_n^*} \quad \text{and} \quad \frac{\partial}{\partial t} = \frac{\partial}{\partial t^*} + \left(W_{lj}(x_j^* - b_j) + \dot{b}_l\right)\frac{\partial}{\partial x_l^*}. \tag{1.19}$$

We insert the transformation rules (1.18) and (1.19) into the equations of balance (1.2) and obtain the equations of balance (1.15) appropriate to a non-inertial frame. The calculation is straightforward; we need only realize that we have to set

$$\frac{\partial R_j}{\partial x_k^*} = W_{jk} \quad \text{and} \quad \frac{\partial R_j}{\partial t^*} = -2W_{jk}\dot{b}_k - W_{jk}^2(x_k^* - b_k) + \dot{W}_{jk}(x_k^* - b_k) + \ddot{b}_j. \tag{1.20}$$

2 Constitutive Theory

2.1 Restrictive Principles

We recall that the system of balance equations (1.2) is closed by the constitutive equations (1.4). Thus if the constitutive functions $\hat{\mathbf{F}}$ and $\hat{\mathbf{S}}$ were known, we would be able, in principle, to calculate thermodynamic processes, that is, solutions of given initial and boundary value problems.

In reality, of course, it is difficult to know the constitutive relations, and therefore the largest part of the work of thermodynamicists consists of attempts to restrict the generality of the constitutive functions. Ideally those functions might be reduced to some scalar coefficients that the experimentalist could measure. Such efforts form the subject of the thermodynamic constitutive theory.

The tools of the constitutive theory are universal physical principles, which we rely on from long experience. The most important of these principles are

i. the entropy principle,

ii. the principle of relativity (material frame indifference), and

iii. the requirement of convexity and causality.

We proceed to formulate and exploit these requirements.

The *entropy principle* provides the balance equation

$$\frac{\partial h}{\partial t} + \frac{\partial h_i}{\partial x_i} = \Sigma \geq 0, \tag{2.1}$$

which must be satisfied for all thermodynamic processes, that is, solutions of the field equations that result from (1.2), (1.4). The entropy density h and the entropy flux h_i are constitutive quantities, and we write generic constitutive relations of the form (1.4), that is,

$$\begin{aligned} h &= \hat{h}\ (F, F_i, F_{ij}, F_{ppi}), \\ h_i &= \hat{h}_i\ (\underline{\qquad\qquad}). \end{aligned} \tag{2.2}$$

The entropy density is non-convective. The entropy flux is decomposed into a convective and a non-convective part,

$$h_i = hv_i + \varphi_i. \tag{2.3}$$

Thus we may write the inequality (2.1) in the form

$$\varrho \left(\frac{h}{\varrho}\right)^{\bullet} + \frac{\partial \varphi_i}{\partial x_i} = \Sigma \geq 0. \tag{2.4}$$

We assume that this inequality holds in every Euclidean frame and conclude that h and Σ are objective scalars while φ_i is an objective vector.

By (1.6) and (2.3) we have a one-to-one correspondence between F, F_i, F_{ij}, F_{ijk}, F_{ppik}, $S_{<ij>}$, S_{ppi}, h, and h_i and the corresponding quantities ϱ, v_i, ϱ_{ij}, ϱ_{ijk}, ϱ_{ppik}, $s_{\langle ij\rangle}$, s_{ppi}, h, and φ_i, which are Galilean tensors—or even objective tensors. We may therefore write the constitutive relations (1.4), (2.2) in the form

$$\begin{aligned} \varrho_{<ijk>} &= \hat{\varrho}_{<ijk>}\ (\varrho, v_i, \varrho_{ij}, \varrho_{ppi}), \\ \varrho_{ppik} &= \hat{\varrho}_{ppik}\ (\underline{\qquad\qquad}), \\ s_{<ij>} &= \hat{s}_{<ij>}\ (\underline{\qquad\qquad}), \\ s_{ppi} &= \hat{s}_{ppi}\ (\underline{\qquad\qquad}), \\ h &= \hat{h}\ (\underline{\qquad\qquad}), \\ \varphi_i &= \hat{\varphi}_i\ (\underline{\qquad\qquad}). \end{aligned} \tag{2.5}$$

These are the constitutive relations in the form appropriate to an observer at rest in an inertial frame. An observer in a non-inertial frame sees different values of the dependent variables, viz. $\varrho^*_{<ijk>}$ through φ^*_i and of the independent variables, viz. $\varrho, v^*_i, \varrho^*_{ij}, \varrho^*_{ppi}$, but the constitutive *functions* $\hat{\varrho}_{<ijk>}$ through $\hat{\varphi}_i$ are unchanged by the change of frame. This assumption is known as the principle of *material frame indifference*. Formally we express this principle by writing

$$
\begin{array}{lll}
\varrho^*_{<ijk>} &= \hat{\varrho}_{<ijk>} & (\varrho, v^*_i, \varrho^*_{ij}, \varrho^*_{ppi}), \\
\varrho^*_{ppik} &= \hat{\varrho}_{ppik} & (\text{———————}), \\
s^*_{<ij>} &= \hat{s}_{<ij>} & (\text{———————}), \\
s^*_{ppi} &= \hat{s}_{ppi} & (\text{———————}), \\
h &= \hat{h} & (\text{———————}), \\
h^*_i &= \hat{\varphi}_i & (\text{———————}).
\end{array}
\tag{2.6}
$$

Note that the constitutive functions $\hat{\varrho}_{<ijk>}$ through $\hat{\varphi}_i$ are the same ones in (2.5) and (2.6).

The *requirement of convexity and causality* demands that $-h$ is a convex function of the densities F, F_i, F_{ij}, F_{ppi} so that the matrix of second derivatives of h is negative definite. This requirement is eminently sensible both mathematically and physically. Mathematically it ensures well-posedness of Cauchy problems for the field equations and physically it guarantees that all wave speeds are finite and that thermodynamic stability holds.

While the principles of entropy and material frame indifference are common ingredients in thermodynamics and continuum mechanics generally, the requirement of convexity and causality is particular to extended thermodynamics. We shall discuss its general consequences in Chapter 6 where we describe the mathematical theory of extended thermodynamics. Here we observe only that for causality, that is, finite speeds, it would have been sufficient to require hyperbolic field equations. Convexity strengthens this demand, because it makes the field equations *symmetric* hyperbolic and, thus, implies well-posedness and thermodynamic stability.

2.2 Exploitation of the Principle of Material Frame Indifference

We recall that the conditions (2.5), (2.6) refer to Euclidean transformations

$$x^*_i = O_{ij}(t)x_j + b_i(t)$$

between an inertial frame and a non-inertial one. For the exploitation we proceed in two steps.

First, we consider translations with $O_{ij} = \delta_{ij}$. In that case the conditions (2.5), (2.6) may be combined to read

$$\hat{C}(\varrho, v_i, \varrho_{ij}, \varrho_{ppi}) = \hat{C}(\varrho, v_i + \dot{b}_i, \varrho_{ij}, \varrho_{ppi}), \tag{2.7}$$

where $\hat{C}$ stands for any one of the constitutive functions. Since this relation must hold for all $\dot{b}_i$, we conclude that none of the constitutive quantities may depend on v_i.

Second, we consider arbitrary Euclidean transformations and make use of the transformation properties of objective tensors to combine the equations (2.5), (2.6) in the form of the following functional relations:

$$
\begin{array}{lllll}
O_{ir}O_{js}O_{kt} & \hat{\varrho}_{<rst>} & (\varrho, \varrho_{ij}, \varrho_{ppi}) &= \hat{\varrho}_{<ijk>} & (\varrho, O_{ir}O_{js}\varrho_{rs}, O_{ir}\varrho_{ppr}), \\
O_{ir}O_{ks} & \hat{\varrho}_{pprs} & (\text{—————}) &= \hat{\varrho}_{ppik} & (\text{——————————}), \\
O_{ir}O_{ks} & \hat{s}_{<rs>} & (\text{—————}) &= \hat{s}_{<ik>} & (\text{——————————}), \\
O_{ir} & \hat{s}_{ppr} & (\text{—————}) &= \hat{s}_{ppi} & (\text{——————————}), \\
 & \hat{h} & (\text{—————}) &= \hat{h} & (\text{——————————}), \\
O_{ir} & \hat{\varphi}_r & (\text{—————}) &= \hat{\varphi}_r & (\text{——————————}).
\end{array}
\tag{2.8}
$$

We express this in words by saying that the constitutive functions are isotropic functions of their variables.

There are representation theorems for isotropic functions (e.g., see [22]). Here we are not interested in those because we shall be content with a linear constitutive theory in which the ϱ's and s's are linear functions of the stress deviator $-\varrho_{<ij>}$ and of the heat flux $\frac{1}{2}\varrho_{ppi}$. The representations read in this case

$$\begin{aligned} \varrho_{<ijk>} &= 0, \\ \varrho_{ppij} &= \alpha\delta_{ij} + \beta\varrho_{<ij>}, \\ s_{<ij>} &= -\frac{1}{\varsigma_t}\varrho_{<ij>}, \\ s_{ppi} &= -\frac{1}{\varsigma_q}\varrho_{ppi}. \end{aligned} \tag{2.9}$$

Note that there is no way to construct an isotropic function for the trace-less tensor $\varrho_{<ijk>}$ from $\varrho_{<ij>}$ and ϱ_{ppi}, if we require linear dependence on these variables. The coefficients α through ς_q in (2.9) may be functions of ϱ and ϱ_{pp}. The entropy principle will impose restrictions on these functions.

2.3 Exploitation of the Entropy Principle

2.3.1 Six Steps

The exploitation of the entropy principle is the cornerstone of the constitutive theory of extended thermodynamics. It requires careful and cumbersome calculations and a certain ingenuity in finding the best, that is, most time-saving path. In order to give the argument some structure, we identify seven steps, viz.

i. Introduction of Lagrange multipliers;
ii. Removal of velocity-dependent terms;
iii. Lagrange multipliers as variables. The scalar potential and the vector potential;
iv. Equilibrium;
v. Determination of near-equilibrium potentials;
vi. Residual inequality; and
vii. Summary of results.

2.3.2 Lagrange Multipliers: Step i

The key to the evaluation of the entropy principle is the statement that the entropy inequality (2.1) must hold for all solutions of the field equations rather than for all fields F, F_i, F_{ij}, F_{ppi}. In a manner of speaking the field equations provide constraints for the fields that satisfy the inequality. Liu has proved in [18] that these constraints may be taken into account by the use of Lagrange multipliers, so the new inequality

$$\begin{aligned} \frac{\partial h}{\partial t} + \frac{\partial h_i}{\partial x_i} - \Lambda & \left(\frac{\partial F}{\partial t} + \frac{\partial F_k}{\partial x_k}\right) \\ - \Lambda_i & \left(\frac{\partial F_i}{\partial t} + \frac{\partial F_{ik}}{\partial x_k}\right) \\ - \Lambda_{ij} & \left(\frac{\partial F_{ij}}{\partial t} + \frac{\partial F_{ijk}}{\partial x_k} - S_{<ij>}\right) \\ - \Lambda_{qqi} & \left(\frac{\partial F_{ppi}}{\partial t} + \frac{\partial F_{ppik}}{\partial x_k} - S_{ppi}\right) \geq 0 \end{aligned} \tag{2.10}$$

must hold for *all* fields F through F_{ppi}. The $\mathbf{\Lambda}$'s are Lagrange multipliers and according to Liu's proof they may be functions of F, F_i, F_{ij}, F_{ppi} whose form depends on material.

We insert the constitutive relations (1.4) and (2.2) for $h, h_i, F_{<ijk>}$, and F_{ppik} and obtain an expression that is explicitly linear in the derivatives

$$\frac{\partial F}{\partial t}, \frac{\partial F_i}{\partial t}, \frac{\partial F_{ij}}{\partial t}, \frac{\partial F_{ppi}}{\partial t}, \frac{\partial F}{\partial x_k}, \frac{\partial F_i}{\partial x_k}, \frac{\partial F_{ij}}{\partial x_k}, \frac{\partial F_{ppi}}{\partial x_k}. \tag{2.11}$$

The inequality must be satisfied for arbitrary values of these derivatives, and therefore their contributions to the inequality must vanish. Thus we obtain

$$\begin{aligned} dh &= \Lambda dF + \Lambda_i dF_i + \Lambda_{ij} dF_{ij} + \Lambda_{ppi} dF_{qqi}, \\ dh_k &= \Lambda dF_k + \Lambda_i dF_{ik} + \Lambda_{ij} dF_{ijk} + \Lambda_{ppi} dF_{qqik}. \end{aligned} \tag{2.12}$$

There remains the residual inequality

$$\Sigma = \Lambda_{ij} S_{<ij>} + \Lambda_{ppi} S_{ppi} \geq 0, \tag{2.13}$$

which represents the non-negative entropy production.

2.3.3 Removal of Velocity-Dependent Terms: Step ii

There is a certain redundance in the conditions (2.12) and (2.13), because by (1.6) and (2.3) the $\mathbf{v}$-dependence of the $\mathbf{F}$'s and of h_i is explicitly known and that knowledge may be used to simplify (2.12), (2.13). After an easy but fairly long calculation we obtain

$$\begin{aligned} dh &= \lambda d\varrho \quad + \lambda_{ij} d\varrho_{ij} + \lambda_{qqi} d\varrho_{ppi}, \\ d\varphi_k &= \quad \lambda_i d\varrho_{ik} + \lambda_{ij} d\varrho_{ijk} + \lambda_{qqi} d\varrho_{ppik}; \end{aligned} \tag{2.14}$$

$$\begin{aligned} \lambda_t \varrho \quad &+ \lambda_{qqi}(2\varrho_{it} + \varrho_{pp}\delta_{it}) = 0, \\ 2\lambda_{tj}\varrho_{jk} &+ \lambda_{qqi}(2\varrho_{itk} + \varrho_{ppk}\delta_{it}) = [h - \lambda\varrho - \lambda_{ij}\varrho_{ij} - \lambda_{qqi}\varrho_{ppi}]\,\delta_{tk}; \end{aligned} \tag{2.15}$$

and

$$\Sigma = \lambda_{ij} s_{<ij>} + \lambda_{qqi} s_{ppi} \geq 0. \tag{2.16}$$

The λ's introduced here are the velocity-independent parts of the Lagrange multipliers. They are related to the previous $\mathbf{\Lambda}$'s by the equations

$$\begin{aligned} \lambda &= \Lambda + \Lambda_i v_i + \Lambda_{ij} v_i v_j + \Lambda_{ppi} v^2 v_i, \\ \lambda_i &= \quad \Lambda_i + 2\Lambda_{ij} v_j + \Lambda_{ppj}(v^2 \delta_{ji} + 2 v_j v_i), \\ \lambda_{ij} &= \quad \Lambda_{ij} + \Lambda_{ppt} 3 v_{(t}\delta_{ij)}, \\ \lambda_{ppi} &= \quad \Lambda_{ppi}. \end{aligned} \tag{2.17}$$

Since by (2.14) the λ's are derivatives of the objective scalar h and of the objective vector φ_k with respect to the objective quantities ϱ, they are themselves objective tensors. For the same reason the λ's must be isotropic functions of $\varrho, \varrho_{ij}, \varrho_{ppi}$, since h and φ_k are such functions.

The conditions (2.15) result from the fact that h and φ_k are both independent of the velocity. Note that the velocity occurs nowhere in (2.14) through (2.16).

This step of removing the velocity dependence in the equations (2.12) through (2.17) is made in a *systematic manner* in Chapter 6, Paragraph 2.3.

2.3.4 Lagrange Multipliers as Variables. The Scalar Potential and the Vector Potential: Step iii

It is not a priori obvious but experience has shown that the further calculations are significantly simplified by replacing $\varrho, \varrho_{ij}, \varrho_{ppi}$ as independent variables by the Lagrange multipliers $\lambda, \lambda_{ij}, \lambda_{ppi}$[2]. We introduce the scalar potential h' and the vector potential h'_k as

$$\begin{aligned} h' &= -h \quad + \lambda\varrho + \lambda_{ij}\varrho_{ij} + \lambda_{qqi}\varrho_{ppi}, \\ h'_k &= -\varphi_k \qquad + \lambda_{ij}\varrho_{ijk} + \lambda_{qqi}\varrho_{ppik} \end{aligned} \tag{2.18}$$

and rewrite (2.14) and $(2.15)_2$ in the alternative forms

$$\begin{aligned} dh' &= \varrho d\lambda + \varrho_{rs} d\lambda_{rs} + \varrho_{ppn} d\lambda_{qqn}, \\ dh'_k &= \left[\lambda_{qqe} a_{ei} \frac{\partial \varrho_{ij}}{\partial \lambda}\right] d\lambda + \left[\varrho_{rsk} + \lambda_{qqe} a_{ei} \frac{\partial \varrho_{ik}}{\partial \lambda_{rs}}\right] d\lambda_{rs} + \left[\varrho_{ppnk} + \lambda_{qqe} a_{ei} \frac{\partial \varrho_{ik}}{\partial \lambda_{qqn}}\right] d\lambda_{qqn}, \end{aligned} \tag{2.19}$$

$$2\lambda_{tj}\varrho_{jk} + \lambda_{qqi}(2\varrho_{itk} + \varrho_{ppk}\delta_{it}) = -h'\delta_{tk}. \tag{2.20}$$

Note that λ_t has been eliminated by use of $(2.15)_1$. This elimination has brought the tensor a_{ei} into $(2.19)_2$, which is defined as

$$a_{ei} = \frac{1}{\varrho}(2\varrho_{ei} + \varrho_{pp}\delta_{ei}). \tag{2.21}$$

We conclude from (2.19) and (2.21) that the knowledge of h' and h'_k as functions of the variables $\lambda, \lambda_{ij}, \lambda_{qqi}$ determines $\varrho, \varrho_{ik}, \varrho_{ppk}, \varrho_{<ijk>}$, and ϱ_{ppik} as functions of these variables. For that reason h' and h'_k are properly called potentials.

Expedient as the Lagrange multipliers may be as variables for the purpose of exploiting the entropy principle, however, they are also non-suggestive and non-measurable except for their equilibrium values. Therefore, eventually we shall have to remove them from the results, but first we calculate their equilibrium values.

2.3.5 Equilibrium: Step iv

Equilibrium is a process in which the productions $s_{<ij>}$ and s_{ppi} vanish. Therefore, by (2.16), the entropy production has a minimum in equilibrium, namely, zero; hence, the following are necessary conditions:

$$\lambda_{<ij>}|_{\mathrm{E}} = 0, \qquad \lambda_{qqi}|_{\mathrm{E}} = 0, \tag{2.22}$$

where E denotes equilibrium.

From $(2.14)_1$ we conclude, with $\varrho_{ii} = 2\varrho\epsilon$ (see $(1.11)_2$),

$$dh_{\mathrm{E}} = \frac{2}{3}\lambda_{ii}|_{\mathrm{E}} d(\varrho\epsilon) + \lambda_{\mathrm{E}} d\varrho. \tag{2.23}$$

Comparison with the Gibbs equation identifies $\lambda_{ii}|_{\mathrm{E}}$ and λ_{E} as

$$\lambda_{ii}|_{\mathrm{E}} = \frac{3}{2}\frac{1}{T} \quad \text{and} \quad \lambda_{\mathrm{E}} = -\frac{g}{T}, \tag{2.24}$$

[2] In the mathematical theory of extended thermodynamics this change of variables appears much less artificial than here. Also there, in Chapter 6, it is shown that the change is always possible.

where T is the absolute temperature and $g = \epsilon - T(h_E/\varrho) + (p_E/\varrho)$ is the specific free enthalpy or Gibbs free energy, sometimes called chemical potential.

Thus by (2.22) and (2.24) we have identified the equilibrium values of the Lagrange multipliers. Before we use this knowledge for the further evaluation of the conditions (2.19), (2.20) we determine the consequences of the Gibbs equation. A combination of (2.23) and (2.24) gives

$$dh_E = \frac{1}{T}d(\varrho\epsilon) - \frac{g}{T}d\varrho \quad \text{or} \quad d\left(\frac{h_E}{\varrho}\right) = \frac{1}{T}\left(d\epsilon - \frac{p_E}{\varrho^2}d\varrho\right). \tag{2.25}$$

As is well known, the Gibbs equation implies the integrability conditions

$$\left(\frac{\partial\epsilon}{\partial\varrho}\right)_T = \frac{1}{\varrho^2}\left(p_E - T\left(\frac{\partial p_E}{\partial T}\right)_\varrho\right), \qquad \left(\frac{\partial p_E}{\partial 1/T}\right)_{g/T} = -\frac{5}{2}p_E T, \qquad \left(\frac{\partial p_E}{\partial g/T}\right)_T = \varrho T. \tag{2.26}$$

For a mon-atomic gas with $p_E = \frac{2}{3}\varrho\epsilon$ the first of these conditions may be integrated to give, with $z = \frac{\varrho}{T^{3/2}}$,

$$p_E = T^{5/2}F(z) \quad \text{and} \quad \frac{h_E}{\varrho} = \frac{3}{2}\int\left(\frac{F'}{z} - \frac{5}{3}\frac{F}{z^2}\right)dz + \varsigma. \tag{2.27}$$

Here $F(z)$ is an arbitrary function and ς is a constant of integration. g/T is a function of z given by

$$\frac{g}{T} = \frac{F}{z} + \int\frac{F}{z^2}dz - \varsigma. \tag{2.28}$$

Note that, by $(2.27)_1$, the thermal equation of state $p_E = p(\varrho, T)$ is determined by a function of a *single* variable. We shall discuss special cases later, see Section 4.

2.3.6 The Potentials near Equilibrium: Step v

The potentials h' and h'_k, being defined by (2.18), are isotropic scalar and vectorial functions of $\lambda, \lambda_{ij}, \lambda_{ppi}$. We write their representations in polynomial form in the neighbourhood of equilibrium by omitting higher-order terms in the variables $\lambda_{<ij>}$ and λ_{qqi}, which vanish in equilibrium.

$$h' = h'_E + h'_1\lambda_{<ij>}\lambda_{<ij>} + h'_2\lambda_{ppi}\lambda_{qqi} + h'_3\lambda_{<in>}\lambda_{<nj>}\lambda_{<ji>} + h'_4\lambda_{<kj>}\lambda_{ppk}\lambda_{qqj} + \mathcal{O}(4), \tag{2.29}$$

$$h'_k = \varphi_1\lambda_{ppk} + \varphi_2\lambda_{<ke>}\lambda_{ppe} + \mathcal{O}(3).$$

All coefficients may be functions of $\lambda_E = -g/T$ and $\lambda_{ii}|_E = \frac{3}{2}\frac{1}{T}$. From the definition $(2.18)_1$ of h' it follows that its equilibrium value is given in terms of the equilibrium pressure by

$$h'_E = -\frac{p_E}{T}. \tag{2.30}$$

We exploit the equation $(2.19)_1$ and with a modicum of calculation obtain

$$\begin{aligned}
\varrho &= \frac{\partial h'}{\partial\lambda} = \varrho_E - \frac{\partial h'_1}{\partial g/T}\lambda_{<ij>}\lambda_{<ij>} - \frac{\partial h'_2}{\partial g/T}\lambda_{ppi}\lambda_{qqi} + \mathcal{O}(3),\\
\frac{1}{3}\varrho_{pp} &= \frac{\partial h'}{\partial\lambda_{pp}} = p_E + \frac{2}{3}\frac{\partial h'_1}{\partial 1/T}\lambda_{<ij>}\lambda_{<ij>} + \frac{2}{3}\frac{\partial h'_2}{\partial 1/T}\lambda_{ppi}\lambda_{qqi} + \mathcal{O}(3),\\
\varrho_{<ij>} &= \frac{\partial h'}{\partial\lambda_{<ij>}} = 2h'_1\lambda_{<ij>} + 3h'_3\left(\lambda_{<ik>}\lambda_{<kj>} - \frac{1}{3}\lambda_{<ln>}\lambda_{<ln>}\delta_{ij}\right) + h'_4\lambda_{qq<i}\lambda_{j>pp} + \mathcal{O}(3),\\
\varrho_{ppi} &= \frac{\partial h'}{\partial\lambda_{ppi}} = 2h'_2\lambda_{ppi} + 2h'_4\lambda_{<ik>}\lambda_{kpp} + \mathcal{O}(3).
\end{aligned} \tag{2.31}$$

We conclude from $(2.31)_{1,2}$ that the density ϱ and the pressure $p = \frac{1}{3}\varrho_{pp}$ are equal to the equilibrium values ϱ_{E} and p_{E} to within second-order terms. In the sequel this fact enables us to drop the distinction.

Next we exploit $(2.19)_2$. We have

$$a_{ei} = \frac{1}{\varrho}\left(4h_1'\lambda_{<ei>} + 5p_{\mathrm{E}}\delta_{ei}\right) + \mathcal{O}(2) \tag{2.32}$$

and obtain

$$\begin{aligned}
0 &= \left(\frac{\partial\varphi_1}{\partial g/T} - 5pT\right)\lambda_{qqk} + \left(\frac{\partial\varphi_2}{\partial g/T} - 10\frac{p}{\varrho}\frac{\partial h_1'}{\partial g/T} - 4h_1'T\right)\lambda_{<kl>}\lambda_{qql} + \mathcal{O}(3),\\
0 &= \varrho_{<rsk>} + \mathcal{O}(2),\\
0 &= \left(\frac{\partial\varphi_1}{\partial 1/T} - h_2' + \frac{25}{2}\frac{p^2T}{\varrho}\right)\lambda_{ppk} + \left(\frac{\partial\varphi_2}{\partial 1/T} - h_4' - 10\frac{p}{\varrho}\frac{\partial h_1'}{\partial 1/T} - 10h_1'\frac{pT}{\varrho}\right)\lambda_{ppl}\lambda_{<lk>} + \mathcal{O}(3),\\
0 &= \left(\varphi_2 - \frac{4}{5}h_2' - 10\frac{p}{\varrho}h_1'\right)\left(\lambda_{qqr}\delta_{ks} + \lambda_{qqs}\delta_{kr} - \frac{2}{3}\lambda_{qqk}\delta_{rs}\right) + \mathcal{O}(2),\\
\varrho_{ppnk} &= \varphi_1\delta_{nk} + \varphi_2\lambda_{<nk>} + \mathcal{O}(2).
\end{aligned} \tag{2.33}$$

Finally we exploit (2.20) and from its trace-less part obtain

$$\begin{aligned}
0 = 2(h_1' + pT)\lambda_{<tk>} &+ (4h_1'T + 3h_3')\left(\lambda_{<tj>}\lambda_{<jk>} - \frac{1}{3}\lambda_{<mn>}\lambda_{<mn>}\delta_{tk}\right)\\
&+ \left(\frac{18}{5}h_2'T + h_4'\right)\lambda_{qq<k}\lambda_{t>pp} + \mathcal{O}(3).
\end{aligned} \tag{2.34}$$

From (2.34) we see that

$$h_1' = -pT \quad \text{and} \quad h_3' = \frac{4}{3}pT^2 \tag{2.35}$$

hold, and therefore the equations $(2.33)_{1,3}$ provide equations for the derivatives of φ_1 and φ_2, viz.

$$\begin{aligned}
\frac{\partial\varphi_1}{\partial g/T} &= 5pT, & \frac{\partial\varphi_1}{\partial 1/T} &= h_2' - \frac{25}{2}\frac{p^2T}{\varrho},\\
\frac{\partial\varphi_2}{\partial g/T} &= -14pT^2, & \frac{\partial\varphi_2}{\partial 1/T} &= h_4' + 45\frac{p^2T^2}{\varrho}.
\end{aligned} \tag{2.36}$$

Again from (2.34) we know that $h_4' = -\frac{18}{5}h_2'T$ holds, and therefore we obtain from $(2.33)_4$

$$h_2' = \frac{5}{4}\left(\varphi_2 + 10\frac{p^2T}{\varrho}\right) \quad \text{and} \quad h_4' = -\frac{9}{2}\left(\varphi_2 + 10\frac{p^2T}{\varrho}\right)T. \tag{2.37}$$

Elimination of h_2' and h_4' between (2.36) and (2.37) provides the equations

$$\frac{\partial\varphi_1}{\partial g/T} = -\frac{5}{14}\frac{\partial\varphi_2/T}{\partial g/T} \quad \text{and} \quad \frac{\partial\varphi_1}{\partial 1/T} = -\frac{5}{14}\frac{\partial\varphi_2/T}{\partial 1/T},$$

which can readily be integrated to give

$$\varphi_1 = -\frac{5}{14}\frac{\varphi_2}{T} + C. \tag{2.38}$$

On the other hand, by (2.36) and (2.27) we have

$$\frac{\partial\varphi_1}{\partial g/T} = 5T^{7/2}F(z), \qquad \text{or, by (2.28),} \qquad \frac{\partial\varphi_1}{\partial z} = 5T^{7/2}\frac{FF'}{z},$$
$$\frac{\partial\varphi_2}{\partial g/T} = -14T^{9/2}F(z) \qquad\qquad\qquad \frac{\partial\varphi_2}{\partial z} = -14T^{9/2}\frac{FF'}{z},$$

where as before z stands for $\varrho/T^{3/2}$. We integrate and obtain

$$\begin{aligned}\varphi_1 &= 5T^{7/2}\left(\int\frac{FF'}{z}dz + c\right) + C,\\ \varphi_2 &= -14T^{9/2}\left(\int\frac{FF'}{z}dz + c\right).\end{aligned} \tag{2.39}$$

With this knowledge we calculate the coefficients h' from (2.35), (2.37) and (2.27)

$$\begin{aligned}h'_1 &= -T^{7/2}F(z),\\ h'_2 &= T^{9/2}\left[\frac{25}{2}\frac{F^2(z)}{z} - \frac{35}{2}\left(\int\frac{FF'}{z}dz + c\right)\right],\\ h'_3 &= \frac{4}{3}T^{9/2}F(z),\\ h'_4 &= T^{11/2}\left[-45\frac{F^2(z)}{z} + 63\left(\int\frac{FF'}{z}dz + c\right)\right].\end{aligned} \tag{2.40}$$

We thus conclude that all coefficients h' and φ that determine the potentials are in turn determined by the function $F(\varrho/T^{3/2})$ that first appeared in the thermal equation of state (2.27).

2.3.7 Residual Inequality: Step vi

Of course as a further consequence of the entropy principle there is the residual inequality (2.16), which entails linear relations between the productions $s_{<ij>}, s_{ppi}$ and the Lagrange multipliers $\lambda_{<ij>}$ and λ_{ppi} of the form

$$s_{<ij>} = s\lambda_{<ij>} \quad \text{and} \quad s_{ppi} = t\lambda_{ppi}, \tag{2.41}$$

where s and t are positive-valued functions of ϱ and T (say)

$$s \geq 0 \quad \text{and} \quad t \geq 0. \tag{2.42}$$

2.3.8 Summary of Results of the Entropy Principle: Step vii

We recall from Section 2.2 that $\varrho_{<ijk>}, \varrho_{ppik}, s_{<ij>}, s_{ppi}$ are the only constitutive quantities that occur in the balance equations. Now we have obtained [see $(2.33)_2, (2.33)_5$ with (2.39) and (2.41)]

$$\begin{aligned}\varrho_{<rsk>} &= \mathcal{O}(2),\\ \varrho_{ppnk} &= \left[5T^{7/2}\left(\int\frac{FF'}{z}dz + c\right) + C\right]\delta_{nk} - 14T^{9/2}\left(\int\frac{FF'}{z}dz + c\right)\lambda_{<nk>} + \mathcal{O}(2),\\ s_{<ij>} &= s\lambda_{<ij>} + \mathcal{O}(2),\\ s_{ppi} &= t\lambda_{ppi} + \mathcal{O}(2).\end{aligned} \tag{2.43}$$

We may use (2.31) and (2.40) to replace the Lagrange multipliers in (2.43) by the original variables $\varrho_{<ij>}$ and ϱ_{ppi}, because we have, to within terms of $\mathcal{O}(2)$

$$\begin{aligned}\lambda_{<ij>} &= -\frac{1}{2pT}\varrho_{<ij>} &&= -\frac{1}{2T^{7/2}F(z)}\varrho_{<ij>},\\ \lambda_{ppi} &= \frac{1}{2h_2'}\varrho_{ppi} &&= \frac{1}{5T^{9/2}\left[5\frac{F^2}{z} - 7\left(\int\frac{FF'}{z}dz + c\right)\right]}\varrho_{ppi}.\end{aligned} \tag{2.44}$$

Therefore we may write the constitutive relations (2.43) in the final forms

$$\begin{aligned}\varrho_{<ijk>} &= \mathcal{O}(2),\\ \varrho_{ppnk} &= \left[5T^{7/2}\left(\int\frac{FF'}{z}dz + c\right) + C\right]\delta_{ij} + 7\frac{T}{F}\left[\int\frac{FF'}{z}dz + c\right]\varrho_{\langle ij\rangle} + \mathcal{O}(2),\\ s_{\langle ij\rangle} &= -\frac{s}{2T^{7/2}F(z)}\varrho_{\langle ij\rangle} + \mathcal{O}(2),\\ s_{ppi} &= \frac{t}{5T^{9/2}\left[5\frac{F^2(z)}{z} - 7\left(\int\frac{FF'}{z}dz + c\right)\right]}\varrho_{ppi} + \mathcal{O}(2).\end{aligned} \tag{2.45}$$

Of course, we may also be interested in the entropy density and the entropy flux. We have by (2.18)

$$h = -h' + \lambda\varrho + \lambda_{ij}\varrho_{ij} + \lambda_{ppi}\varrho_{qqi} \quad \text{and} \quad \varphi_k = -\varphi_k' + \lambda_{ij}\varrho_{ijk} + \lambda_{ppi}\varrho_{ppik}. \tag{2.46}$$

From these equations, the representations (2.29) with (2.27) and with coefficients (2.39), (2.40) we obtain, to within terms of order 3,

$$\begin{aligned}h &= \varrho\left[\frac{3}{2}\int\left(\frac{F'}{z} - \frac{5}{3}\frac{F}{z^2}\right)dz + \varsigma\right] - \frac{\varrho_{\langle ij\rangle}\varrho_{\langle ij\rangle}}{4T^{7/2}F(z)} - \frac{\varrho_{ppi}\varrho_{qqi}}{10T^{7/2}F(z)\left[7\frac{T}{F(z)}\left(\int\frac{FF'}{z}dz + c\right) - 5T\frac{F(z)}{z}\right]},\\ \varphi_k &= \frac{1}{T}\frac{1}{2}\varrho_{ppk} - \frac{2}{5}\frac{1}{T^{7/2}F(z)}\varrho_{\langle kj\rangle}\frac{1}{2}\varrho_{ppj}.\end{aligned} \tag{2.47}$$

The equations (2.45), (2.47) are the final results obtained from the entropy principle. We conclude that $\varrho_{\langle ijk\rangle}$, ϱ_{ppnk}, h, and φ_k are fully explicit to within the constants c, C, and ς, if only the function $F(z)$ is known. That function determines the thermal equation of state and, thus, is known either from experiment or from statistical mechanics.

The productions $s_{\langle ij\rangle}$ and s_{ppi} are determined to within two positive-valued functions of ϱ and T.

These amazingly restrictive results have first been derived by Liu and Müller [15] in a slightly different and more laborious manner.

2.4 Exploitation of the Requirement of Convexity and Causality

In Chapter 6, within the framework of the formal theory of extended thermodynamics, we shall prove that the requirement of convexity of $-h$ with respect to the densities F, F_i, F_{ij}, F_{ppi} implies that $-h$ is also a convex function of the densities $\varrho, \varrho_{pp}, \varrho_{\langle ij\rangle}$, and ϱ_{ppi}. Hence, it follows by inspection of $(2.47)_1$ that we must have

$$F(z) > 0 \quad \text{and} \quad 7\left(\int\frac{FF'}{z}dz + c\right) - 5\frac{F^2}{z} > 0, \tag{2.48}$$

but this is not all. The matrix of second derivatives of $-h_E$ with respect to ϱ and ϱ_{ll} must be positive definite. Hence we have

$$\left\| \begin{matrix} -\left(\dfrac{\partial^2 h_E}{\partial \varrho^2}\right)_{\varrho_{pp}} & -\dfrac{\partial^2 h_E}{\partial \varrho \partial \varrho_{pp}} \\ -\dfrac{\partial^2 h_E}{\partial \varrho \partial \varrho_{pp}} & -\left(\dfrac{\partial^2 h_E}{\partial \varrho_{pp}^2}\right)_{\varrho} \end{matrix} \right\| = \left\| \begin{matrix} \left(\dfrac{\partial \frac{g}{T}}{\partial \varrho}\right)_{\varrho_{pp}} & -\dfrac{1}{2}\left(\dfrac{\partial \frac{1}{T}}{\partial \varrho}\right)_{\varrho_{pp}} \\ -\dfrac{1}{2}\left(\dfrac{\partial \frac{1}{T}}{\partial \varrho}\right)_{\varrho_{pp}} & -\dfrac{1}{2}\left(\dfrac{\partial \frac{1}{T}}{\partial \varrho_{pp}}\right)_{\varrho} \end{matrix} \right\|$$

$$= \left\| \begin{matrix} \dfrac{1}{T^{3/2}} \dfrac{\frac{5}{3}\frac{FF'}{z^2}}{\frac{5}{3}\frac{F}{z} - F'} & -\dfrac{1}{\varrho T} \dfrac{\frac{1}{3}F'}{\frac{5}{3}\frac{F}{z} - F'} \\ -\dfrac{1}{\varrho T} \dfrac{\frac{1}{3}F'}{\frac{5}{3}\frac{F}{z} - F'} & \dfrac{1}{\varrho T^2} \dfrac{\frac{1}{9}}{\frac{5}{3}\frac{F}{z} - F'} \end{matrix} \right\| \text{ — pos. def.} \tag{2.49}$$

The first equation follows from (2.25), the second from (2.27). The condition is satisfied, if we have

$$0 < F' < \frac{5}{3}\frac{F}{z}. \tag{2.50}$$

It is clear that the above exploitation of the requirement of convexity is appropriate to the present near-equilibrium approximation. In Chapter 7 we shall prove that it is possible to determine the region of hyperbolicity around equilibrium.

3 Field Equations and the Thermodynamic Limit

3.1 Field Equations

In the equations (1.15) we replace the $\mathbf{F}^*$'s by ϱ^*'s as dictated by (1.6) and eliminate the explicit dependence of all quantities on the velocity in the manner that was described, for Galilean systems, in Paragraph 1.3. Thus we obtain the balance laws of extended thermodynamics in the form

$$\begin{aligned} &\dot{\varrho} + \varrho \frac{\partial v_i^*}{\partial x_i^*} = 0, \\ &\varrho\left[\dot{v}_i^* - i_i^0 - 2W_{ik}v_k^*\right] + \frac{\partial \varrho_{ij}^*}{\partial x_j^*} = 0, \\ &\varrho\left(\frac{\varrho_{pp}}{\varrho}\right)^{\bullet} + \frac{\partial \varrho_{ppi}^*}{\partial x_j^*} + 2(\varrho_{\langle ij\rangle}^* + \frac{1}{3}\varrho_{pp}\delta_{ij})\frac{\partial v_i^*}{\partial x_j^*} = 0, \\ &\varrho\left[\left(\frac{\varrho_{\langle ik\rangle}^*}{\varrho}\right)^{\bullet} + 2\varrho_{l\langle i}^*\frac{\partial v_{k\rangle}^*}{\partial x_l} - 4\varrho_{l\langle i}^* W_{k\rangle l}\right] + \frac{\partial \varrho_{\langle ik\rangle j}^*}{\partial x_j^*} = s_{\langle ik\rangle}^*, \\ &\varrho\left[\left(\frac{\varrho_{ppi}^*}{\varrho}\right)^{\bullet} + \frac{\varrho_{ppk}^*}{\varrho}\frac{\partial v_i^*}{\partial x_k^*} - 2\varrho_{ppk}^* W_{ik}\right] + \frac{\partial \varrho_{ppik}^*}{\partial x_k^*} \\ &\qquad + 2\varrho_{ijk}^*\frac{\partial v_j^*}{\partial x_k^*} - 2\frac{\varrho_{ij}^*}{\varrho}\frac{\partial \varrho_{jk}^*}{\partial x_k^*} - \frac{\varrho_{pp}}{\varrho}\frac{\partial \varrho_{ik}^*}{\partial x_k^*} = s_{ppi}^*. \end{aligned} \tag{3.1}$$

For simplicity in notation we drop the stars that characterize quantities referred to non-inertial frames. The presence of the inertial terms with i_i^0 and W_{ik} must suffice as a reminder that the frame is non-inertial. Also we introduce the suggestive notation (1.11), (1.12) for ϱ_{ij}, ϱ_{pp}, and

ϱ_{ppi}. Furthermore we insert the explicit expressions (2.45) for $\varrho_{\langle ijk\rangle}, \varrho_{ppij}, s_{\langle ij\rangle}$, and s_{ppi} that are dictated by the entropy principle. In this manner we obtain from (3.1), always with $z = \varrho/T^{3/2}$,

$$
\begin{aligned}
&\dot{\varrho} + \varrho\frac{\partial v_i}{\partial x_i} = 0, \\
&\varrho\left[\dot{v}_i - i_i^0 - 2W_{ik}v_k\right] - \frac{\partial(t_{\langle ij\rangle} - T^{5/2}F(z)\delta_{ij})}{\partial x_j} = 0, \\
&\varrho\frac{3}{2}\left(T\frac{F(z)}{z}\right)^{\bullet} + \frac{\partial q_j}{\partial x_j} - \left(t_{\langle ij\rangle} - T^{5/2}F(z)\delta_{ij}\right)\frac{\partial v_i}{\partial x_j} = 0, \\
&\varrho\left[\left(\frac{t_{\langle ik\rangle}}{\varrho}\right)^{\bullet} + 2\frac{t_{l\langle i}}{\varrho}\frac{\partial v_{k\rangle}}{\partial x_l} - 4\frac{t_{l\langle i}}{\varrho}W_{k\rangle l}\right] - \frac{4}{5}\frac{\partial q_{\langle i}}{\partial x_{k\rangle}} = -\frac{1}{\tau_t}t_{\langle ik\rangle}, \\
&\varrho\left[\left(\frac{q_i}{\varrho}\right)^{\bullet} + \frac{q_k}{\varrho}\frac{\partial v_i}{\partial x_k} - 2\frac{q_k}{\varrho}W_{ik}\right] - \frac{\partial\left\{\frac{7}{2}\left(\frac{T}{F(z)}t_{\langle ij\rangle} - \frac{5}{7}T^{7/2}\delta_{ij}\right)\left(\int\frac{FF'}{z}dz + c\right)\right\}}{\partial x_j} \\
&\qquad + \frac{2}{5}q_i\frac{\partial v_k}{\partial x_k} + \frac{4}{5}q_k\frac{\partial v_{(i}}{\partial x_{k)}} - \frac{\left(t_{\langle ij\rangle} - T^{5/2}F(z)\delta_{ij}\right)}{\varrho}\frac{\partial\left(t_{\langle jk\rangle} - T^{5/2}F(z)\delta_{jk}\right)}{\partial x_k} \\
&\qquad + \frac{3}{2}\frac{T^{5/2}F(z)}{\varrho}\frac{\partial\left(t_{\langle ik\rangle} - T^{5/2}F(z)\delta_{ik}\right)}{\partial x_k} = -\frac{1}{\tau_q}q_i.
\end{aligned}
\tag{3.2}
$$

We have introduced

$$
\frac{1}{\tau_t} \quad \text{for} \quad \frac{s}{2T^{7/2}F(z)} \qquad \text{and} \qquad \frac{1}{\tau_q} \quad \text{for} \quad \frac{t}{T^{9/2}\left[7\left(\int\frac{FF'}{z}dz + c\right) - 5\frac{F^2}{z}\right]}. \tag{3.3}
$$

The τ's may be considered as relaxation times for stress and heat flux respectively.

Inspection of (3.2) shows that this system of equations for the determination of the thirteen variables

$$\varrho, \ v_i, \ T, \ t_{\langle ij\rangle}, \ q_i$$

is quite explicit. What remains to be determined is:

the constant c,
the function $F(z)$ from the thermal equation of state, and
the coefficients $\tau_t(\varrho, T)$ and $\tau_q(\varrho, T)$.

The functions $\tau_t(\varrho, T)$ and $\tau_q(\varrho, T)$ are related to the viscosity and the heat conductivity of the gas. We proceed to show this in Paragraph 3.2.

The equations $(3.2)_{2,4,5}$ contain inertial terms, and yet, as they stand, the equations hold in all Euclidean frames because the expressions

$$
\dot{v}_i - i_i^0 - 2W_{ik}v_k, \qquad \left(\frac{t_{\langle ik\rangle}}{\varrho}\right)^{\bullet} + 2\frac{t_{l\langle i}}{\varrho}\frac{\partial v_{k\rangle}}{\partial x_l} - 4\frac{t_{l\langle i}}{\varrho}W_{k\rangle l}, \qquad \left(\frac{q_i}{\varrho}\right)^{\bullet} + \frac{q_k}{\varrho}\frac{\partial v_i}{\partial x_k} - 2\frac{q_k}{\varrho}W_{ik}, \tag{3.4}
$$

while frame-dependent, are objective vectors and tensors.

3.2 The Thermodynamic Limit

A natural question to be asked from extended thermodynamics is how it relates to ordinary thermodynamics. In ordinary thermodynamics of viscous, heat-conducting mon-atomic gases the stress-deviator and the heat flux are given by the constitutive relations of Navier–Stokes and Fourier, viz.

$$
\begin{aligned}
t_{\langle ij\rangle} &= 2\mu\,\frac{\partial v_{\langle i}}{\partial x_{j\rangle}}, \\
q_i &= -\kappa\,\frac{\partial T}{\partial x_i},
\end{aligned}
\tag{3.5}
$$

with μ and κ being viscosity and heat conductivity respectively. In extended thermodynamics we have replaced that simple set of equations for $t_{\langle ij\rangle}, q_i$ by the complicated set of differential equations $(3.2)_{4,5}$; however, there is a relation between the two.

This relation is provided by a formal iterative scheme, by which the equations (3.5) emerge from $(3.2)_{4,5}$ as first iterates. The scheme is called Maxwellian iteration, because Maxwell invented it when faced with a similar problem in the kinetic theory of gases (see [23]). The Maxwellian iteration proceeds as follows:

In the first step we calculate first iterates $t^{(1)}_{\langle ij\rangle}$ and $q^{(1)}_i$ from $(3.2)_{4,5}$ by using the zeroth iterates $t^{(0)}_{\langle ij\rangle} = 0$ and $q^{(0)}_i = 0$ on the left-hand sides of the equations. (The zeroth iterates are the equilibrium values, of course.) In the second step we calculate second iterates $t^{(2)}_{\langle ij\rangle}$ and $q^{(2)}_i$ by insertion of $t^{(1)}_{\langle ij\rangle}$ and $q^{(1)}_i$ into the left-hand sides of $(3.2)_{4,5}$; etc. In this manner we obtain as first iterates, after a brief calculation,

$$
\begin{aligned}
t^{(1)}_{\langle ij\rangle} &= 2T^{5/2}F(z)\tau_t\frac{\partial v_{\langle i}}{\partial x_{j\rangle}}, \\
q^{(1)}_i &= -\frac{5}{4}T^{5/2}\left\{5\frac{F^2}{z} - 7\left(\int\frac{FF'}{z}dz + c\right)\right\}\tau_q\frac{\partial T}{\partial x_i}.
\end{aligned}
\tag{3.6}
$$

We conclude that the classical laws of Navier–Stokes and Fourier result from the equations of extended thermodynamics in an approximate manner as first iterates in the Maxwellian scheme. The viscosity and the heat conductivity of the gas result from the comparison of (3.5) and (3.6)

$$
\mu = 2T^{5/2}F(z)\tau_t, \qquad \kappa = \frac{5}{4}T^{5/2}\left\{5\frac{F^2}{z} - 7\left(\int\frac{FF'}{z}dz + c\right)\right\}\tau_q. \tag{3.7}
$$

Since μ and κ may be measured as functions of ϱ and T—and infact are tabulated—we see that the coefficients τ_t and τ_q may be determined.

Insertion of (3.6) on the left-hand side of $(3.2)_{4,5}$ furnishes the second iterates $t^{(2)}_{\langle ij\rangle}$ and $q^{(2)}_i$, viz.

$$
\begin{aligned}
t^{(2)}_{\langle ij\rangle} &= 2p\tau_t\frac{\partial v_{\langle i}}{\partial x_{j\rangle}} - \varrho\tau_t\left[\left(\frac{2p\tau_t}{\varrho}\frac{\partial v_{\langle i}}{\partial x_{j\rangle}}\right)^{\bullet} + \frac{4p\tau_t}{\varrho}\frac{\partial v_{\langle l}}{\partial x_{\langle i\rangle}}\frac{\partial v_{k\rangle}}{\partial x_l} - \frac{8p\tau_t}{\varrho}\frac{\partial v_{\langle l}}{\partial x_{\langle i\rangle}}W_{k\rangle l}\right] - \frac{4}{5}\tau_t\frac{\partial\left(\pi\tau_q\frac{\partial T}{\partial x_{\langle i}}\right)}{\partial x_{j\rangle}}, \\
q^{(2)}_i &= -\pi\tau_q\frac{\partial T}{\partial x_i} + \varrho\tau_q\left[\left(\frac{\pi\tau_q}{\varrho}\frac{\partial T}{\partial x_i}\right)^{\bullet} + \frac{\pi\tau_q}{\varrho}\frac{\partial T}{\partial x_k}\frac{\partial v_i}{\partial x_k} - 2\frac{\pi\tau_q}{\varrho}\frac{\partial T}{\partial x_k}W_{ik}\right] \\
&\quad + \tau_q\frac{\partial\left\{7T^{7/2}\tau_t\frac{\partial v_{\langle i}}{\partial x_{j\rangle}}\left(\int\frac{FF'}{z}dz + c\right)\right\}}{\partial x_j} \\
&\quad + \frac{2}{5}\pi\tau_q^2\frac{\partial T}{\partial x_i}\frac{\partial v_k}{\partial x_k} + \frac{4}{5}\pi\tau_q^2\frac{\partial T}{\partial x_k}\frac{\partial v_{(i}}{\partial x_{k)}} + 2\frac{p}{\varrho}\tau_q\tau_t\frac{\partial v_{\langle i}}{\partial x_{j\rangle}}\frac{\partial\left(2p\tau_t\frac{\partial v_{j\rangle}}{\partial x_{k\rangle}}\right)}{\partial x_k} \\
&\quad - 2\frac{p}{\varrho}\tau_q\tau_t\frac{\partial v_{\langle i}}{\partial x_{j\rangle}}\frac{\partial p}{\partial x_j} - \frac{p}{\varrho}\tau_q\frac{\partial\left(2p\tau_t\frac{\partial v_{\langle i}}{\partial x_{k\rangle}}\right)}{\partial x_k} - \frac{3}{2}\frac{p}{\varrho}\tau_q\frac{\partial\left(2p\tau_t\frac{\partial v_{\langle i}}{\partial x_{k\rangle}}\right)}{\partial x_k}.
\end{aligned}
\tag{3.8}
$$

Here p, of course, is equal to $T^{5/2}F(z)$ and π has been introduced as an abbreviation for

$$\frac{5}{4}T^{5/2}\left\{5\frac{F^2}{z} - 7\left(\int \frac{FF'}{z}dz + c\right)\right\}.$$

While we shall not discuss the second iterates in any detail, we point to three features that seem worthy of comment. First, we observe that the first iterates are linear in τ_t and τ_q, while the contributions of the second iterative step are quadratic or cubic in the relaxation times. It is easy to extrapolate and conclude that the contributions of the nth iterative step are of nth order at least in τ_t and τ_q. Thus the iteration may be considered as a kind of expansion in powers of the relaxation times, and since these are expected to be small, one may thus justify the break-off of the iteration. Of course, this is far from a rigorous argument.

Second, we see that in the second step the stress depends on the temperature gradient as well as the velocity gradient. Similarly the heat flux depends on the velocity gradient. That observation was already made by Maxwell in the context of the kinetic theory of gases. To my knowledge such dependencies have never been observed because they are too small.

Third, we conclude from (3.8) that the stress $t^{(2)}_{\langle ij\rangle}$ and the heat flux $q^{(2)}_i$ are affected by the angular velocity $\mathbf{W}$ of the frame. This fact may be the most noteworthy aspect of the second iterates and shall be investigated in a separate paragraph. The effect was first described by Müller [24] in the kinetic theory of gases.

3.3 The Frame-Dependence of the Heat Flux

The effect of the frame dependent terms in (3.8) is hard to evaluate in general, because the equations are too complicated. In order to understand the nature and origin of the frame dependence we shall therefore concentrate on the heat flux and consider the special case of a rigid rotation in which the fluid elements have constant temperature. That case is characterized by

$$\frac{\partial v_{(i}}{\partial x_{j)}} = 0 \quad \text{and} \quad \dot{T} = 0. \tag{3.9}$$

It is easy to see that in such a case the equation $(3.8)_2$ reduces to the simple form

$$q_i = -\pi\tau_q\left[\frac{\partial T}{\partial x_i} + 2\tau_q\left(W_{ik} - \frac{\partial v_{[k}}{\partial x_{i]}}\right)\frac{\partial T}{\partial x_k}\right]. \tag{3.10}$$

This is a remarkable formula, because it says that in general the heat flux is not parallel to the temperature gradient. Indeed, if the gas rotates with respect to an inertial frame, the heat flux contains the term $2\pi\tau_q^2\frac{\partial v_{[k}}{\partial x_{i]}}\frac{\partial T}{\partial x_k}$, which is perpendicular to the temperature gradient; or when the gas is at rest in a non-inertial frame there is the contribution $-2\pi\tau_q^2 W_{ik}\frac{\partial T}{\partial x_k}$ to the heat flux. In general the term perpendicular to the temperature gradient reads

$$-2\pi\tau_q^2\left(W_{ik} - \frac{\partial v_{[k}}{\partial x_{i]}}\right)\frac{\partial T}{\partial x_k}, \tag{3.11}$$

which is a frame-dependent objective vector.

In order to understand the nature of the frame-dependent term we consider a gas at rest in a cylinder with a radial temperature gradient as depicted in Figure 1(a). We concentrate on a small volume element of the linear dimensions of the mean free path of the atoms. First we consider the gas at rest in an inertial frame. In that case the free paths of the atoms are straight lines, and because of the temperature gradient, the atoms flying from bottom to top carry a bigger energy upward across the plane $s-s$ than is carried downward by the atoms moving down; see Figure 1(b). A net flux of energy therefore accompanies the passage of a pair of atoms through $s-s$, and that flux is proportional to the temperature gradient and opposite to it just as predicted by (3.10) for $W_{ik}=0$ and $\partial v_{[k}/\partial x_{i]}=0$. Next we consider the same situation for a gas at rest in a non-inertial frame. Now the paths of free flight are curved by the Coriolis force and there is a flux of energy through the plane $H-H$ as well as through the plane $s-s$, see Figure 1(c). Thus the flux has a component perpendicular to the gradient of temperature as predicted by (3.10) in the case $W_{ik}\neq 0$, $\partial v_{[k}/\partial x_{i]}=0$.

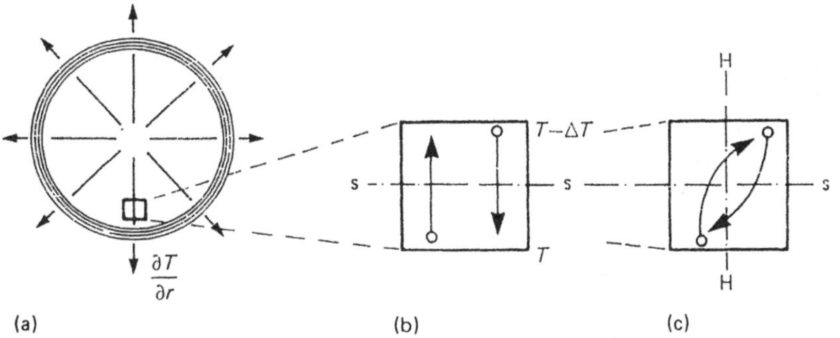

Figure 1. On the frame dependence of the heat flux.

It is clear from the above argument that the frame-dependent term in the heat flux is small, because it depends on the extent to which the Coriolis force can bend the free path of an atom. Also formally the smallness is apparent from equation (3.10), since the components of $\tau_q\mathbf{W}$ are quotients of the relaxation time τ_q and the period of rotation of the frame. If we consider τ_q to be of the order of a mean time of free flight of an atom, this quotient is very small indeed for practical frequencies of revolution.

The above suggestive explanation is akin to the explanation of the Hall effect in a conductor under a magnetic field. Indeed, the Lorentz force on the freely moving electrons in the conductor bends their paths in much the same manner as the Coriolis force bends the paths of the atoms. The Hall effect is greater than the Coriolis effect, because the electrons have a much smaller mass than an atom; also, the magnetic flux density can easily be made big, while the Coriolis force is limited since a number of revolutions of the frame in the order 10^6 Hz is impractical. Therefore, the Hall effect is well documented whereas the Coriolis effect has never been observed. Müller [25] has treated both effects jointly.

According to $(3.8)_1$ there is also a frame-dependent term in the stress. That term can also be explained suggestively much in the same manner as the term in the heat flux, but we do not do this here.

4 Thermal Equations of State and Specific Equation for Ideal Gases

4.1 The Classical Ideal Gas

As we have seen it remains to insert into (3.2) the function $F(z)$, which determines the thermal equation of state, because $p(\varrho, T) = T^{5/2}F(z)$ with $z = \varrho/T^{3/2}$. The thermal equation of state of ideal gases has been found empirically in the early history of thermodynamics and it reads

$$p = \varrho \frac{k}{m} T, \tag{4.1}$$

where m is the atomic mass and $k = 1,38 \cdot 10^{-23} \frac{J}{K}$ is the Boltzmann constant. It follows that in a classical ideal gas we must have

$$F(z) = \frac{k}{m} z. \tag{4.2}$$

Therefore, the field equations (3.2) of extended thermodynamics for a classical ideal gas read

$$\begin{aligned}
&\dot{\varrho} + \varrho \frac{\partial v_i}{\partial x_i} = 0, \\
&\varrho \left[\dot{v}_i - i_i^0 - 2W_{ik}v_k \right] - \frac{\partial (t_{\langle ij \rangle} - p\delta_{ij})}{\partial x_j} = 0, \\
&\varrho \frac{3}{2} \frac{k}{m} \dot{T} + \frac{\partial q_j}{\partial x_j} - (t_{\langle ij \rangle} - p\delta_{ij}) \frac{\partial v_i}{\partial x_j} = 0, \\
&t_{\langle ij \rangle} = 2p\tau_t \left\{ \frac{\partial v_{\langle i}}{\partial x_{j\rangle}} - \frac{1}{2p} \left[\dot{t}_{\langle ij \rangle} + 2t_{\langle n \langle i \rangle} \frac{\partial v_{j\rangle}}{\partial n_n} - 4t_{\langle n \langle i \rangle} W_{j\rangle n} \right] + \frac{2}{5p} \frac{\partial q_{\langle i}}{\partial x_{j\rangle}} \right\} - \tau_t t_{\langle ij \rangle} \frac{\partial v_n}{\partial x_n}, \\
&q_i = -\left(\frac{5}{2} \frac{k}{m} p + \frac{35}{4} T^{5/2} c \right) \tau_q \left\{ \frac{\partial T}{\partial x_i} + \frac{1}{\frac{5}{2}\frac{k}{m}p + \frac{35}{4}T^{5/2}c} \left[\dot{q}_i + q_k \frac{\partial v_i}{\partial x_k} - 2q_k W_{ik} \right] \right. \\
&\qquad \left. - \frac{\frac{k}{m}T + \frac{7}{2}c\frac{m}{k}\frac{T^{3/2}}{\varrho}}{\frac{5}{2}\frac{k}{m}p + \frac{35}{4}T^{5/2}c} \frac{\partial t_{\langle ij \rangle}}{\partial x_j} \right\} \\
&\qquad - \tau_q \left(\frac{7}{5} q_i \frac{\partial v_k}{\partial x_k} + \frac{4}{5} q_k \frac{\partial v_i}{\partial x_k} - \frac{7}{2} \frac{k}{m} t_{\langle ij \rangle} \frac{\partial T(1 + (\frac{m}{k})^2 c \frac{T^{3/2}}{\varrho})}{\partial x_j} - \frac{t_{\langle ij \rangle}}{\varrho} \frac{\partial t_{\langle jk \rangle}}{\partial x_k} + \frac{t_{\langle ij \rangle}}{\varrho} \frac{\partial p}{\partial x_j} \right).
\end{aligned} \tag{4.3}$$

The relaxation times τ_t and τ_q are related to the viscosity μ and the heat conductivity κ by the equations (3.7), which for a classical ideal gas read

$$\mu = \varrho \frac{k}{m} T \tau_t \quad \text{and} \quad \kappa = \left(\frac{5}{2} \varrho \left(\frac{k}{m} \right)^2 T + \frac{35}{4} T^{5/2} c \right) \tau_q. \tag{4.4}$$

4.2 Comparison with the Kinetic Theory

We compare the equations $(4.3)_{4,5}$ with the corresponding equations (1.3.34) of the kinetic theory of gases and find complete agreement, provided the constant c in (4.3) is set equal to zero. Henceforth we shall assume this to be the case.

Actually there is another minor discrepancy between the equations $(4.3)_{4,5}$ and (1.3.34), and that concerns the relaxation times. While τ_t and τ_q are independent, the corresponding values in the kinetic theory are related by the factor $\frac{2}{3}$. This is due to the assumption of a Maxwellian interatomic potential in the derivation of (1.3.34). For other, more realistic, potentials there is no such relation, even though τ_t/τ_q is close to 2/3 for most mon-atomic gases.

The fact that the phenomenological theory of extended thermodynamics has led to the same results as the kinetic theory of gases in the Grad approximation is truly amazing. It shows how little of the complicated apparatus of the kinetic theory is needed for these results and that they follow from a few macroscopic phenomenological principles like Galilei and Euclidean invariance, the entropy principle, and the balance type of the field equations.

Nor does the agreement between extended thermodynamics and the kinetic theory stop with the 13 field equations. The entropy and its flux agree as well. Indeed, from (2.47) we obtain with (4.2)

$$\frac{h}{\varrho} = \left[\frac{k}{m} ln \frac{T^{3/2}}{\varrho} + \varsigma\right] - \frac{t_{\langle ij\rangle} t_{\langle ij\rangle}}{4p\varrho T} - \frac{q_i q_i}{5p^2 T},$$
$$\varphi_i = \frac{q_i}{T} + \frac{2}{5pT} t_{\langle ij\rangle} q_j, \tag{4.5}$$

provided we set $c = 0$. These formulae agree completely with (1.3.29) and (1.3.30) from the kinetic theory.

4.3 Comparison with Extended TIP

Comparison of $(4.3)_{4,5}$ with the "phenomenological relations" (1.2.13) of extended TIP shows some agreement but also differences. The differences are the same as those between the kinetic theory and extended TIP, namely,

i. λ in (1.2.13) must be set equal to zero as is appropriate for an ideal gas that does not exhibit bulk viscosity.

ii. The time derivatives $\dot{q}_i$ and $\dot{t}_{\langle ij\rangle}$ in (1.2.13) must be replaced by

$$\dot{q}_i + q_k \frac{\partial v_i}{\partial x_k} - 2q_k W_{ik} \quad \text{and} \quad \dot{t}_{\langle ij\rangle} + 2t_{\langle n\langle i\rangle} \frac{\partial v_{j\rangle}}{\partial x_n} - 4t_{\langle n\langle i\rangle} W_{j\rangle n}.$$

iii. Non-linear terms of the type

$$q_i \frac{\partial v_k}{\partial x_k} \quad \text{or} \quad t_{\langle ij\rangle} \frac{\partial t_{\langle ik\rangle}}{\partial x_k} \quad \text{or} \quad t_{\langle ij\rangle} \frac{\partial p}{\partial x_i}$$

in $(4.3)_{4,5}$ must be ignored.

iv. The coefficients a, b, and K in (1.2.13) are now specific. We have

$$a = -\frac{1}{5p^2 T}, \qquad b = -\frac{1}{4p\varrho T}, \qquad k = \frac{2}{5pT}.$$

All these points together show to what extent the present systematic version of extended thermodynamics is superior to extended TIP. Indeed, in extended TIP there is no way to calculate the coefficients a, b, and K so that the theory has five unknown coefficients, viz. κ, μ, a, b, and K in the case of mon-atomic ideal gases, whereas the present theory has only two, namely, κ and μ.

4.4 Degenerate Ideal Gases

The thermal equation of state (4.1) is an empirical relation originally, but it can also be derived from molecular considerations in statistical mechanics. That theory reveals that $p = \varrho \frac{k}{m} T$ holds only for high temperatures and small densities where quantum effects may be neglected. At temperatures and densities where such effects are important we say that the gas is degenerate and the particles must be distinguished as Bosons and Fermions. We may then calculate the thermal equation of state by means of statistical mechanics of equilibrium.

The equilibrium phase density f_{E} is then not given by the Maxwell distribution (1.3.23), rather it reads (e.g., see [15])

$$f_{\mathrm{E}} = \frac{1}{e^{\alpha + (m(\mathbf{c}-\mathbf{v})^2/2kT)} \mp 1} \qquad \text{for} \quad \begin{array}{l}\text{Bosons,}\\ \text{Fermions.}\end{array} \tag{4.6}$$

Thus the density $\varrho = \int mf\, d\mathbf{c}$ and the pressure $p = \frac{1}{3}\int mC^2 f\, d\mathbf{c}$ are given by

$$\varrho = 4\pi m \sqrt{\frac{2kT}{m}}^{3} i_2^{\mp}(\alpha) \quad \text{and} \quad p = \frac{4\pi}{3} m \sqrt{\frac{2kT}{m}}^{5} i_4^{\mp}(\alpha), \tag{4.7}$$

where the i's stand for

$$i_n^{\mp}(\alpha) = \int_0^{\infty} \frac{x^n}{e^{\alpha + x^2} \mp 1} dx. \tag{4.8}$$

It is easy to show that the i's satisfy the recurrence relation

$$\frac{di_n(\alpha)}{d\alpha} = -\frac{n-1}{2} i_{n-2}(\alpha). \tag{4.9}$$

Note that by $(4.7)_1$ α is a function of $z = \varrho/T^{3/2}$ so that $(4.2)_2$ is a thermal equation of state of the form

$$p = T^{5/2} F(z) \qquad \text{with } F(z) = \frac{4\pi}{3} m \sqrt{\frac{2k}{m}}^{5} i_4^{\mp}(\alpha). \tag{4.10}$$

Since we are unable in general to solve $(4.7)_1$ for α in terms of $z = \varrho/T^{3/2}$, it is impossible to calculate the thermal equation of state $p = p(\varrho, T)$ explicitly, except in the following limiting cases:

i. $\alpha \gg 1$, non-degenerate gas.
In this case the i's assume the values

$$i_n^{\mp}(\alpha) = e^{-\alpha} \int_0^{\infty} x^n e^{-x^2} dx = e^{-\alpha} \frac{1}{2} \Gamma\left(\frac{n+1}{2}\right) \tag{4.11}$$

and the equilibrium phase density (4.6) tends to the Maxwell distribution

$$f_{\mathrm{E}} = \frac{\varrho}{m} \sqrt{\frac{m}{2\pi kT}}^{3} e^{-m(\mathbf{c}-\mathbf{v})^2/2kT}. \tag{4.12}$$

$\Gamma(n)$ in (4.11) is the Γ-function. We obtain $F(z) = \frac{k}{m} z$. This case has already been discussed in Paragraphs 4.1. through 4.3.

ii. $\alpha \ll -1$, strongly degenerate Fermi gas.
The i's may be calculated to give

$$i_n^{+}(\alpha) = \int_0^{\sqrt{-\alpha}} x^n dx = \frac{1}{n+1} (\sqrt{-\alpha})^{n+1}. \tag{4.13}$$

Thus by (4.10) we have

$$F(z) = \frac{1}{5} \sqrt{\frac{3}{4\pi m}}^{3} \left(\frac{\varrho}{T^{3/2}}\right)^{5/3}, \quad \text{hence,} \quad p = \frac{1}{5} \sqrt{\frac{3}{4\pi m}}^{3} \varrho^{5/3}. \tag{4.14}$$

In the limit of the strongly degenerate Fermi gas the pressure is thus independent of T; it depends only on ϱ.

iii. $0 < \alpha \ll 1$, strongly degenerate Bose gas.
The i's assume the form

$$i_n^-(0) = \int_0^\infty \frac{x^n}{e^{x^2}-1} = \frac{1}{2}\Gamma\left(\frac{n+1}{2}\right)\zeta_{(n+1)/2},$$

where Γ denotes the Γ-function and $\zeta_n = \sum_{r=1}^{\infty}(1/r^n)$. The values of this sum for different values of n may be found in mathematical handbooks, for example, see [26]. It follows that $F(z)$ and $p = p(\varrho, T)$ have the forms

$$F(z) = \frac{1}{3}4\pi m\sqrt{\frac{2k}{m}}^{-5} i_4^-(0) \quad \text{and} \quad p = \frac{1}{3}4\pi m\sqrt{\frac{2k}{m}}^{-5} i_4^-(0)T^{5/2}. \tag{4.15}$$

In the strongly degenerate Bose gas the pressure is independent of ϱ, it depends only on T.

iv. $\alpha = 0$, Bose-Einstein condensation.
In this case f_{E} has a singularity for $\mathbf{c} - \mathbf{v} = \mathbf{0}$. The singularity corresponds to an indefinite number of atoms in the state of zero energy. We must therefore replace the equation $(4.7)_1$ by

$$\varrho = \varrho_0 + 4\pi m\sqrt{\frac{2kT}{m}}^{-3} i_2^+(0), \tag{4.16}$$

where ϱ_0 is the density of the "condensate," which is formed by the atoms in the zero-energy state. $F(z)$ and p are still given by (4.15) in this case, however.
For $\alpha = 0$ the function F is a constant given by $(4.15)_1$, and it follows from $(2.47)_1$ that the specific entropy has the form

$$s = \frac{5}{6}4\pi m\sqrt{\frac{2k}{m}}^{-5} i_4^+(0)\frac{T^{3/2}}{\varrho} + \varsigma$$

or, by (4.16),

$$s = \frac{5}{3}\frac{k}{m}\frac{i_4^+(0)}{i_2^+(0)}(1-\zeta) + \varsigma, \tag{4.17}$$

where ζ is the fraction of atoms in the state of zero energy. Note that for the fully condensed gas the entropy s is equal to the constant ς, which is usually taken to be zero.

The field equations for degenerate gases result from the insertion of (4.10) into the equations (3.2). We do not write out these results. The limiting cases characterized heretofore lead to several degeneracies in the field equations. It is also obvious from (2.50) that these limiting cases are singular. Indeed, for the strongly degenerate Fermi gas we have $F' = \frac{5}{3}\frac{F}{z}$, while for the strongly degenerate Bose gas we have $F' = 0$. Thus the limiting cases violate the requirement of convexity, and hence the corresponding degeneracies in the field equations are not surprising.

Chapter 3

Thermodynamics of Mixtures of Fluids

The paradoxa of heat conduction and shear wave propagation described in Chapter 1 are not alone. A similar phenomenon occurs in diffusion as we shall demonstrate in this chapter. This fact calls for the formulation of an extended thermodynamic theory of mixtures.

The extension of the ordinary thermodynamic theory of mixtures is conceptually simpler than the extension of the theory of heat conduction and shear stresses that was described in Chapter 2. The reason for this simplicity is that no new—unconventional—balance laws need to be postulated. Indeed, the extension consists of considering the velocities of all constituents as basic fields, and the only new balance laws are the equations of balance of the momenta of the constituents.

This chapter presents ordinary thermodynamics of mixtures of fluids and extended thermodynamics of non-viscous simple mixtures. The particular case so characterized is chosen for demonstrative purposes: Onsager relations for diffusion and thermal diffusion can be proved in this case, and the speeds of first and second sound can be calculated. Liquid Helium II turns out to be a special case of non-viscous simple mixtures.

The inertial contributions that have appeared in the new balance equations of extended thermodynamics will also appear in this theory of mixtures. They represent the inertial forces on the relative motion of the constituents.

1 Ordinary Thermodynamics of Mixtures (TIP)

1.1 Constitutive Equations

We consider a homogeneous mixture of ν fluids in which all volume elements are occupied by particles of all constituents. The objective of thermodynamics of such mixtures is the determination of the $4\nu + 1$ fields of

$$\begin{array}{lll} \text{mass densities} & \varrho_\alpha & \\ \text{velocities} & v_i^\alpha & (\alpha = 1, 2, \dots \nu), \\ \text{temperature} & T. & \end{array} \tag{1.1}$$

In thermodynamics of irreversible processes (TIP) the necessary equations are based on the equations of balance of masses of the constituents and of momentum and energy of the mixture, viz.

$$\begin{aligned} \frac{\partial \varrho_\alpha}{\partial t} + \frac{\partial \varrho_\alpha v_j^\alpha}{\partial x_j} &= \tau_\alpha \\ \frac{\partial \varrho v_i}{\partial t} + \frac{\partial (\varrho v_i v_j - t_{ij})}{\partial x_j} &= 0 \\ \frac{\partial \varrho \varepsilon}{\partial t} + \frac{\partial (\varrho \varepsilon v_j + q_j)}{\partial x_j} &= t_{ij} \frac{\partial v_i}{\partial x_j} \end{aligned} \quad \text{or} \quad \begin{aligned} \dot{\varrho} + \varrho \frac{\partial v_i}{\partial x_i} &= 0. \\ \varrho \dot{c}_\alpha + \frac{\partial J_j^\alpha}{\partial x_j} &= \tau_\alpha \\ \varrho \dot{v}_i + \frac{\partial t_{ij}}{\partial x_j} &= 0 \\ \varrho \dot{\varepsilon} + \frac{\partial q_j}{\partial x_j} &= t_{ij} \frac{\partial v_i}{\partial x_j}. \end{aligned} \tag{1.2}$$

The dot denotes the rate of change seen by the observer who moves with the velocity $v_i = \sum_{\alpha=1}^{\nu} \frac{\varrho_\alpha}{\varrho} v_i^\alpha$ of the mixture; $c_\alpha = \frac{\varrho_\alpha}{\varrho}$ is the concentration, and $J_i^\alpha = \varrho_\alpha(v_i^\alpha - v_i)$ is the diffusion flux of constituent α. τ_α is its mass-production due to chemical reactions. Among the ν productions there are only n independent ones, where n is the number of independent chemical reactions. Indeed, we may express the τ_α's in terms of n reaction rate densities Λ^a by the relations

$$\tau_\alpha = \sum_{a=1}^{n} \gamma_\alpha^a m_\alpha \Lambda^a. \tag{1.3}$$

γ_α^a is the stoichiometric coefficient of constituent α in reaction a, and m_α is the molecular mass of constituent α. Note that $\sum_{\alpha=1}^{\nu} \tau_\alpha = 0$ by virtue of the equation $\sum_{\alpha=1}^{\nu} \gamma_\alpha^a m_\alpha = 0$, which represents the conservation of mass in every molecular reaction.

In order to close the system (1.2) we need constitutive equations for

$$\begin{array}{ll} \text{diffusion fluxes} & J_i^\alpha, \\ \text{reaction rate densities} & \Lambda^a, \\ \text{stress tensor} & t_{ij}, \\ \text{internal energy} & \epsilon, \\ \text{heat flux} & q_i. \end{array} \tag{1.4}$$

In TIP such relations are derived from the Gibbs equation for mixtures, viz.

$$(\varrho s)^\bullet = \frac{1}{T}\left((\varrho\epsilon)^\bullet - \sum_{\alpha=1}^{\nu} \mu_\alpha \dot{\varrho}_\alpha\right) \quad \text{or} \quad \dot{s} = \frac{1}{T}\left(\dot{\epsilon} - \frac{p}{\varrho^2}\dot{\varrho} - \sum_{\alpha=1}^{\nu-1} (\mu_\alpha - \mu_\nu)\dot{c}_\alpha\right). \tag{1.5}$$

Here s is the specific entropy, p is the pressure of the mixture in equilibrium, and μ_α are the chemical potentials. ϵ, p, and μ_α are considered to be known functions of ϱ, c_α, T from measurements in mixtures in equilibrium.

Elimination of $\dot{\varepsilon}, \dot{\varrho}, \dot{c}_\alpha$ between (1.5) and (1.2) and some rearrangement leads to the equation

$$\begin{aligned}\varrho\dot{s} &+ \frac{\partial}{\partial x_j}\left(\frac{q_j - \sum_{\alpha=1}^{\nu-1}(\mu_\alpha - \mu_\nu)J_j^\alpha}{T}\right) \\ &= \frac{1}{T}\sum_{a=1}^{n}\left(\sum_{\alpha=1}^{\nu-1}(\mu_\alpha - \mu_\nu)\gamma_\alpha^a m_\alpha\right)\Lambda^a + q_j\frac{\partial\frac{1}{T}}{\partial x_j} - \sum_{\alpha=1}^{\nu-1}J_j^\alpha\frac{\partial\frac{\mu_\alpha-\mu_\nu}{T}}{\partial x_j} \\ &+ \frac{1}{T}t_{\langle ij\rangle}\frac{\partial v_{\langle i}}{\partial x_{j\rangle}} + \frac{1}{T}\left(\frac{1}{3}t_{ii} + p\right)\frac{\partial v_i}{\partial x_i},\end{aligned} \tag{1.6}$$

which may be interpreted as a balance of entropy, if we identify

$$\begin{aligned}\varphi_i &= \frac{1}{T}\left(q_i - \sum_{\alpha=1}^{\nu-1}(\mu_\alpha - \mu_\nu)J_i^\alpha\right), \\ \Sigma &= q_j\frac{\partial\frac{1}{T}}{\partial x_j} - \sum_{\alpha=1}^{\nu-1}J_i^\alpha\frac{\partial\frac{\mu_\alpha-\mu_\nu}{T}}{\partial x_i} + \frac{1}{T}t_{\langle ij\rangle}\frac{\partial v_{\langle i}}{\partial x_{j\rangle}} + \frac{1}{T}\left(\frac{1}{3}t_{ii}+p\right)\frac{\partial v_i}{\partial x_i} + \frac{1}{T}\sum_{a=1}^{n}\left(\sum_{\alpha=1}^{\nu-1}(\mu_\alpha-\mu_\nu)\gamma_\alpha^a m_\alpha\right)\Lambda^a\end{aligned} \tag{1.7}$$

as the entropy flux and entropy production, respectively.

We observe that the entropy production is a sum of products of

thermodynamic fluxes	and	thermodynamic forces
heat flux q_i		temperature gradient $\partial\frac{1}{T}/\partial x_i$
diffusion fluxes J_i^α		chemical potential gradients $\partial\frac{\mu_\alpha-\mu_\nu}{T}/\partial x_i$
stress deviator $t_{\langle ij\rangle}$		deviatoric velocity gradient $\partial v_{\langle i}/\partial x_{j\rangle}$
dynamic pressure $\pi = \frac{1}{3}t_{ii} + p$		divergence of velocity $\partial v_i/\partial x_i$
reaction rate densities Λ^a		chemical "force" $\sum_{\alpha=1}^{\nu-1}(\mu_\alpha - \mu_\nu)\gamma_\alpha^a m_\alpha$.

The entropy production must be non-negative. Assuming linear relations between forces and fluxes TIP ensures this inequality by constitutive equations of the type

$$\begin{aligned}\Lambda^a &= \sum_{b=1}^{n}L^{ab}\left(\sum_{\alpha=1}^{\nu-1}(\mu_\alpha-\mu_\nu)\gamma_\alpha^b m_\alpha\right) + L^a\frac{\partial v_i}{\partial x_i}, \\ \frac{1}{3}t_{ii} + p &= \sum_{b=1}^{n}\tilde{L}^b\left(\sum_{\alpha=1}^{\nu-1}(\mu_\alpha-\mu_\nu)\gamma_\alpha^b m_\alpha\right) + \lambda\,\frac{\partial v_i}{\partial x_i};\end{aligned} \tag{1.8}$$

$$\begin{aligned}q_i &= L\,\frac{\partial\frac{1}{T}}{\partial x_i} + \sum_{\beta=1}^{\nu-1}L_\beta\left(-\frac{\partial\frac{\mu_\beta-\mu_\nu}{T}}{\partial x_i}\right), \\ J_i^\alpha &= \tilde{L}_\alpha\frac{\partial\frac{1}{T}}{\partial x_i} + \sum_{\beta=1}^{\nu-1}L_{\alpha\beta}\left(-\frac{\partial\frac{\mu_\beta-\mu_\nu}{T}}{\partial x_i}\right),\end{aligned} \tag{1.9}$$

$$t_{\langle ij\rangle} = 2\mu\frac{\partial v_{\langle i}}{\partial x_{j\rangle}}; \tag{1.10}$$

where the *phenomenological coefficients* L^{ab} through μ must satisfy the following inequalities:

$$\begin{bmatrix} L^{ab} & L^a \\ \tilde{L}^b & \lambda \end{bmatrix} \quad \text{— positive definite,} \tag{1.11}$$

$$\begin{bmatrix} L & L_\beta \\ \tilde{L}_\alpha & L_{\alpha\beta} \end{bmatrix} \quad \text{— positive definite,} \tag{1.12}$$

$$\mu \geq 0. \tag{1.13}$$

The equations (1.9) are generalizations of the traditional laws of Fourier and Fick according to which the heat flux and the diffusion flux depend on the gradients of temperature and concentrations, respectively, *but not on both.* TIP permits the temperature gradient to influence the diffusion fluxes and concentration gradients to influence the heat flux; both effects are indeed observed, and they are called thermo-diffusion effect and diffusion-thermo effect, respectively.

μ and λ are shear and bulk viscosities, respectively. The coefficients $L, L_\beta, \tilde{L}_\alpha$, and $L_{\alpha\beta}$ are the transport coefficients of heat conduction and diffusion, while $\tilde{L}^b$ is called the chemical viscosity.

If the coefficients L and λ, μ are known, the balance equations (1.2) and the constitutive relations (1.8) through (1.10) combine to provide a set of field equations for the fields $\varrho_\alpha, v_i^\alpha$, and T.

In conclusion of this short description of TIP in mixtures we briefly mention Onsager relations. These are symmetry or antisymmetry relations for the matrices of phenomenological coefficients. As generic notation for fluxes and forces we write J_A and X_A so that the constitutive relations (1.8) through (1.10) assume the schematic form

$$J_A = \sum_B L_{AB} X_B. \tag{1.14}$$

Onsager has postulated that either $L_{AB} = L_{BA}$ or $L_{AB} = -L_{BA}$ holds depending on the transformation property of the forces X_A and X_B under time reversal $t^* = -t$. If

$$X_C^* = \varepsilon_C X_C \tag{1.15}$$

holds, with $\varepsilon_C = \pm 1$ for even and odd forces under time reversal, the Onsager relations read

$$L_{AB} = \varepsilon_A \varepsilon_B L_{BA}. \tag{1.16}$$

In the present case we have

$$\sum_{\alpha=1}^{\nu} \mu_\alpha \gamma_\alpha^a m_\alpha, \quad \frac{\partial \frac{1}{T}}{\partial x_i}, \quad -\frac{\partial \frac{\mu_\alpha - \mu_\nu}{T}}{\partial x_i}$$

as even forces and

$$\frac{\partial v_i}{\partial x_i}, \quad \frac{\partial v_{<i}}{\partial x_{j>}}$$

as odd forces, so that (1.16) implies

$$\begin{aligned} L^{ab} &= L^{ba}, & \tilde{L}^b &= -L^b, \\ L_{\alpha\beta} &= L_{\beta\alpha}, & \tilde{L}_\beta &= +L_\beta. \end{aligned} \tag{1.17}$$

There is no good proof for the Onsager postulate, but we shall see that $(1.17)_2$ can be proved in extended thermodynamics of simple mixtures.

1.2 Paradox of Diffusion

We investigate the special case of a non-reacting binary mixture at rest, that is, with $v_i \equiv 0$ in which the temperature is uniform. In that case the only equations of interest among (1.2) and (1.9) are

$$\varrho \frac{\partial c}{\partial t} + \frac{\partial J_i}{\partial x_i} = 0 \quad \text{with } J_i = -L_{11} \frac{\mu_c}{T} \frac{\partial c}{\partial x_i}, \tag{1.18}$$

where c and J_i stand for c_1 and J_i^1 and μ_c is an abbreviation for $\partial(\mu^1 - \mu^2)/\partial c$. Elimination of J_i provides a differential equation for c, viz.

$$\frac{\partial c}{\partial t} = L_{11} \frac{\mu_c}{\varrho T} \frac{\partial^2 c}{\partial x_i^2}. \tag{1.19}$$

This equation is called the *diffusion equation*, it is a differential equation of parabolic type analogous to the equations (1.1.9) of heat conduction and shear diffusion. Consequently it predicts the propagation for disturbances in concentration with infinite speed. This is the same kind of paradox that was discussed in Chapter 1 in the context of heat conduction and shear diffusion.

We proceed to develop extended thermodynamics of mixtures in which no such paradox arises.

2 Extended Thermodynamics of Non-Viscous Simple Mixtures

2.1 Balance Equations

Again we consider a homogeneous mixture of ν constituents. The equations of balance of mass, momentum, and energy of the constituents read as follows:

$$\begin{aligned}
&\frac{\partial \varrho_\alpha}{\partial t} + \frac{\partial}{\partial x_j}(\varrho_\alpha v_j^\alpha) = \tau_\alpha, \\
&\frac{\partial \varrho_\alpha v_i^\alpha}{\partial t} + \frac{\partial}{\partial x_j}(\varrho_\alpha v_i^\alpha v_j^\alpha - t_{ij}^\alpha) = m_i^\alpha, \\
&\frac{\partial \varrho_\alpha(\varepsilon_\alpha + \frac{1}{2}v_\alpha^2)}{\partial t} + \frac{\partial}{\partial x_j}\left(\varrho_\alpha(\varepsilon_\alpha + \frac{1}{2}v_\alpha^2)v_j^\alpha - t_{ij}^\alpha v_i^\alpha + q_j^\alpha\right) = e_\alpha \qquad (\alpha = 1, 2, \ldots, \nu).
\end{aligned} \tag{2.1}$$

They have the same form as the corresponding equations of balance for a single body, except for the non-zero right-hand sides, which represent the production densities of mass, momentum, and energy of constituent α; these terms are due to interaction with the other constituents. Of course, since the mass, momentum, and energy of the mixture is conserved, we must have

$$\sum_{\alpha=1}^{\nu} \tau_\alpha = 0, \qquad \sum_{\alpha=1}^{\nu} m_i^\alpha = 0, \qquad \sum_{\alpha=1}^{\nu} e_\alpha = 0. \tag{2.2}$$

If we sum the equations $(2.1)_1$ over all constituents and introduce

$$\text{the density:} \quad \varrho = \sum_{\alpha=1}^{\nu} \varrho_\alpha \qquad \text{and} \qquad \text{velocity:} \quad v_i = \sum_{\alpha=1}^{\nu} \frac{\varrho_\alpha}{\varrho} v_i^\alpha \tag{2.3$_1$}$$

of the mixture, we obtain the mass balance of the mixture, viz.

$$\frac{\partial \varrho}{\partial t} + \frac{\partial}{\partial x_j}(\varrho v_j) = 0. \tag{2.4$_1$}$$

It is convenient to introduce the diffusion velocity u_i^α and the diffusion flux J_i^α by

$$u_i^\alpha = v_i^\alpha - v_i, \qquad J_i^\alpha = \varrho_\alpha(v_i^\alpha - v_i). \tag{2.5}$$

With $(2.5)_1$ and the definition of

$$\text{the stress:} \qquad t_{ij} = \sum_{\alpha=1}^{\nu} (t_{ij}^\alpha - \varrho_\alpha u_i^\alpha u_j^\alpha) \tag{2.3$_2$}$$

of the mixture we obtain, by summing the equations $(2.1)_2$,

$$\frac{\partial \varrho v_i}{\partial t} + \frac{\partial}{\partial x_j}(\varrho v_i v_j - t_{ij}) = 0. \qquad (2.4)_2$$

This is the balance equation of momentum of the mixture. Summing over all equations $(2.1)_3$ and introducing

$$\begin{aligned} &\text{the internal energy density:} && \varrho\varepsilon = \sum_{\alpha=1}^{\nu} \varrho_\alpha(\varepsilon_\alpha + \frac{1}{2}u_\alpha^2), \\ &\text{the flux of internal energy:} && q_i = \sum_{\alpha=1}^{\nu}(q_i^\alpha + \varrho_\alpha(\varepsilon_\alpha + \frac{1}{2}u_\alpha^2)u_i^\alpha - t_{ij}^\alpha u_j^\alpha) \end{aligned} \qquad (2.3)_3$$

we obtain the balance of energy of the mixture

$$\frac{\partial \varrho(\varepsilon + \frac{1}{2}v^2)}{\partial t} + \frac{\partial}{\partial x_j}\left(\varrho(\varepsilon + \frac{1}{2}v^2)v_j - t_{ij}v_i + q_j\right) = 0. \qquad (2.4)_3$$

Note that the equations $(2.4)_{1,2,3}$ for the mixture have the same form as those for a single fluid. It was in order to achieve that structure that the definitions $(2.3)_{1,2,3}$ were chosen. This choice goes back to Truesdell [27]; it seems to be the most rational procedure to follow.

Of course, $(2.4)_{2,3}$ imply a balance of internal energy, which we may write in the form

$$\frac{\partial \varrho\varepsilon}{\partial t} + \frac{\partial}{\partial x_j}(\varrho\varepsilon v_j + q_j) = t_{ij}\frac{\partial v_i}{\partial x_j}. \qquad (2.6)$$

2.2 Thermodynamic Processes

As stated before, the objective of thermodynamics of mixtures of fluids is the determination of the $4\nu + 1$ fields of

$$\begin{array}{lll} \text{mass densities} & \varrho_\alpha & \\ \text{velocities} & v_i^\alpha & (\alpha = 1, 2, \ldots, \nu), \\ \text{temperature} & T. & \end{array} \qquad (2.7)$$

For the determination of these fields we need the appropriate number of field equations. They are based on the equations of balance of masses and momenta of the constituents and of the energy—or internal energy—of the mixture, viz.

$$\begin{array}{llll} \text{masses} & \dfrac{\partial \varrho_\alpha}{\partial t} + \dfrac{\partial \varrho_\alpha v_j^\alpha}{\partial x_j} & = \tau_\alpha, & \\ \text{momenta} & \dfrac{\partial \varrho_\alpha v_i^\alpha}{\partial t} + \dfrac{\partial(\varrho_\alpha v_i^\alpha v_j^\alpha - t_{ij}^\alpha)}{\partial x_j} & = m_i^\alpha, & \\ \text{(internal) energy} & \dfrac{\partial \varrho\varepsilon}{\partial t} + \dfrac{\partial(\varrho\varepsilon v_j + q_j)}{\partial x_j} & = t_{ij}\dfrac{\partial v_i}{\partial x_j}. & \end{array} \qquad (2.8)$$

Among the mass productions τ_α there are only n independent ones as we have argued in Section 1. Indeed we have

$$\tau_\alpha = \sum_{a=1}^{n} \gamma_\alpha^a m_\alpha \Lambda^a, \qquad (2.9)$$

where, as before, the Λ^a's are the reaction rate densities of the n independent reactions.

The momentum productions m_i^α are not objective vectors, because the momentum of the chemically produced mass contributes to them. The combinations $m_i^\alpha - \tau_\alpha v_i^\alpha$, however, are objective vectors. We call these expressions the interaction forces and denote them by

$$M_i^\alpha = m_i^\alpha - \tau_\alpha v_i^\alpha. \tag{2.10}$$

By $(2.3)_3$ the specific internal energy ε and its flux q_i contain explicit expressions in terms of ϱ_α and v_i^α, viz.

$$\sum_{\alpha=1}^{\nu} \frac{1}{2}\frac{\varrho_\alpha}{\varrho} u_\alpha^2 \quad \text{and} \quad \sum_{\alpha=1}^{\nu} \frac{1}{2}\varrho_\alpha u_\alpha^2 u_i^\alpha. \tag{2.11}$$

It is convenient to split off these expressions and work with the intrinsic values ε^{I} and q_i^{I} of the internal energy and its flux, viz.

$$\varrho\varepsilon^{\mathrm{I}} = \sum_{\alpha=1}^{\nu} \varrho_\alpha \varepsilon_\alpha \quad \text{and} \quad q_i^{\mathrm{I}} = \sum_{\alpha=1}^{\nu} \left(q_i^\alpha + \varrho_\alpha \varepsilon_\alpha u_i^\alpha - t_{ij}^\alpha u_j^\alpha\right). \tag{2.12}$$

After these remarks we may close the system (2.8) of equations of balance by constitutive equations for $\Lambda^a, t_{ij}^\alpha, M_i^\alpha, \varepsilon^{\mathrm{I}}$, and q_i^{I}. In extended thermodynamics the constitutive quantities depend only on the local and present values of the fields, here ρ_α, v_i^α, and T. Therefore the generic form of the constitutive relations in extended thermodynamics of mixtures reads

$$C = \hat{C}(\varrho_\alpha, v_i^\alpha, T). \tag{2.13}$$

We call fluid mixtures that are characterized by such equations *non-viscous* and *simple*. Viscous and non-simple mixtures would require gradients of v_i^α and ϱ_α among the variables, or else they would require the consideration of higher moments as in Chapter 2.

If the constitutive functions $\hat{C}$ were known explicitly, between (2.8) and (2.13) we should have a system of explicit field equations for the fields $\varrho_\alpha, v_i^\alpha, T$. Every solution of that system is called a *thermodynamic process.*

2.3 Constitutive Theory

As always in extended thermodynamics we adopt the principle of material frame indifference, which states that the constitutive functions $\hat{C}$ are invariant under Euclidean transformations. In the present case this principle requires that the velocities v_i^α cannot appear separately among the variables but only as relative velocities V_i^α defined by

$$V_i^\alpha = v_i^\alpha - v_i^\nu.[1] \tag{2.14}$$

If we restrict the constitutive quantities Λ^a through q_i^{I} to be at most linearly dependent on V_i^α, we have

$$\begin{aligned}
\Lambda^a &= \Lambda^a(\varrho_\gamma, T),\\
t_{ij}^\alpha &= -p^\alpha(\varrho_\gamma, T)\delta_{ij},\\
M_i^\alpha &= \sum_{\beta=1}^{\nu-1} M_V^{\alpha\beta}(\varrho_\gamma, T) V_i^\beta,\\
\varepsilon_{\mathrm{I}} &= \varepsilon_{\mathrm{I}}(\varrho_\gamma, T),\\
q_i^{\mathrm{I}} &= \sum_{\beta=1}^{\nu-1} q_V^\beta(\varrho_\gamma, T) V_i^\beta.
\end{aligned} \tag{2.15}$$

Note that the intrinsic flux of internal energy is determined by the relative velocities V_i^β. Therefore heat conduction in a simple, non-viscous mixture is due to the diffusion of the constituents.

[1] It is convenient to single out one constituent—here ν—and refer the motion of the others to it.

The entropy principle will provide restrictions on the functions $\Lambda^a(\varrho_\gamma, T)$ through $q_V^\beta(\varrho_\gamma, T)$ that appear in (2.15). The entropy principle requires that the inequality

$$\frac{\partial \varrho s}{\partial t} + \frac{\partial(\varrho s v_j + \varphi_j)}{\partial x_j} \geq 0 \tag{2.16}$$

must hold for *all thermodynamic processes.* Here s is the specific entropy and φ_i is the (non-convective) entropy flux. Both are considered as constitutive quantities of the type (2.13), subject to the principle of material frame indifference and at most linearly dependent on V_i^α. Thus we have

$$\begin{aligned} s &= s(\varrho_\gamma, T), \\ \varphi_i &= \sum_{\alpha=1}^{\nu-1} \varphi_V^\alpha(\varrho_\gamma, T) V_i^\alpha. \end{aligned} \tag{2.17}$$

The exploitation of the inequality (2.16) makes use of the method of Lagrange multipliers, see [18]. We form the new inequality

$$\begin{aligned} \frac{\partial \varrho s}{\partial t} + \frac{\partial(\varrho s v_i + \varphi_i)}{\partial x_i} &- \sum_{\alpha=1}^{\nu} \Lambda^{\varrho_\alpha} \left(\frac{\partial \varrho_\alpha}{\partial t} + \frac{\partial \varrho_\alpha v_j^\alpha}{\partial x_j} - \tau_\alpha \right) \\ &- \sum_{\alpha=1}^{\nu} \Lambda^{v_i^\alpha} \left(\frac{\partial \varrho_\alpha v_i^\alpha}{\partial t} + \frac{\partial(\varrho_\alpha v_i^\alpha v_j^\alpha - t_{ij}^\alpha)}{\partial x_j} - m_i^\alpha \right) \\ &- \Lambda^\varepsilon \left(\frac{\partial \varrho \varepsilon}{\partial t} + \frac{\partial(\varrho \varepsilon v_j + q_j)}{\partial x_j} - t_{ij} \frac{\partial v_i}{\partial x_j} \right) \geq 0, \end{aligned} \tag{2.18}$$

which must hold for *all fields* $\varrho_\alpha, v_i^\alpha, T$. The Λ's are the Lagrange multipliers; they may be functions of $\varrho_\alpha, v_i^\alpha, T$.

The key to the exploitation of the inequality (2.18) is that, after insertion of the constitutive equations and after all indicated differentiations have been carried out, its left-hand side is explicitly linear in the derivatives

$$\frac{\partial \varrho_\beta}{\partial t}, \frac{\partial v_i^\beta}{\partial t}, \frac{\partial T}{\partial t}, \frac{\partial \varrho_\beta}{\partial x_k}, \frac{\partial v_i^\beta}{\partial x_k}, \frac{\partial T}{\partial x_k}. \tag{2.19}$$

The factors of these derivatives must then vanish lest the inequality not be violated for some choice of the values (2.19). This requirement leads to the following conditions:

$$\begin{aligned} &\frac{\partial s}{\partial \varrho_\beta} - \Lambda^\varepsilon \frac{\partial \varepsilon}{\partial \varrho_\beta} + \frac{1}{\varrho}(s - \Lambda^\varepsilon \varepsilon) - \frac{1}{\varrho} \Lambda^{\varrho_\beta} - \frac{1}{\varrho} v_k^\beta \Lambda^{v_k^\beta} = 0, \\ &\frac{\partial s}{\partial V_i^\beta} - \Lambda^\varepsilon \frac{\partial \varepsilon}{\partial V_i^\beta} - \frac{\varrho_\beta}{\varrho} \Lambda^{v_i^\beta} = 0 \qquad (\beta = 1, 2, \ldots, \nu - 1), \\ &-\sum_{\beta=1}^{\nu-1} \left(\frac{\partial s}{\partial V_i^\beta} - \Lambda^\varepsilon \frac{\partial \varepsilon}{\partial V_i^\beta} \right) - \frac{\varrho_\nu}{\varrho} \Lambda^{v_i^\nu} = 0, \\ &\frac{\partial s}{\partial T} - \Lambda^\varepsilon \frac{\partial \varepsilon}{\partial T} = 0, \\ &\frac{\partial \varphi_k}{\partial \varrho_\beta} - \Lambda^\varepsilon \frac{\partial q_k}{\partial \varrho_\beta} - \varrho \left(\frac{\partial s}{\partial \varrho_\beta} - \Lambda^\varepsilon \frac{\partial \varepsilon}{\partial \varrho_\beta} \right) u_k^\beta - \sum_{\alpha=1}^{\nu} \Lambda^{v_k^\alpha} \frac{\partial p^\alpha}{\partial \varrho_\beta} + \Lambda^\varepsilon t_{ik} \frac{1}{\varrho} u_i^\beta = 0, \\ &\left(\frac{\partial s}{\partial \varrho_\beta} - \Lambda^\varepsilon \frac{\partial \varepsilon}{\partial \varrho_\beta} \right) \delta_{ik} - \frac{1}{\varrho \varrho_\beta} \left(\frac{\partial \varphi_k}{\partial V_i^\beta} - \Lambda^\varepsilon \frac{\partial q_k}{\partial V_i^\beta} \right) + \frac{1}{\varrho} \Lambda^{v_i^\beta} u_k^\beta - \frac{1}{\varrho^2} \Lambda^\varepsilon t_{ik} = 0 \quad (\beta = 1, 2, \ldots, \nu - 1), \end{aligned} \tag{2.20}$$

$$\left(\frac{\partial s}{\partial \varrho_\nu} - \Lambda^\varepsilon \frac{\partial \varepsilon}{\partial \varrho_\nu}\right)\delta_{ik} + \sum_{\beta=1}^{\nu-1} \frac{1}{\varrho\varrho_\nu}\left(\frac{\partial \varphi_k}{\partial V_i^\beta} - \Lambda^\varepsilon \frac{\partial q_k}{\partial V_i^\beta}\right) - \frac{1}{\varrho}\Lambda^{v_i^\nu} u_k^\nu - \frac{1}{\varrho^2}\Lambda^\varepsilon t_{ik} = 0,$$

$$\left(\frac{\partial \varphi_k}{\partial T} - \Lambda^\varepsilon \frac{\partial q_k}{\partial T} - \sum_{\alpha=1}^{\nu} \Lambda^{v_k^\alpha} \frac{\partial p_\alpha}{\partial T}\right) = 0.$$

There remains the residual inequality

$$\sum_{\alpha=1}^{\nu} \Lambda^{v_i^\alpha} m_i^\alpha + \sum_{a=1}^{n}\left(\sum_{\alpha=1}^{\nu} \Lambda^{\varrho_\alpha} \gamma_\alpha^a m_\alpha\right)\Lambda^a \geq 0. \tag{2.21}$$

We proceed to evaluate the conditions (2.20). First, we list two useful identities that allow us to switch between the relative velocities V_i^α and the diffusion fluxes $J_i^\alpha = \varrho_\alpha u_i^\alpha$. They read

$$\varrho_\beta u_i^\beta = \sum_{\gamma=1}^{\nu-1} \underbrace{\left(\varrho_\beta \delta_{\beta\gamma} - \frac{\varrho_\beta \varrho_\gamma}{\varrho}\right)}_{\overset{-1}{F}_{\beta\gamma}} V_i^\gamma \quad \Longleftrightarrow \quad V_i^\gamma = \sum_{\beta=1}^{\nu-1} \underbrace{\left(\frac{1}{\varrho_\beta}\delta_{\gamma\beta} + \frac{1}{\varrho_\nu}\right)}_{F_{\beta\gamma}} \varrho_\beta u_i^\beta. \tag{2.22}$$

Now we recall that s is independent of V_i^γ and that ε depends on V_i^γ only through the explicit term $\sum_{\alpha=1}^{\nu} \frac{1}{2}\frac{\varrho_\alpha}{\varrho} u_\alpha^2$. We find

$$\frac{\partial \varepsilon}{\partial V_i^\beta} = -\frac{\varrho_\beta}{\varrho} u_i^\beta$$

and thus obtain from $(2.20)_{2,3}$

$$\Lambda^{v_i^\beta} = -\Lambda^\varepsilon u_i^\beta. \tag{2.23}$$

The equations $(2.20)_{4,1}$ may be written in the forms

$$\frac{\partial s}{\partial T} = \Lambda^\varepsilon \frac{\partial \varepsilon_I}{\partial T},$$

$$\frac{\partial \varrho s}{\partial \varrho_\beta} = \Lambda^\varepsilon \frac{\partial \varrho \varepsilon_I}{\partial \varrho_\beta} + \underbrace{\Lambda^{\varrho_\beta} + \Lambda^\varepsilon\left(\frac{1}{2}u_\beta^2 - v_k^\beta u_k^\beta\right)}_{\Lambda_I^{\varrho_\beta}},$$

which we may combine to read

$$d(\varrho s) = \Lambda^\varepsilon d(\varrho \varepsilon_I) + \sum_{\beta=1}^{\nu} \Lambda_I^{\varrho_\beta} d\varrho_\beta. \tag{2.24}$$

$\Lambda_I^{\varrho_\beta}$ is the velocity independent part of the Lagrange multiplier Λ^{ϱ_β}. Comparing equation (2.24) with the Gibbs relation of thermostatics [see $(1.5)_1$] we conclude that

$$\Lambda^\varepsilon = \frac{1}{T} \quad \text{and} \quad \Lambda_I^{\varrho_\beta} = -\frac{\mu_\beta}{T} \tag{2.25}$$

hold. Thus all Lagrange multipliers have been identified. T and μ_β are the absolute temperature of the mixture and the chemical potentials of its constituents. For future reference we rewrite the Gibbs equation in two forms, viz.

$$d(\varrho s) = \frac{1}{T}\left(d(\varrho\varepsilon_I) - \sum_{\alpha=1}^{\nu} \mu_\alpha d\varrho_\alpha\right) \quad \text{or} \quad d\left[\varrho(\varepsilon_I - Ts)\right] = -s dT + \sum_{\alpha=1}^{\nu} \mu_\alpha d\varrho_\alpha. \tag{2.26}$$

From $(2.20)_6$ and $(2.20)_7$ we obtain, by summation and subtraction, respectively,

$$\varrho(\varepsilon_I - Ts) + p = \sum_{\alpha=1}^{\nu} \varrho_\alpha \mu_\alpha \tag{2.27}$$

and

$$\varphi_j = \frac{q_j^I - \sum_{\alpha=1}^{\nu} \varrho_\alpha \mu_\alpha u_j^\alpha}{T}, \tag{2.28}$$

where $p = \sum_{\alpha=1}^{\nu} p_\alpha$ has been introduced. It remains to exploit the conditions $(2.20)_5$ and $(2.20)_8$. From $(2.20)_5$ we obtain after some rearrangement

$$\frac{\partial \mu_\alpha}{\partial \varrho_\beta} = \frac{1}{\varrho_\alpha} \frac{\partial p_\alpha}{\partial \varrho_\beta}, \tag{2.29}$$

a relation that has a severely restrictive consequence. Indeed, the integrability condition implied by (2.29) reads

$$\frac{\partial p_\alpha}{\partial \varrho_\beta} = 0 \;\; (\alpha \neq \beta), \quad \text{hence by (2.29)} \;\; \frac{\partial \mu_\alpha}{\partial \varrho_\beta} = 0 \;\; (\alpha \neq \beta). \tag{2.30}$$

It follows that a mixture characterized by generic constitutive relations of the form (2.13) is simple in the sense that the partial pressure p_α and the chemical μ_α of constituent α depend only on their own density ϱ_α—and on T, of course. Another aspect of this simplicity reveals itself, if we make use of the relation

$$\mu_\alpha = \frac{\partial \left[\varrho(\varepsilon_I - Ts)\right]}{\partial \varrho_\alpha}, \tag{2.31}$$

which follows from $(2.26)_2$. Insertion into $(2.30)_2$ gives

$$\frac{\partial^2 \varrho[\varepsilon_I - Ts]}{\partial \varrho_\alpha \partial \varrho_\beta} = 0 \quad \text{whence follows} \quad \varrho[\varepsilon_I - Ts] = \sum_{\alpha=1}^{\nu} \varrho_\alpha \psi_\alpha(\varrho_\alpha, T) \tag{2.32}$$

so that the free energy density $\varrho[\varepsilon_I - Ts]$ is a sum of functions of which each one depends only on one density. Obviously, by (2.26) this same summational character then holds for ϱs and, hence, for $\varrho \varepsilon_I$, namely,

$$\varrho s = \sum_{\alpha=1}^{\nu} \varrho_\alpha s_\alpha(\varrho_\alpha, T), \qquad \varrho \varepsilon_I = \sum_{\alpha=1}^{\nu} \varrho_\alpha \varepsilon_\alpha(\varrho_\alpha, T). \tag{2.33}$$

Clearly $s_\alpha(\varrho_\alpha, T)$ and $\varepsilon_\alpha(\varrho_\alpha, T)$ must be interpreted as the specific entropy and internal energy of the constituent α. Also we may conclude that the relation (2.27) and the Gibbs equation (2.26) may be decomposed into expressions for the single constituents, viz.

$$\varrho_\alpha(\varepsilon_\alpha - Ts_\alpha) + p_\alpha = \varrho_\alpha \mu_\alpha, \tag{2.34}$$

$$ds_\alpha = \frac{1}{T}\left(d\varepsilon_\alpha - \frac{p_\alpha}{\varrho_\alpha^2} d\varrho_\alpha\right). \tag{2.35}$$

The only relation in (2.20) yet to be exploited is $(2.20)_8$. We insert whatever we have learned so far and after a little calculation obtain

$$q_k^I = \sum_{\alpha=1}^{\nu} \left(\varepsilon_\alpha + \frac{p_\alpha}{\varrho_\alpha}\right) \varrho_\alpha u_k^\alpha. \tag{2.36}$$

By comparison with $(2.3)_3$ we conclude that the intrinsic flux of internal energy represents only the convection of the partial enthalpy $\varepsilon_\alpha + p_\alpha/\varrho_\alpha$ with the diffusive motion; in particular, in a simple mixture with the generic constitutive relations (2.13) the partial fluxes q_i^α are all zero.

It remains to exploit the residual inequality (2.21). An easy calculation shows that this inequality may be written in the form

$$-\sum_{\alpha,\beta=1}^{\nu-1} M_V^{\alpha\beta} V_i^\alpha V_i^\beta - \sum_{a=1}^{n} \left(\sum_{\alpha=1}^{\nu-1} \gamma_\alpha^a m_\alpha (\mu_\alpha - \mu_\nu + \frac{1}{2} V_\alpha^2) \right) \Lambda^a \geq 0. \tag{2.37}$$

We conclude that in equilibrium, where V_i^α and Λ^a all vanish, we must have

$$M_V^{\alpha\beta}|_E \text{ —positive definite} \quad \text{and} \quad \sum_{\alpha=1}^{\nu} \gamma_\alpha^a m_\alpha \mu_\alpha |_E = 0. \tag{2.38}$$

The second relation is the law of mass action, which at a given temperature provides n constraints on the equilibrium values of the densities ϱ_β—as many as there are independent reactions. If the law of mass action is not satisfied, reactions will proceed and in a linear theory we must set

$$\Lambda^a = \sum_{b=1}^{n} L_{ab} \left(\sum_{\alpha=1}^{\nu} \gamma_\alpha^b m_\alpha \mu_\alpha \right), \tag{2.39}$$

where the coefficients L_{ab} may depend on $\nu - n$ densities and T.

2.4 Summary of Results

We summarize the results of the previous two paragraphs. These results are applicable to non-viscous simple mixtures.

For the determination of the $4\nu + 1$ fields $\varrho_\beta, v_i^\beta, T$ we have the equations of balance

$$\frac{\partial \varrho_\alpha}{\partial t} + \frac{\partial \varrho_\alpha v_j^\alpha}{\partial x_j} = \tau_\alpha \quad \text{with } \tau_\alpha = \sum_{a=1}^{n} \gamma_\alpha^a m_\alpha \Lambda^a; \tag{2.40}$$

$$\frac{\partial \varrho_\alpha v_i^\alpha}{\partial t} + \frac{\partial (\varrho_\alpha v_i^\alpha v_j^\alpha - t_{ij}^\alpha)}{\partial x_j} = m_i^\alpha \quad \text{with } t_{ij}^\alpha = -p_\alpha(\varrho_\alpha, T)\delta_{ij}, \tag{2.41}$$

$$m_i^\alpha = \tau_\alpha v_i^\alpha + \sum_{\beta=1}^{\nu-1} M_V^{\alpha\beta}(\varrho_\gamma, T) V_i^\beta;$$

$$\frac{\partial \varrho \varepsilon}{\partial t} + \frac{\partial (\varrho \varepsilon v_j + q_j)}{\partial x_j} = t_{ij} \frac{\partial v_i}{\partial x_j} \quad \text{with } \varrho = \sum_{\alpha=1}^{\nu} \varrho_\alpha, \quad v_i = \sum_{\alpha=1}^{\nu} \frac{\varrho_\alpha}{\varrho} v_i^\alpha, \tag{2.42}$$

$$t_{ij} = -\sum_{\alpha=1}^{\nu} \left(p_\alpha(\varrho_\alpha T)\delta_{ij} + \varrho_\alpha u_i^\alpha u_j^\alpha \right),$$

$$\varepsilon = \sum_{\alpha=1}^{\nu} \frac{\varrho_\alpha}{\varrho} \left(\varepsilon_\alpha(\varrho_\alpha, T) + \frac{1}{2} u_\alpha^2 \right)$$

$$= \varepsilon_I + \sum_{\alpha=1}^{\nu} \frac{1}{2} \frac{\varrho_\alpha}{\varrho} u_\alpha^2,$$

$$q_i = \sum_{\alpha=1}^{\nu} \varrho_\alpha \left(\varepsilon_\alpha + \frac{p_\alpha}{\varrho_\alpha} + \frac{1}{2} u_\alpha^2 \right) u_i^\alpha.$$

The specific entropy s and the entropy flux φ_i are given in the forms

$$s = \sum_{\alpha=1}^{\nu} \frac{\varrho_\alpha}{\varrho} s_\alpha(\varrho_\alpha, T), \tag{2.43}$$

$$\varphi_i = \frac{\sum_{\alpha=1}^{\nu} \varrho_\alpha \left(T s_\alpha + \frac{1}{2} u_\alpha^2\right) u_j^\alpha}{T}. \tag{2.44}$$

Gibbs equations hold for the mixture and for all constituents

$$d(\varrho s) = \frac{1}{T}\left(d(\varrho \varepsilon_I) - \sum_{\alpha=1}^{\nu} \mu_\alpha d\varrho_\alpha\right), \tag{2.45}$$

$$ds_\alpha = \frac{1}{T}\left(d\varepsilon_\alpha - \frac{p_\alpha}{\varrho_\alpha^2} d\varrho_\alpha\right), \qquad \text{hence,} \quad \frac{\partial \varepsilon_\alpha}{\partial \varrho_\alpha} = -\frac{1}{\varrho_\alpha^2} T^2 \frac{\partial (p_\alpha/T)}{\partial T}, \tag{2.46}$$

and the chemical potential of constituent α may be written as

$$\mu_\alpha(\varrho_\alpha, T) = \varepsilon_\alpha - T s_\alpha + \frac{p_\alpha}{\varrho_\alpha}, \quad \text{hence,} \quad \frac{\partial \mu_\alpha}{\partial \varrho_\alpha} = \frac{1}{\varrho_\alpha} \frac{\partial p_\alpha}{\partial \varrho_\alpha}. \tag{2.47}$$

The coefficients $M_V^{\alpha\beta}$ form a positive-definite matrix and in equilibrium the densities ϱ_α and temperature T are related by the law of mass action, viz.

$$\sum_{\alpha=1}^{\nu} \gamma_\alpha^a m_\alpha \mu_\alpha \big|_E = 0. \tag{2.48}$$

In non-equilibrium the linear expression for the reaction rate densities reads

$$\Lambda^a = \sum_{b=1}^{n} L^{ab} \left(\sum_{\alpha=1}^{\nu} \gamma_\alpha^b m_\alpha \mu_\alpha\right). \tag{2.49}$$

2.5 Wave Propagation in a Non-Reacting Binary Mixture

In this section we wish to demonstrate explicitly that all wave speeds are finite in mixtures according to extended thermodynamics. As an illustrative special case we treat only the case of a non-reacting binary mixture. In such a mixture it is appropriate for intuitive reasons to replace the variables $\varrho_1, \varrho_2, v_i^1, v_i^2, T$ by the equivalent set

$$\varrho = \varrho_1 + \varrho_2, \qquad v_i = \frac{\varrho_1}{\varrho} v_i^1 + \frac{\varrho_2}{\varrho} v_i^2, \qquad T, \qquad c = \frac{\varrho_1}{\varrho}, \qquad J_i = \varrho_1 (v_i^1 - v_i). \tag{2.50}$$

Here c is called the concentration of constituent 1 and J_i is its diffusion flux. It is then proper to replace the governing balance laws by the set

$$\begin{aligned}
&\frac{\partial \varrho}{\partial t} + \frac{\partial \varrho v_j}{\partial x_j} = 0, \\
&\frac{\partial \varrho v_i}{\partial t} + \frac{\partial (\varrho v_i v_j - t_{ij})}{\partial x_j} = 0, \\
&\frac{\partial \varrho \varepsilon}{\partial t} + \frac{\partial (\varrho \varepsilon v_j + q_j)}{\partial x_j} = t_{ij} \frac{\partial v_i}{\partial x_j}, \\
&\varrho \left(\frac{\partial c}{\partial t} + v_j \frac{\partial c}{\partial x_j}\right) + \frac{\partial J_j}{\partial x_j} = 0, \\
&\frac{\partial J_i}{\partial t} + \frac{\partial}{\partial x_j}\left(J_i v_j - \frac{1}{\varrho c} J_i J_j - (t_{ij}^1 - c t_{ij})\right) = t_{ij} \frac{\partial c}{\partial x_j} - J_j \frac{\partial v_i}{\partial x_j} + m_i^1.
\end{aligned} \tag{2.51}$$

The constitutive relations are still the same ones as those summarized in Paragraph 2.4, but the new choice of variables makes some of the restrictive conditions inappropriate. In particular, the Gibbs equations (2.45) should now be written in the new form

$$ds = \frac{1}{T}\left(d\varepsilon_I - \frac{p}{\varrho^2}d\varrho - \mu dc\right), \tag{2.52}$$

where μ stands for $\mu_1 - \mu_2$. The integrability conditions implied by (2.52) read

$$\frac{\partial \varepsilon_I}{\partial \varrho} = -\frac{1}{\varrho^2}T^2\frac{\partial(p/T)}{\partial T}, \qquad \frac{\partial \varepsilon_I}{\partial c} = -T^2\frac{\partial(\mu/T)}{\partial T}, \qquad \frac{\partial p}{\partial c} = \varrho^2\frac{\partial \mu}{\partial \varrho}. \tag{2.53}$$

We linearize the equations in all deviations $\bar{\varrho}, \bar{v}_i, \bar{T}, \bar{c}, \bar{J}_i$ from a homogeneous and constant rest state $\tilde{\varrho}, \tilde{v}_i = 0, \tilde{T}, \tilde{c}, \tilde{J}_i = 0$ and obtain

$$\begin{bmatrix} \frac{\partial}{\partial t} & \tilde{\varrho}\frac{\partial}{\partial x_i} & 0 & 0 & 0 \\ \tilde{p}_\varrho\frac{\partial}{\partial x_i} & \tilde{\varrho}\frac{\partial}{\partial t} & \tilde{p}_\mathrm{T}\frac{\partial}{\partial x_i} & \tilde{p}_c\frac{\partial}{\partial x_i} & 0 \\ -\frac{\tilde{T}}{\tilde{\varrho}^2}\tilde{p}_\mathrm{T}\frac{\partial}{\partial t} & 0 & \tilde{\varepsilon}_\mathrm{T}\frac{\partial}{\partial t} & 0 & \frac{\tilde{T}}{\tilde{\varrho}}\left(\frac{\tilde{p}^1_\mathrm{T}}{\tilde{\varrho}_1} - \frac{\tilde{p}^2_\mathrm{T}}{\tilde{\varrho}_2}\right)\frac{\partial}{\partial x_i} \\ 0 & 0 & 0 & \tilde{\varrho}\frac{\partial}{\partial t} & \frac{\partial}{\partial x_i} \\ -\frac{\tilde{\varrho}_1\tilde{\varrho}_2}{\tilde{\varrho}}\tilde{\mu}_\varrho\frac{\partial}{\partial x_i} & 0 & -\frac{\tilde{\varrho}_1\tilde{\varrho}_2}{\tilde{\varrho}}\left(\frac{\tilde{p}^1_\mathrm{T}}{\tilde{\varrho}_1} - \frac{\tilde{p}^2_\mathrm{T}}{\tilde{\varrho}_2}\right)\frac{\partial}{\partial x_i} & -\frac{\tilde{\varrho}_1\tilde{\varrho}_2}{\tilde{\varrho}}\tilde{\mu}_c\frac{\partial}{\partial x_i} & \frac{\partial}{\partial t} - M_V^{11}\frac{\tilde{\varrho}}{\tilde{\varrho}_1\tilde{\varrho}_2} \end{bmatrix} \begin{bmatrix} \bar{\varrho} \\ \bar{v}_i \\ \bar{T} \\ \bar{c} \\ \bar{J}_i \end{bmatrix} = 0. \tag{2.54}$$

The quantities p_ϱ, ε_T, μ_c, etc., denote derivatives of p, ε, and μ with respect to the indicated variables.

We eliminate v_i, T, and J_i from this linear first-order system of partial differential equations and obtain a coupled set of second order differential equations for $\bar{\varrho}$ and $\bar{c}$, viz.

$$\begin{aligned} &\frac{\partial^2 \bar{\varrho}}{\partial t^2} - \left[p_\varrho + \frac{Tp_T^2}{\varrho^2\varepsilon_T}\right]\Delta\bar{\varrho} - \left[p_c + \frac{T}{c_T}p_T\left(\frac{p_T^1}{\varrho_1} - \frac{p_T^2}{\varrho_2}\right)\right]\Delta\bar{c} = 0, \\ &\frac{\partial^2 \bar{c}}{\partial t^2} - M_V^{11}\frac{\varrho}{\varrho_1\varrho_2}\frac{\partial \bar{c}}{\partial t} - \frac{\varrho_1\varrho_2}{\varrho^2}\left[\mu_c + \frac{T}{\varepsilon_T}\left(\frac{p_T^1}{\varrho_1} - \frac{p_T^2}{\varrho_2}\right)^2\right]\Delta\bar{c} - \frac{\varrho_1\varrho_2}{\varrho^4}\left[p_c + \frac{T}{\varepsilon_T}p_T\left(\frac{p_T^1}{\varrho_1} - \frac{p_T^2}{\varrho_2}\right)\right]\Delta\bar{\varrho} = 0. \end{aligned} \tag{2.55}$$

The tilde has been dropped from the coefficients for convenience.

The coupling of the fields $\bar{\varrho}$ and $\bar{c}$ is provided by the last term in each equation with the coefficient

$$W = p_c + \frac{T}{\varepsilon_T}p_T\left(\frac{p_T^1}{\varrho_1} - \frac{p_T^2}{\varrho_2}\right). \tag{2.56}$$

Let us suppose for a moment that that term were absent. In that case the first equation is the usual equation for the propagation of sound with the speed

$$V_\varrho^2 = p_\varrho + \frac{Tp_T^2}{\varrho^2\varepsilon_T}. \tag{2.57}$$

The second equation is the telegraph equation—always assuming $W = 0$—and it describes the propagation of what has been called *second sound*, a strongly damped mode of sound propagation because of the term with M_V^{11}. According to $(2.55)_2$ the speed of the second sound is given by

$$V_c^2 = \frac{\varrho_1\varrho_2}{\varrho^2}\left(\mu_c + \frac{T}{\varepsilon_T}\left(\frac{p_T^1}{\varrho_1} - \frac{p_T^2}{\varrho_2}\right)^2\right) = \frac{\varrho_1\varrho_2}{\varrho^2}\left(\mu_c - \frac{p_c}{p_T}\left(\frac{p_T^1}{\varrho_1} - \frac{p_T^2}{\varrho_2}\right)\right), \tag{2.58}$$

where $(2.58)_2$ only holds for $W = 0$.

We continue to consider the case of no interaction and note that without the term $\partial J_i/\partial t$ in (2.51), which describes the acceleration of the relative motion of the two constituents, the second equation in (2.55) would read

$$M_V^{11}\frac{\partial \bar{c}}{\partial t} + \frac{\varrho_1^2\varrho_2^2}{\varrho^3}\left[\mu_c + \frac{T}{\varepsilon_T}\left(\frac{p_T^1}{\varrho_1} - \frac{p_T^2}{\varrho_2}\right)^2\right]\Delta\bar{c} = 0. \tag{2.59}$$

This is a parabolic equation. It provides infinite speeds of propagation of disturbances in concentration. We recall the arguments of Section 1 about the paradox of diffusion and conclude that no such paradox arises, if we account for the inertia of the diffusive motion.

Now, with coupling, that is, with $W \neq 0$, we proceed to eliminate $\bar{c}$ between the equations (2.55) and obtain a single equation of fourth order in time, viz.

$$\frac{\partial^4\bar{c}}{\partial t^4} - M_V^{11}\frac{\varrho}{\varrho_1\varrho_2}\frac{\varrho^3\bar{c}}{\partial t^3} - (V_\varrho^2 + V_c^2)\frac{\partial^2\Delta\bar{c}}{\partial t^2} + M_V^{11}\frac{\varrho}{\varrho_1\varrho_2}V_\varrho^2\frac{\partial\Delta\bar{c}}{\partial t} + \left(V_\varrho^2V_c^2 - \frac{\varrho_1\varrho_2}{\varrho^4}W^2\right)\Delta\Delta c = 0. \tag{2.60}$$

This is again a hyperbolic equation. It predicts the propagation of the first and second sound with the speeds

$$V_\pm = \sqrt{2}\frac{\sqrt{\frac{V_\varrho^2V_c^2 - (\varrho_1\varrho_2/\varrho^4)W^2}{V_\varrho^2 + V_c^2}}}{\sqrt{\left\{1 \pm \sqrt{1 - 4\frac{V_\varrho^2V_c^2 - (\varrho_1\varrho_2/\varrho^4)W^2}{(V_\varrho^2+V_c^2)^2}}\right\}}}. \tag{2.61}$$

Additional insight into the nature of first and second sound is gained by the consideration of plane harmonic waves of small amplitude. We introduce

$$\bar{\varrho} = \hat{\varrho}E, \qquad \bar{v}_i = \hat{v}_iE, \qquad \bar{T} = \hat{T}E, \qquad \bar{c} = \hat{c}E, \qquad \bar{J}_i = \hat{J}_iE \tag{2.62}$$

into the linearized equations (2.54). E stands for

$$E = e^{i(\omega t - k\mathbf{n}\cdot\mathbf{x})} \tag{2.63}$$

and $\hat{\varrho}$ through $\hat{J}_i$ are complex amplitudes. We obtain with $u = \omega/k$

$$\begin{bmatrix} u & -\varrho n_i & 0 & 0 & 0 \\ -p_\varrho n_i & \varrho u & -p_T n_i & -p_c n_i & 0 \\ -\frac{T}{\varrho^2}p_T u & 0 & \varepsilon_T u & 0 & -\frac{T}{\varrho}\left(\frac{p_T^1}{\varrho_1} - \frac{p_T^2}{\varrho_2}\right)n_i \\ 0 & 0 & 0 & \varrho u & -n_i \\ \frac{\varrho_1\varrho_2}{\varrho}\mu_\varrho n_i & 0 & \frac{\varrho_1\varrho_2}{\varrho}\left(\frac{p_T^1}{\varrho_1} - \frac{p_T^2}{\varrho_2}\right)n_i & \frac{\varrho_1\varrho_2}{\varrho}\mu_c n_i & u\left(1 + iM_V^{11}\frac{\varrho}{\varrho_1\varrho_2}\frac{1}{\omega}\right) \end{bmatrix} \begin{bmatrix} \hat{\varrho} \\ \hat{v}_i \\ \hat{T} \\ \hat{c} \\ \hat{J}_i \end{bmatrix} = 0. \tag{2.64}$$

Inspection of $(2.64)_{2,5}$ shows that both v_i and J_i are longitudinal waves. Without loss of generality we may then take n_i, v_i, and J_i to point in the x_1-direction. Elimination of $\hat{\varrho}, \hat{T}$, and $\hat{c}$ provides a linear system of equations for $\hat{v}$ and $\hat{J}$, viz. with $\pi = (p_T^1/\varrho_1) - (p_T^2/\varrho_2)$

$$\begin{aligned}
&-\varrho\left(p_\varrho - \frac{Tp_T^2}{\varrho^2\varepsilon_T} - u^2\right)\hat{v} - \frac{1}{\varrho}\left(p_c - T\frac{p_T}{\varepsilon_T}\pi\right)\hat{J} = 0,\\
&-\varrho_1\varrho_2\left(\mu_\varrho - \frac{1}{p_T}\pi(p_\varrho - u^2)\right)\hat{v} - \left(\frac{\varrho_1\varrho_2}{\varrho^2}\left(\mu_c - \frac{p_c}{p_T}\pi\right) - u^2\left(1 + iM_V^{11}\frac{\varrho}{\varrho_1\varrho_2}\frac{1}{\omega}\right)\right)\hat{J} = 0.
\end{aligned} \tag{2.65}$$

We investigate the uncoupled case in which W, defined by (2.56), vanishes and obtain

$$\begin{aligned}
&\left(p_\varrho + \frac{Tp_T^2}{\varrho^2\varepsilon_T} - u^2\right)\hat{v} = 0,\\
&\frac{\varrho_1\varrho_2}{p_T}\pi\left(p_\varrho + \frac{Tp_T^2}{\varrho^2\varepsilon_T} - u^2\right)\hat{v} - \left(\frac{\varrho_1\varrho_2}{\varrho^2}\left(\mu_c - \frac{p_c}{p_T}\pi\right) - u^2\left(1 + iM_V^{11}\frac{\varrho}{\varrho_1\varrho_2}\frac{1}{\omega}\right)\right)\hat{J} = 0.
\end{aligned} \tag{2.66}$$

For $\omega \to \infty$ the solutions $u_{1,2}$ are the pulse speeds of disturbances. From (2.66) we, therefore, conclude from

i. Either $\hat{v} = 0$ holds, in which case $\hat{J}$ may be unequal to zero and pulses propagate with the speed

$$u = V_c \quad \text{where } V_c \text{ is given by (2.58)},$$

ii. or $\hat{J} = 0$ holds, in which case $\hat{v}$ may be unequal to zero and pulses propagate with the speed

$$u = V_\varrho \quad \text{where } V_\varrho \text{ is given by (2.57)}.$$

In the first case we speak of the second sound. The two constituents oscillate with respect to each other and the mixture as a whole is at rest. We have $\hat{\varrho} = 0$, while $\hat{c} \neq 0$. The second case corresponds to the propagation of the first sound. The two constituents are at rest with respect to each other while the mixture as a whole oscillates. We have $\hat{c} = 0$ and $\hat{\varrho} \neq 0$. In both cases the temperature oscillates; indeed, from $(2.64)_3$ we conclude that

$$\hat{T} = \frac{1}{p_T}\left((u^2 - p_\varrho)\hat{\varrho} - p_c\hat{c}\right) \tag{2.67}$$

holds, always for the uncoupled case.

Note that for the second sound u is complex, so the second sound is damped. In fact, the damping is so strong that the second sound has never been observed in mixtures of gases.

2.6 Landau Equations: First and Second Sound in He II

The only case in which the second sound has been observed in a fluid is the case of liquid He II. This, of course, is not a mixture in the proper sense of the word, because it consists only of Helium atoms. We may consider He II as a *phase* mixture consisting of the superfluid component s whose atoms are in the state of Bose-Einstein condensation and the normal fluid n whose atoms have non-zero energies. We have given a brief description of a gas in the state of Bose-Einstein condensation in Chapter 2 (see Paragraph 2.4.4), and we recall that the entropy of the fully condensed gas vanishes. Thus in the theory of mixtures between the constituents s and n we have to set

$$s_s = 0. \tag{2.68}$$

The most striking observation that can be made in He II is that there is no dissipation accompanying the diffusion and the transition between the phases s and n; that is, there is no entropy production. We recall from (2.37) that in a binary mixture there are two "mechanisms" of entropy production, namely, diffusion and "chemical reaction," that is, here, transition between the phases s and n. Since there is only one "reaction" and since the stoichiometric coefficients are 1 and -1, we may write (2.37) in the form

$$-M_V^{ss} V_s^2 - m\left(\mu_s - \mu_n + \frac{1}{2}V_s^2\right)\Lambda \geq 0. \tag{2.69}$$

The only way that the left-hand side can vanish in the presence of diffusion and phase transition is that $M_V^{ss} = 0$ and $\mu_s - \mu_n + \frac{1}{2}V_s^2 = 0$ hold. In other words, the interaction force between the diffusing phases must vanish and, ignoring the non-linear term V_s^2, the chemical potentials of the phases must always be equal. We may therefore write

$$M_V^{11} = 0 \quad \text{and} \quad \mu_s(\varrho_s, T) = \mu_n(\varrho_n, T). \tag{2.70}$$

We proceed to find the special forms of the equations (2.40) through (2.42) appropriate for He II. First of all we recall (2.39), or (2.49), which in the present case reads

$$\Lambda = Lm(\mu_s - \mu_n). \tag{2.71}$$

Since $\mu_s = \mu_n$ holds and Λ does not vanish, L must be infinite. Thus the equations (2.40) have an indeterminate right-hand side and the only useful consequence is given by their sum, viz. in linearized form

$$\boxed{\frac{\partial \varrho}{\partial t} + \tilde{\varrho}\frac{\partial v_i}{\partial x_i} = 0}. \tag{2.72}$$

From (2.41) we derive the balance of total momentum by summation

$$\boxed{\tilde{\varrho}\frac{\partial v_i}{\partial t} + \frac{\partial p}{\partial x_i} = 0} \tag{2.73}$$

and the balance of relative momentum by subtraction

$$\frac{\partial V_i^s}{\partial t} + \frac{1}{\varrho_s}\frac{\partial p^s}{\partial x_i} - \frac{1}{\varrho_n}\frac{\partial p^n}{\partial x_i} = 0. \tag{2.74}$$

We rewrite (2.74) by using (2.29). Hence we have

$$\begin{aligned}
\frac{1}{\varrho_s}\frac{\partial p^s}{\partial x_i} - \frac{1}{\varrho_n}\frac{\partial p^n}{\partial x_i} &= \frac{\partial(\mu^s - \mu^n)}{\partial x_i} - \left(\frac{\partial(\mu^s - (p_s/\varrho_s))}{\partial T} - \frac{\partial\mu^n - (p_n/\varrho_n))}{\partial T}\right)\frac{\partial T}{\partial x_i} \quad \text{with (2.70), (2.47):}\\
&= -\left[\left(\frac{\partial \varepsilon_s}{\partial T} - T\frac{\partial s_s}{\partial T} - s_s\right) - \left(\frac{\partial \varepsilon_n}{\partial T} - T\frac{\partial s_n}{\partial T} - s_n\right)\right]\frac{\partial T}{\partial x_i} \quad \text{with (2.68), (2.46):}\\
&= -s_n\frac{\partial T}{\partial x_i} \quad \text{with } \varrho s = \varrho_n s_n \text{ [see (2.68)]:}\\
&= -\frac{\varrho}{\varrho_n}s\frac{\partial T}{\partial x_i}.
\end{aligned}$$

Thus (2.74) may be written in the form

$$\boxed{\frac{\tilde{\varrho}_n}{\tilde{\varrho}}\frac{\partial V_i^s}{\partial t} - \tilde{s}\frac{\partial T}{\partial x_i} = 0}. \tag{2.75}$$

There remains the energy equation (2.42), which, when linearized, assumes the form

$$\tilde{\varrho}\frac{\partial \varepsilon}{\partial t} + \frac{\tilde{\varrho}_s\tilde{\varrho}_n}{\tilde{\varrho}}\left(\tilde{\varepsilon}_s + \frac{\tilde{p}_s}{\tilde{\varrho}_s} - \tilde{\varepsilon}_n - \frac{\tilde{p}_n}{\tilde{\varrho}_n}\right)\frac{\partial V_i^s}{\partial x_i} + \tilde{p}\frac{\partial v_i}{\partial x_i} = 0. \tag{2.76}$$

We use (2.52) and (2.51)$_4$ to write

$$\begin{aligned}\tilde{\varrho}\frac{\partial \varepsilon}{\partial t} + \tilde{p}\frac{\partial v_i}{\partial x_i} &= \tilde{\varrho}\tilde{T}\frac{\partial s}{\partial t} + \tilde{\varrho}(\tilde{\mu}^s - \tilde{\mu}^n)\frac{\partial c^s}{\partial t} \\ &= \tilde{\varrho}\tilde{T}\frac{\partial s}{\partial t} + (\tilde{\mu}^s - \tilde{\mu}^n)\frac{\tilde{\varrho}_s\tilde{\varrho}_n}{\tilde{\varrho}}\frac{\partial V_i^s}{\partial x_i}.\end{aligned}$$

Hence follows an alternative of (2.76) with (2.47) and (2.68), viz.

$$\boxed{\frac{\partial s}{\partial t} - \frac{\tilde{\varrho}_s}{\tilde{\varrho}}\tilde{s}\frac{\partial V_i^s}{\partial x_i} = 0}. \tag{2.77}$$

Some of the preceding reformulations may seem arbitrary; they have been made so as to arrive at the framed equations that are those that Landau postulated for He II. We conclude that the Landau bi-fluid theory is a special case of a non-viscous simple mixture. The special character is provided by the assumptions (2.68) and (2.70). The recognition of this fact has long been delayed. The issue was confused, because one of the mass balance equations is missing from the set of Landau equations. We have seen that that equation is not needed for the calculation of the fields $\varrho_s, \varrho_n, v_i^s, v_i^n$, and T, because by (2.70)$_2$ there is an algebraic equation between ϱ_s, ϱ_n, and T. Once the fields $\varrho_s, \varrho_n, v_i^s, v_i^n$, and T have been calculated from (2.70)$_2$ and the Landau equations, we may use the "missing mass balance" to calculate the indeterminate reaction rate density Λ.

We conclude this section with a treatment of first and second sound in He II. We choose p, T and v_i^s, v_i^n as variables and write

$$\frac{\partial \varrho}{\partial t} = \left(\frac{\partial \varrho}{\partial p}\right)_T\frac{\partial p}{\partial t} + \left(\frac{\partial \varrho}{\partial T}\right)_p\frac{\partial T}{\partial t}, \qquad \frac{\partial s}{\partial t} = \left(\frac{\partial s}{\partial p}\right)_T\frac{\partial p}{\partial t} + \left(\frac{\partial s}{\partial T}\right)_p\frac{\partial T}{\partial t}. \tag{2.78}$$

We recall that in an arbitrary binary mixture ϱ and s should depend on three variables, e.g., on p, T, and c^s, but in Helium only two of those are independent because of the constraint $\mu_s = \mu_n$ [see (2.70)$_2$].

We assume that there is no thermal expansion so that

$$\left(\frac{\partial \varrho}{\partial T}\right)_p = 0 \quad \text{and} \quad \left(\frac{\partial s}{\partial p}\right)_T = \frac{1}{\varrho^2}\left(\frac{\partial \varrho}{\partial T}\right)_p = 0,$$

and obtain the Landau equations in the form

$$\begin{bmatrix} 0 & \left(\frac{\partial \varrho}{\partial p}\right)_T\frac{\partial}{\partial t} & \varrho_s\frac{\partial}{\partial x_i} & \varrho_n\frac{\partial}{\partial x_i} \\ 0 & \frac{\partial}{\partial x_i} & \varrho_s\frac{\partial}{\partial t} & \varrho_n\frac{\partial}{\partial t} \\ -\varrho s\frac{\partial}{\partial x_i} & 0 & \varrho_n\frac{\partial}{\partial t} & -\varrho_n\frac{\partial}{\partial t} \\ \frac{1}{\varrho T}c_p\frac{\partial}{\partial t} & 0 & -\varrho_s s\frac{\partial}{\partial x_i} & \varrho_s s\frac{\partial}{\partial x_i} \end{bmatrix}\begin{bmatrix} T \\ p \\ v_i^s \\ v_i^n \end{bmatrix} = \begin{bmatrix} 0 \\ 0 \\ 0 \\ 0 \end{bmatrix}, \tag{2.79}$$

where $c_p = T(\partial s/\partial T)_p$ is the specific heat. We investigate plane harmonic waves and introduce

$$T = \tilde{T} + \hat{T}E, \qquad p = \tilde{p} + \hat{p}E, \qquad v_i^s = \hat{v}_i^s E, \qquad v_i^n = \hat{v}_i^n E, \tag{2.80}$$

where $E = \exp\left[i(\omega t - k\mathbf{nx})\right]$. It turns out that the waves are longitudinal; that is, v_i^s and v_i^n point in the direction of propagation $\mathbf{n}$, which we choose to be the x_1-direction. We insert (2.80) into (2.79) and obtain with $u = \omega/k$ as the phase speed

$$\begin{bmatrix} 0 & \left(\frac{\partial\varrho}{\partial p}\right)_T u & -\varrho_s & -\varrho_n \\ 0 & -1 & \varrho_s u & \varrho_n u \\ \varrho s & 0 & \varrho_n u & -\varrho_n u \\ \frac{c_p}{\varrho T} u & 0 & \varrho_s s & -\varrho_s s \end{bmatrix} \begin{bmatrix} \hat{T} \\ \hat{p} \\ \hat{v}_1^s \\ \hat{v}_i^n \end{bmatrix} = \begin{bmatrix} 0 \\ 0 \\ 0 \\ 0 \end{bmatrix}. \tag{2.81}$$

Elimination of $\hat{T}$ and $\hat{p}$ between those four equations results in the equations

$$\begin{aligned} &\left(\frac{\varrho_s}{\varrho_n}\frac{s^2 T}{c_p} - u^2\right)(v_1^s - v_1^n) = 0, \\ &\left(\left(\frac{\partial p}{\partial \varrho}\right)_T - u^2\right)\left(\frac{\varrho_s}{\varrho}v_1^s + \frac{\varrho_n}{\varrho}v_1^n\right) = 0. \end{aligned} \tag{2.82}$$

We conclude that there are two modes of propagation:

i. Either $V_i^s = v_1^s - v_1^n = 0$ holds, in which case $v_1 = (\varrho_s/\varrho)v_1^s + (\varrho_n/\varrho)v_1^n$ may be unequal to zero. The phase speed is

$$u = \sqrt{\left(\frac{\partial p}{\partial \varrho}\right)_T}, \tag{2.83}$$

ii. or $v_1 = 0$ holds, in which case V_1^s may be unequal to zero and the wave moves with the speed

$$u = \sqrt{\frac{\varrho_s}{\varrho_n}\frac{s^2 T}{c_p}}. \tag{2.84}$$

These two cases correspond to the first and second sound respectively in He II. In the first sound the two phases move jointly and $(2.81)_{1,4}$ show that the pressure oscillates while the temperature remains constant. In the second sound the two phases oscillate with respect to each other but the mixture as a whole does not move. We see from $(2.81)_{1,4}$ that now the pressure is constant while the temperature oscillates. For that reason the second sound must be generated by temperature oscillations rather than pressure oscillations.

3 Ordinary and Extended Thermodynamics of Mixtures

3.1 The Laws of Fick and Fourier in Extended Thermodynamics

We recall the equations (1.9), which are the generalizations offered by TIP of the laws of Fourier and Fick. In extended thermodynamics we have a rather different equation for q_i [see (2.42)], while J_i^α or V_i^α is a variable rather than a constitutive quantity. The question arises how the two theories compare and, in particular, what the status of the laws of Fourier and Fick is in extended thermodynamics. For simplicity we answer that question for a non-reacting mixture.

It turns out that Fick's law results from a truncation of the balance equations of momenta. In order to see this we recall $(2.41)_1$ and, for generality, we refer it to a non-inertial frame so that there are inertial forces. We have

$$\dot{v}_i^\alpha + u_j^\alpha \frac{\partial v_i^\alpha}{\partial x_j} - \ddot{i}_i^0 - 2W_{ik}v_k^\alpha - \frac{1}{\varrho_\alpha}\frac{\partial t_{ij}^\alpha}{\partial x_j} = \frac{1}{\varrho_\alpha}m_i^\alpha \qquad (\alpha = 1, 2, \ldots, \nu), \tag{3.1}$$

where $\dot{i}_i^0$ is the velocity-independent part of the inertial acceleration and $\mathbf{W}$ is the angular velocity of the frame. We subtract the νth equation from all others and obtain with $(2.41)_{2,3}$ and $(2.47)_2$

$$\left[\dot{V}_i^\alpha + u_j^\alpha \frac{\partial v_i^\alpha}{\partial x_j} - u_j^\nu \frac{\partial v_i^\nu}{\partial x_j} - 2W_{ik}V_k^\alpha\right] + T\frac{\partial \frac{\mu_\alpha - \mu_\nu}{T}}{\partial x_i} + T^2\left(T\frac{\partial \frac{\mu_\alpha - \mu_\nu}{T}}{\partial T} - \frac{\partial\left(\frac{p_\alpha}{\varrho_\alpha} - \frac{p_\nu}{\varrho_\nu}\right)}{\partial T}\right)\frac{\partial \frac{1}{T}}{\partial x_i} = \sum_{\beta=1}^{\nu-1}(\mathbf{F}\mathbf{M}_V\mathbf{F})_{\alpha\beta}J_i^\beta, \tag{3.2}$$

where $\mathbf{M}_V$ is the matrix consisting of the $M_V^{\alpha\beta}$'s and $\mathbf{F}$ is the matrix defined by $(2.22)_2$. The equations (3.2) represent the equations of relative motion in the mixture and they are exact within the present theory of non-viscous simple mixtures.

In order to obtain from (3.2) the constitutive equation for J_i^β, which TIP requires, we rely on the Maxwellian iteration again, which was explained, in a different context, in Paragraph 2.3.2: On the left-hand side of (3.2) we replace V_i^α by its equilibrium value, namely, zero, and use the equation to calculate $\overset{(1)}{J_i^\beta}$, the first iterate. This, or the corresponding $\overset{(1)}{V_i^\beta}$, is then used on the left-hand side to give $\overset{(2)}{J_i^\beta}$, etc. Here we shall be content with the first iterate because it gives us Fick's law. Indeed with $V_i^\alpha \equiv 0$ on the left-hand side of (3.2) these equations may be written in the form

$$\boxed{J_i^\alpha = \sum_{\beta=1}^{\nu-1}\underbrace{\left\{-T\left[\mathbf{A}\right]_{\alpha\beta}\right\}}_{L_{\alpha\beta}}\frac{\partial\left(-\frac{\mu^{\beta\nu}}{T}\right)}{\partial x_i} + \underbrace{\left\{\sum_{\beta=1}^{\nu-1}T^2\left[\mathbf{A}\right]_{\alpha\beta}\left(T\frac{\partial\frac{\mu^{\beta\nu}}{T}}{\partial T} - \frac{\partial \pi^{\beta\nu}}{\partial T}\right)\right\}}_{\tilde{L}_\alpha}\frac{\partial\frac{1}{T}}{\partial x_i},} \tag{3.3}$$

where $\mu^{\beta\nu}$ stands for $\mu^\beta - \mu^\nu$, $\pi^{\beta\nu}$ for $p_\beta/\varrho_\beta - p_\nu/\varrho_\nu$, and $\mathbf{A}$ for $\left(\overset{-1}{\mathbf{F}}\overset{-1}{\mathbf{M}_V}\overset{-1}{\mathbf{F}}\right)$. This equation is equivalent to TIP's equation $(1.9)_2$. Comparison of the coefficients in the two equations leads to the identification of $L_{\alpha\beta}$ and $\tilde{L}_\alpha$ as indicated in (3.3).

Fourier's law, or its generalization $(1.9)_1$, can now also be derived from the arguments of extended thermodynamics. By (2.42) we have, if we ignore a non-linear term

$$q_i = \sum_{\alpha=1}^{\nu-1}\left(\varepsilon_\alpha + \frac{p_\alpha}{\varrho_\alpha} - \varepsilon_\nu - \frac{p_\nu}{\varrho_\nu}\right)J_i^\alpha \quad \text{or by } (2.22)_1 \text{ and with} \quad h_\alpha = \varepsilon_\alpha + \frac{p_\alpha}{\varrho_\alpha},$$

$$\boxed{q_i = \sum_{\beta=1}^{\nu-1}\underbrace{\left\{\sum_{\alpha=1}^{\nu-1}h_{\alpha\nu} - T\left[\mathbf{A}\right]_{\alpha\beta}\right\}}_{L_\beta}\frac{\partial\left(-\frac{\mu^{\beta\nu}}{T}\right)}{\partial x_i} + \underbrace{\left\{\sum_{\alpha,\beta=1}^{\nu-1}h_{\alpha\nu}T^2\left[\mathbf{A}\right]_{\alpha\beta}\left(T\frac{\partial\frac{\mu^{\beta\nu}}{T}}{\partial T} - \frac{\partial \pi^{\beta\nu}}{\partial T}\right)\right\}}_{L}\frac{\partial\frac{1}{T}}{\partial x_i},} \tag{3.4}$$

where $h_{\alpha\nu}$ stands for $h_\alpha - h_\nu$. This equation is analogous to TIP's equation $(1.9)_1$, if we identify the coefficients L_β and L as indicated.

We conclude that the constitutive laws of Fick and Fourier—or TIP's generalizations (1.9) of these classical laws—arise in extended thermodynamics in an approximate manner as a result of a formal iterative scheme. Roughly speaking Fick's law for the diffusion fluxes J_i^α comes out by ignoring the acceleration of the relative motions of the constituents. The flux of internal energy in extended thermodynamics is solely due to the convection of energy with the diffusive motion. Thus Fourier's law is merely a corollary of Fick's law. This of course is a peculiarity of simple mixtures.

3.2 Onsager Relations

It is true that we treat a rather special case of mixtures in extended thermodynamics. Not only are they non-viscous, which means that the partial stresses are isotropic tensors, but also in some sense they are not heat-conducting, because the flux of internal energy is due solely to the convection of energy with the diffusive motion; yet the laws of Fick and Fourier, viz. (3.3) and (3.4), have the same general form here and in ordinary thermodynamics, except that we now have some information about the coefficients $L_{\alpha\beta}, \tilde{L}_{\alpha}$, and L_{β}. We proceed to show that we may use this knowledge to prove the validity of the Onsager relations $(1.17)_2$ for diffusion and thermal diffusion.

Inspection of (3.3) shows that $L_{\alpha\beta}$ is symmetric, if $M_V^{\alpha\beta}$ is. Truesdell [28] has made some eminently reasonable assumptions under which the symmetry of $M_V^{\alpha\beta}$ can be proved. We follow his argument. First of all we introduce the new coefficient $M_V^{\alpha\nu}$ by $M_V^{\alpha\nu} = -\sum_{\beta=1}^{\nu-1} M_V^{\alpha\beta}$ and rewrite the constitutive equation $(2.15)_3$ in the form

$$m_i^\alpha = -\sum_{\beta=1}^{\nu} M_V^{\alpha\beta}(v_i^\alpha - v_i^\beta). \tag{3.5}$$

Chemical reactions are ignored for this argument, so $\sum_{\alpha=1}^{\nu} M_V^{\alpha\beta} = 0$ holds. The advantage of the form (3.5) is that the coefficients $M_V^{\alpha\beta}$ may now be interpreted suggestively as coefficients that determine how the force on constituent α is influenced by the relative motion of that constituent and constituent β. It therefore seems reasonable to assume that the following two properties hold:

i. $M_V^{\alpha\beta}$ is independent of ϱ_γ for $\gamma \neq \alpha, \beta$; and

ii. $M_V^{\alpha\beta}$ tends to zero for $\varrho_\beta \to 0$.

We say that the mixture experiences binary drags only if i. is true, and ii. states that there is no drag on constituent α from constituent β, if the latter is absent. We proceed to prove the symmetry of $M_V^{\alpha\beta}$ under these conditions.

Because of conservation of momentum, we must have

$$\sum_{\alpha,\beta=1}^{\nu} M_V^{\alpha\beta}(v_i^\alpha - v_i^\beta) = 0 \quad \text{or} \quad \sum_{\alpha,\beta=1}^{\nu} M_V^{[\alpha\beta]} v_i^\alpha = 0, \tag{3.6}$$

where the square bracket denotes the antisymmetric part of $M_V^{\alpha\beta}$. Since (3.6) must hold for all v_i^α, we have

$$\sum_{\beta=1}^{\nu} M_V^{[\alpha\beta]} = 0.$$

This means that the sum over all elements in every row and every column must vanish. In particular, in a binary mixture we have

$$M_V^{[\alpha\beta]} = \begin{bmatrix} 0 & \alpha \\ -\alpha & 0 \end{bmatrix},$$

and since all rows and columns must add up to zero, we must conclude that $\alpha = 0$ holds. In a mixture of three constituents we have

$$M_V^{[\alpha\beta]} = \begin{bmatrix} 0 & \alpha & -\alpha \\ -\alpha & 0 & \alpha \\ \alpha & -\alpha & 0 \end{bmatrix}.$$

Now, if the above conditions i. and ii. hold for $M_V^{\alpha\beta}$, they must also hold for its symmetric and antisymmetric parts. Thus $M_V^{[12]}$ is independent of ϱ_3, $M_V^{[31]}$ is independent of ϱ_2, and $M_V^{[23]}$ is independent of ϱ_1; but all of these components are equal to α. Hence α is a constant and, by condition ii., that constant is zero, so $M_V^{[\alpha\beta]} = 0$ holds for a ternary mixture. The corresponding proof for more than three constituents can easily be reduced to the case of three constituents by condition ii. Thus $M_V^{[\alpha\beta]} = 0$ is generally true for our mixtures and we have proved the symmetry of $M_V^{\alpha\beta}$, whence follows the symmetry of $L_{\alpha\beta}$ by (3.3).

We continue to establish the validity of the Onsager relations $(1.17)_2$ and proceed to show that by (3.3) and (3.4) the coefficients $\tilde{L}_\alpha$ and L_α are equal. Indeed, by (2.47) we have with $\partial\varepsilon_\alpha/\partial T = T(\partial s_\alpha/\partial T)$

$$T\frac{\partial\frac{\mu_\beta-\mu_\nu}{T}}{\partial T} - \frac{\partial\left(\frac{p_\beta}{\varrho_\beta}-\frac{p_\nu}{\varrho_\nu}\right)}{\partial T} = -\frac{1}{T}(h_\beta - h_\nu)$$

from which we conclude that $\tilde{L}_\alpha = L_\alpha$ holds. This was first proved by Müller [29].

The important conclusion to be drawn here is that Onsager relations hold as a consequence not of microscopic considerations—as in Onsager's motivations, for example, see [30]—but as a consequence of the macroscopic structure of the coefficients $L_{\alpha\beta}$, L_α, and $\tilde{L}_\alpha$, a structure that is revealed by extended thermodynamics of simple mixtures. It is true, though, that this structure is not maintained for non-simple mixtures and properly heat-conducting mixtures; it may therefore be that in such more complicated, and more realistic, mixtures Onsager relations do not hold.

3.3 Inertial Contribution to the Laws of Diffusion

We return to the consideration of the equations (3.2) which represent the equations of relative motion of the constituents. For the present argument we restrict our attention to a binary mixture with a uniform temperature in which case there is only one equation, namely, with $\mu = \mu^1 - \mu^2$

$$\left[\dot{V}_i^1 + V_j^1\frac{\partial v_i}{\partial x_j} - 2W_{ji}V_j^1\right] + \underline{u_j^1\frac{\partial u_i^1}{\partial x_j} - u_j^2\frac{\partial u_i^2}{\partial x_j}} + \frac{\partial \mu}{\partial x_i} - M_V^{11}\frac{\varrho}{\varrho_1\varrho_2}V_i^1 . \tag{3.7}$$

We ignore the underlined terms that are non-linear in the diffusion velocities and apply the Maxwellian iteration to (3.7). In the first step we obtain

$$\overset{(1)}{V_i^1} = \tau\frac{\partial\mu}{\partial x_i} \quad \text{where } \tau \equiv \frac{1}{M_V^{11}}\frac{\varrho_1\varrho_2}{\varrho}, \tag{3.8}$$

and in the second step we obtain

$$\overset{(2)}{V_i^1} = \tau\frac{\partial\mu}{\partial x_i} + \tau^2\left[\frac{1}{\tau}\left(\tau\frac{\partial\mu}{\partial x_i}\right)^{\bullet} + \frac{\partial v_i}{\partial x_j}\frac{\partial\mu}{\partial x_j} - 2W_{ji}\frac{\partial\mu}{\partial x_j}\right]. \tag{3.9}$$

Note that the terms in square brackets in (3.7) and (3.9) are objective vectors but frame-dependent ones, since they contain the matrix $\mathbf{W}$ of angular velocity of the frame.

We simplify (3.9) by looking at the special case of a rigid rotation with $\dot{\mu} = 0$. In that case we obtain

$$\overset{(2)}{V_i^1} = \tau\left\{\frac{\partial\mu}{\partial x_i} - 2\tau\left(\frac{\partial v_{[j}}{\partial x_{i]}} - W_{ji}\right)\frac{\partial\mu}{\partial x_j}\right\}. \tag{3.10}$$

It follows that the diffusion flux created by a chemical potential gradient in a mixture at rest in a non-inertial frame has a component perpendicular to $\partial\mu/\partial x_i$. Equation (3.10) is analogous to the equation (2.3.10) for heat conduction and indeed, the interpretation of the phenomena is very similar as we proceed to explain.

Consider a binary mixture of gases with $v_i = 0$ in a cylinder with a radial gradient of μ as depicted in Figure 1(a). The radial gradient of μ may be created by a gradient of the concentration. We focus attention on a small volume element of the linear dimensions of the mean free paths of the atoms, and first we consider a gas at rest in an inertial frame. In that case the free paths of the atoms are straight lines and, because of the concentration gradient, there are more atoms of constituent 1 (say) flying from top to bottom than in the other direction; see Figure 1(b). Therefore we observe a diffusion flux, or a non-zero vector V_i^1, which is proportional to grad μ as it should be according to (3.10) in an inertial frame and when the mixture is at rest. Next we consider the same situation for a mixture at rest in a non-inertial frame. Now the free paths between collisions are curved by the Coriolis force as shown in Figure 1(c). Consequently there is now not only a radial mass flux but also a flux perpendicular to grad μ and to the angular velocity of the frame as predicted by (3.10).

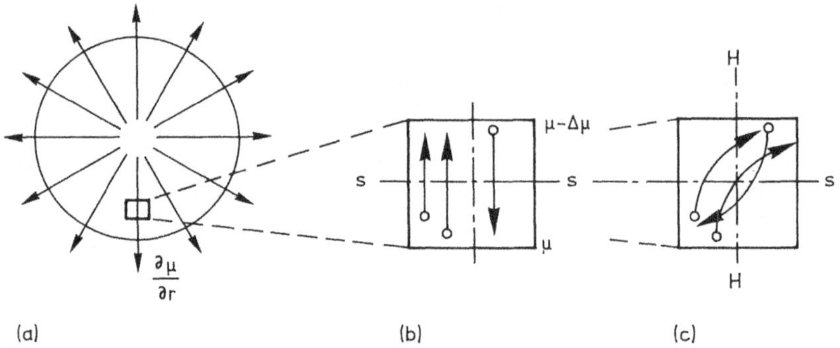

Figure 1. On the frame dependence of the diffusion flux.

This argument about the nature of the frame dependence is entirely analogous to the corresponding argument about the heat flux in Chapter 2. It seems convincing in the sense that nobody will doubt the effect of the Coriolis force on the relative motion of two constituents of a mixture. In contrast to that, the corresponding argument about heat flux (see Paragraph 2.3.3) attracted a lot of criticism at the time when it was presented; see [31], [32]. That criticism was always unfounded, and the present argument confirms this fact.

Chapter 4

Relativistic Extended Thermodynamics of Mon-Atomic Gases

Relativists have shown a keen interest in the development of extended thermodynamics, because the paradoxa of infinite speeds implied by the parabolic equations of ordinary thermodynamics disturbed them acutely. Indeed immediately after the formulation of the early version of extended thermodynamics Müller [5] formulated the theory relativistically, but this work was not properly published until Israel [33] presented a very similar theory. These theories had all the shortcomings of extended TIP that were discussed in Chapter 1.

After the development of the systematic version of extended thermodynamics—the theory presented in Chapter 2, relativistic thermodynamics could be revisited and properly formulated. This was done by Liu, Müller, and Ruggeri [34], and this chapter gives a detailed account of their work (see also [35], [36]). The basic fields are the particle flux vector and the energy-momentum tensor—14 fields altogether. The necessary field equations are based on the conservation laws of particles, energy, and momentum and on a balance of fluxes. Constitutive equations are needed for the flux tensor and the flux production. The constitutive theory makes use of the principle of relativity, the entropy principle, and the requirement of convexity and causality, and it derives explicit field equations in which only four functions must be determined experimentally or from statistical mechanics, viz. the thermal equation of state, the heat conductivity, and the shear and bulk viscosities. Specific results are given for some special cases, namely, the non-degenerate relativistic gas, the degenerate non-relativistic gas, and the ultrarelativistic gas.

Two applications are discussed. The famous treatment of white dwarfs by Chandrasekhar [37] and the speeds of discontinuity waves calculated by Seccia and Strumia [38].

It is not obvious how relativistic extended thermodynamics reduces to the non-relativistic theory. The easiest way to investigate this question is by reference to the kinetic theory. We therefore give a brief account of the relativistic version of the kinetic theory of gases and investigate its non-relativistic limit. It turns out that in that limit the 14 variables of the relativistic theory reduce to 13; the dynamic pressure drops out because it is $\mathcal{O}(1/c^2)$.

1 Balance Equations and Constitutive Restrictions

1.1 Thermodynamic Processes

The principal objective of relativistic extended thermodynamics is the determination of the 14 fields

$$\begin{array}{ll} \text{particle flux vector} & A^A, \\ \text{energy-momentum tensor} & A^{AB} \end{array} \tag{1.1}$$

in all events x^D. The energy-momentum tensor is assumed symmetric so that it has 10 independent components.

For the determination of the fields (1.1) we need field equations and these are formed by the conservation laws of particle number and energy-momentum, viz.

$$A^A{}_{,A} = 0, \qquad (1.2)_1$$

$$A^{AB}{}_{,B} = 0 \qquad (1.2)_2$$

and by the equations of balance of fluxes

$$A^{ABC}{}_{,C} = I^{AB}. \qquad (1.2)_3$$

A^{ABC} is a completely symmetric tensor of fluxes and I^{AB} is its production density. The equations (1.2) are 15 in number for only 14 fields. We therefore assume

$$I^A{}_A = 0 \quad \text{and} \quad A^{AC}{}_C \sim A^C, \tag{1.3}$$

so that the trace of $(1.2)_3$ reduces to $(1.2)_1$. We are then left with only 14 independent equations (1.2).

The components of A^A and A^{AB} have the following interpretations:

$$\begin{aligned} A^0 &= c \cdot \text{rest mass density}, \\ A^a &= \text{flux of rest mass}, \\ A^{00} &= \text{energy density}, \\ A^{0a} &= \frac{1}{c} \cdot \text{energy flux}, \\ A^{a0} &= c \cdot \text{momentum density}, \\ A^{ab} &= \text{momentum flux}. \end{aligned} \tag{1.4}$$

The motivation for the choice of the equations (1.2) and, in particular, for the equation $(1.2)_3$ stems from the kinetic theory of mon-atomic gases of which a brief account is given in Section 6. Indeed, A^A and A^{AB} are the first two moments in the kinetic theory and $A^A{}_{,A} = 0$ and $A^{AB}{}_{,B} = 0$ are the first two equations of transfer. It is then reasonable to take the further equations from the equation of transfer for the third moment A^{ABC} and these have the form $(1.2)_3$. In the kinetic theory the two conditions (1.3) are satisfied and the factor of proportionality in $(1.3)_2$ turns out to be c^2, see (6.11). We therefore rewrite (1.3) in the form

$$I^A{}_A = 0 \quad \text{and} \quad A^{AC}{}_C = c^2 A^A. \tag{1.5}$$

The kinetic theory makes it also clear that the assumed symmetry of A^{ABC} is characteristic for mon-atomic gases. The present theory is therefore restricted to that case. This is not much of a restriction, because molecular compounds cannot exist at temperatures that are high enough for relativistic effects to matter.

Of course, the set of equations (1.2) is not a set of field equations for the basic fields (1.1), because two new quantities appear in them, namely, the flux tensor A^{ABC} and the flux production I^{AB}. In this situation we let ourselves be guided by the arguments of non-relativistic continuum mechanics and thermodynamics and assume that A^{ABC} and I^{AB} are constitutive quantities. Such quantities are related to the basic fields in a materially dependent manner by constitutive functions. In particular, when the constitutive relations are of the general form

$$\begin{aligned} A^{ABC} &= \hat{A}^{ABC}\,(A^M, A^{MN}), \\ I^{AB} &= \hat{I}^{AB}\quad(A^M, A^{MN}), \end{aligned} \tag{1.6}$$

we say that they characterize a viscous, heat conducting gas.

If the constitutive functions $\hat{A}^{ABC}$ and $\hat{I}^{AB}$ were known, we could eliminate A^{ABC} and I^{AB} between (1.2) and (1.6) and come up with an explicit set of 14 field equations for the fourteen fields A^A and A^{AB}. Each solution of this set of equations is called a *thermodynamic process.*

1.2 Principles of the Constitutive Theory

In reality, of course, the constitutive functions $\hat{A}^{ABC}$ and $\hat{I}^{AB}$ are not known, and therefore the thermodynamicist tries to determine them or at least to restrict their generality. In these efforts he is guided by universal physical principles to which the constitutive functions must conform. The most important such principles are

i. the entropy principle,
ii. the principle of relativity, and
iii. the requirement of causality and convexity.

The entropy principle postulates the existence of the entropy flux vector h^A, which is a constitutive quantity, so in a viscous, heat conducting gas we have

$$h^A = \hat{h}^A(A^M, A^{MN}). \tag{1.7}$$

Moreover, the principle requires that h^A satisfy the entropy inequality

$$h^A{}_{,A} \geq 0 \tag{1.8}$$

for all thermodynamic processes.

The principle of relativity assumes that the field equation and the entropy inequality are invariant under space-time transformations

$$\overset{*}{x}{}^A = \overset{*}{x}{}^A\,(x^B). \tag{1.9}$$

Since the balance equations (1.2) are tensor equations, which are naturally invariant, the principle in effect requires that the constitutive functions be invariant. If C and C^* are generic expressions for the constitutive quantities A^{ABC}, I^{AB}, and h^A in the two frames connected by the transformation (1.9), the principle of relativity can be stated as

$$C = \hat{C}(A^M, A^{MN}) \quad \text{and} \quad C^* = \hat{C}(A^{*M}, A^{*MN}). \tag{1.10}$$

Note that $\hat{C}$ is the *same* function in both equations.

Causality and convexity is a condition on the character of the field equations. It requires that these equations be symmetric hyperbolic. This property guarantees that Cauchy problems are well-posed and that all wave speeds are finite.

2 Constitutive Theory

2.1 Scope and Structure

The exploitation of the above restrictive principles is the subject of the thermodynamic constitutive theory. The details of this theory are cumbersome and require much work; so as to maintain the clarity of the argument we proceed in simple steps as follows:

i. Lagrange multipliers and the vector potential;
ii. Linear representations;
iii. Suggestive variables;
iv. Equilibrium: Fugacity and absolute temperature;
v. Linear expressions for the Lagrange multipliers;
vi. The linear flux tensor;
vii. The entropy flux vector;
viii. Residual inequality;
ix. Causality and convexity; and
x. Summary of results.

2.2 Lagrange Multipliers and the Vector Potential: Step i

The key to the exploitation of the entropy principle is the observation that the entropy inequality (1.8) must not hold for arbitrary fields $A^M(x^D), A^{MN}(x^D)$ but only for thermodynamic processes, that is, solutions of the field equations. One may say that the field equations form constraints on the fields that satisfy (1.8). By a lemma of Liu [18] we may take care of these constraints by use of Lagrange multipliers. Application of the Liu lemma in this case shows that the new inequality

$$h^A{}_{,A} + \Lambda A^A{}_{,A} + \Lambda_B A^{BA}{}_{,A} + \Lambda_{BC}(A^{BCA}{}_{,A} - I^{BC}) \geq 0 \tag{2.1}$$

must hold for arbitrary fields A^A, A^{BA}. The $\mathbf{\Lambda}$'s in (2.1) are Lagrange multipliers, which may depend on the same variables as A^{BCA}, I^{BC}, and h^A, that is, on A^M and A^{MN}. The mulitiplier Λ_{BC} may be considered trace-less, because its factor in (2.1) is trace-less according to (1.5) Obviously the inequality (2.1) may be written in the form

$$h'^A{}_{,A} - A^A \Lambda_{,A} - A^{AB}\Lambda_{B,A} - A^{BCA}\Lambda_{BC,A} - \Lambda_{BC} I^{BC} \geq 0, \tag{2.2}$$

where h'^A has been introduced as an abbreviation for

$$h'^A = h^A + \Lambda A^A + \Lambda_B A^{BA} + \Lambda_{BC} A^{BCA}. \tag{2.3}$$

The inequality (2.2) is most easily evaluated after a change of independent variables

$$\text{from} \quad A^A, A^{AB} \quad \text{to} \quad \Lambda, \Lambda_B, \Lambda_{BC}. \tag{2.4}$$

In Chapter 6 we prove that this transformation is globally invertible for reasons of convexity.

With this change of variables A^A and A^{AB} become constitutive quantities just as h^A, or h'^A, and A^{BCA} and I^{BC}. If the chain rule is used on $h'^A(\Lambda, \Lambda_B, \Lambda_{BC})$ in (2.2), we obtain

$$\left(\frac{\partial h'^A}{\partial \Lambda} - A^A\right)\Lambda_{,A} + \left(\frac{\partial h'^A}{\partial \Lambda_B} - A^{AB}\right)\Lambda_{B,A} + \left(\frac{\partial h'^A}{\partial \Lambda_{BC}} - A^{BCA}\right)\Lambda_{BC,A} - \Lambda_{BC} I^{BC} \geq 0. \tag{2.5}$$

This inequality must hold for all fields Λ, Λ_B, and Λ_{BC} and, therefore, in particular, for arbitrary values of the derivatives $\Lambda_{,A}$, $\Lambda_{B,A}$, and $\Lambda_{BC,A}$ in one event; but these derivatives occur linearly on the left-hand side of (2.5); the inequality could thus be violated, if the brackets were contributing to its left-hand side. We therefore must have

$$A^A = \frac{\partial h'^A}{\partial \Lambda}, \qquad A^{BA} = \frac{\partial h'^A}{\partial \Lambda_B}, \qquad A^{\langle BC \rangle A} = \frac{\partial h'^A}{\partial \Lambda_{BC}} - \frac{1}{4} g^{BC} g_{MN} \frac{\partial h'^A}{\partial \Lambda_{MN}}. \tag{2.6}$$

The last condition is complicated by the fact that Λ_{BC} is trace-less. g is the metric tensor with the signature $(1, -1, -1, -1)$.

There remains the residual inequality on the entropy production Σ

$$\Sigma = -\Lambda_{BC} I^{BC} \geq 0. \tag{2.7}$$

We conclude from (2.6) that the dependence of A^A, A^{BC}, and A^{BCA} on the Lagrange multipliers $\Lambda, \Lambda_B, \Lambda_{BC}$ is determined by the derivatives of the single-vector function $h'^A(\Lambda, \Lambda_B, \Lambda_{BC})$. We therefore call this function the *vector potential*. The equations (2.6) represent the main result of the entropy principle and, rather obviously, they imply a great number of integrability conditions. Those integrability conditions lead to all the restrictive conditions, which we are about to derive.

From the residual inequality we may deduce a condition concerning equilibrium. Equilibrium is defined as a process in which all productions I^{AB} vanish. Therefore inspection of (2.7) shows that the entropy production, which may be considered as a function of I^{AB}, has its minimum in equilibrium, namely, zero. Of necessity then we must have

$$\frac{\partial \Sigma}{\partial I^{BC}} \Big|_E = -\Lambda_{BC} \Big|_E = 0. \tag{2.8}$$

Thus the Lagrange multiplier Λ_{BC} vanishes in equilibrium.

2.3 Principle of Relativity and Linear Representations: Step ii

We recall from (2.6) that the vector h'^A, a constitutive quantity, dependent on Λ, Λ_B, and Λ_{BC}, plays the dominant role in the determination of A^A, A^{AB}, and A^{ABC}. The principle of relativity requires that we have

$$h'^A = \hat{h}'^A(\Lambda, \Lambda^M, \Lambda^{MN}) \quad \text{and} \quad \overset{*}{h'^A} = \hat{h}'^A(\Lambda^*, \Lambda^{*M}, \Lambda^{*MN}). \tag{2.9}$$

Since h'^A and $\Lambda, \Lambda_M, \Lambda_{MN}$ are tensors of rank zero, one, and two, the equations (2.9) may be combined to read

$$\frac{\partial x^{*A}}{\partial x^B} \hat{h}'^B(\Lambda, \Lambda^M, \Lambda^{MN}) = \hat{h}'^A\left(\Lambda, \frac{\partial x^{*M}}{\partial x^P}\Lambda^P, \frac{\partial x^{*M}}{\partial x^P}\frac{\partial x^{*N}}{\partial x^Q}\Lambda^{PQ}\right), \tag{2.10}$$

which means that $\hat{h}'^B$ is an isotropic vector function.

In non-relativistic continuum mechanics there are representation theorems for isotropic functions (e.g., see [22]), and Pennisi [39] showed that these carry over to the present case. Thus the most general form of h'^A that satisfies (2.10) reads

$$h'_A = \gamma_0 \Lambda_A + \gamma_1 \Lambda_{AB} \Lambda^B + \gamma_2 \Lambda^2_{AB} \Lambda^B + \gamma_3 \Lambda^3_{AB} \Lambda^B, \tag{2.11}$$

where the γ's may depend on the following 8 scalar combinations of variables:

$$
\begin{aligned}
H_0 &= \Lambda, & G_0 &= \Lambda^A\Lambda_A, \\
H_2 &= tr(\Lambda^2_{AB}), & G_1 &= \Lambda^A\Lambda_{AB}\Lambda^B, \\
H_3 &= tr(\Lambda^3_{AB}), & G_2 &= \Lambda^A\Lambda^2_{AB}\Lambda^B, \\
H_4 &= tr(\Lambda^4_{AB}), & G_3 &= \Lambda^A\Lambda^3_{AB}\Lambda^B.
\end{aligned} \tag{2.12}
$$

The indices on H and G indicate the order to which these scalars depend on the components of Λ_{AB}.

In this book we describe a *linear* theory so that A^A, A^{BA}, and A^{BCA} are linear functions of the multiplier Λ_{BC}, which by (2.8) vanishes in equilibrium. Inspection of (2.6) shows that we need h'^A to within third-order terms in Λ_{BC}. In this approximation we may ignore γ_3 in the representation (2.11) and set

$$
h'^A = \left[\Gamma_0 + \Gamma_{01}G_1 + \Gamma_{02}G_1^2 + \Gamma_{03}G_2 + \Gamma_{04}H_2\right]\Lambda_A + \left[\Gamma_1 + \Gamma_{11}G_1\right]\Lambda_{AB}\Lambda^B + \Gamma_2\Lambda^2_{AB}\Lambda^B, \tag{2.13}
$$

where the Γ's may depend only on Λ and G_0.

In order to obtain linear representations for A^A, A^{BA}, and A^{BCA} we insert (2.13) into (2.6) and ignore non-linear terms after the differentiation. In this manner we obtain expressions for A^A, A^{AB}, and A^{ABC} that do not automatically satisfy the requirement that the tensors A^{AB} and A^{BCA} be symmetric, nor is the trace condition (1.5) satisfied, of course. These conditions can be fulfilled, however, by a proper choice of the coefficients (2.13).

We must have

$$
\begin{aligned}
h'_A = &\left[\Gamma_0 + \frac{\partial\Gamma_1}{\partial G_0}G_1 + \frac{1}{2}\frac{\partial^2\Gamma_2}{\partial G_0^2}G_1^2 + \frac{\partial\Gamma_2}{\partial G_0}G_2 + \frac{1}{4}\Gamma_2 H_2\right]\Lambda_A \\
&+ \left[\Gamma_1 + \frac{\partial\Gamma_2}{\partial G_0}G_1\right]\Lambda_{AB}\Lambda^B + \left[\Gamma_2\right]\Lambda^2_{AB}\Lambda^B.
\end{aligned} \tag{2.14}
$$

Insertion into (2.6) now provides

$$
A^A = \left(\frac{\partial\Gamma_0}{\partial\Lambda} + \frac{\partial^2\Gamma_1}{\partial\Lambda\partial G_0}G_1\right)\Lambda^A + \frac{\partial\Gamma_1}{\partial\Lambda}\Lambda^{AB}\Lambda_B, \tag{2.15}
$$

$$
\begin{aligned}
A^{BA} = &\left(\Gamma_0 + \frac{\partial\Gamma_1}{\partial G_0}G_1\right)g^{BA} + \Gamma_1\Lambda^{BA} + 2\left(\frac{\partial\Gamma_0}{\partial G_0} + \frac{\partial^2\Gamma_1}{\partial G_0^2}G_1\right)\Lambda^B\Lambda^A \\
&+ 2\frac{\partial\Gamma_1}{\partial G_0}\left(\Lambda^B\Lambda^{AC}\Lambda_C + \Lambda^A\Lambda^{BC}\Lambda_C\right),
\end{aligned} \tag{2.16}
$$

$$
\begin{aligned}
A^{BCA} = &\frac{1}{2}\left(\Gamma_1 + \frac{\partial\Gamma_2}{\partial G_0}G_1\right)\left(g^{BC}\Lambda^A + g^{CA}\Lambda^B + g^{AB}\Lambda^C\right) \\
&+ \frac{1}{2}\Gamma_2\left(g^{BC}\Lambda^{AD}\Lambda_D + g^{CA}\Lambda^{BD}\Lambda_D + g^{AB}\Lambda^{CD}\Lambda_D\right. \\
&\qquad\left. + \Lambda^{BC}\Lambda^A + \Lambda^{CA}\Lambda^B + \Lambda^{AB}\Lambda^C\right) \\
&+ \left(\frac{\partial\Gamma_1}{\partial G_0} + \frac{\partial^2\Gamma_2}{\partial G_0^2}G_1\right)\Lambda^B\Lambda^C\Lambda^A \\
&+ \frac{\partial\Gamma_2}{\partial G_0}\left(\Lambda^B\Lambda^C\Lambda^{AD}\Lambda_D + \Lambda^C\Lambda^A\Lambda^{BD}\Lambda_D + \Lambda^A\Lambda^B\Lambda^{CD}\Lambda_C\right).
\end{aligned} \tag{2.17}
$$

The tensors (2.16) and (2.17) are fully symmetric and the trace condition (1.5) will be satisfied, if we have

$$
\frac{\partial G_0^3\Gamma_1}{\partial G_0} = \frac{\partial c^2 G_0^2\Gamma_0}{\partial\Lambda} \quad \text{and} \quad \frac{\partial G_0^4\Gamma_2}{\partial G_0} = \frac{\partial c^2 G_0^3\Gamma_1}{\partial\Lambda}. \tag{2.18}
$$

A linear expression for the entropy flux vector follows from (2.3), (2.14) and from the above linear equations for A^A, A^{BA}, and A^{BCA}. We get

$$h^A = -\left[\Lambda\frac{\partial\Gamma_0}{\partial\Lambda} + 2\frac{\partial\Gamma_0}{\partial G_0}G_0 + \left(\Lambda\frac{\partial^2\Gamma_1}{\partial\Lambda\partial G_0} + 2G_0\frac{\partial^2\Gamma_1}{\partial G_0^2} + 3\frac{\partial\Gamma_1}{\partial G_0}\right)G_1\right]\Lambda^A - \left[\Gamma_1 + \Lambda\frac{\partial\Gamma_1}{\partial\Lambda} + 2G_0\frac{\partial\Gamma_1}{\partial G_0}\right]\Lambda^{AB}\Lambda_B. \quad (2.19)$$

From (2.18) we conclude that, if only $\Gamma_0(\Lambda, G_0)$ is known, we may calculate Γ_1 and Γ_2 by integration to within two functions of Λ alone; therefore, the knowledge of the single scalar function $\Gamma_0(\Lambda, G_0)$ goes a long way in determining the linear constitutive quantities A^A, A^{BA}, A^{BCA}, and h^A in (2.15), (2.16), (2.17), and (2.19), albeit only in terms of the Lagrange multipliers Λ, Λ_B, and Λ_{BC}.

This last-mentioned fact is a drawback, because the Lagrange multipliers have no suggestive meaning at all—at least not at this stage. It is therefore imperative that we determine the Λ's in terms of meaningful, suggestive, and measurable variables.

2.4 Stress Deviator, Heat Flux, and Dynamic Pressure: Step iii

By use of the four-velocity U^A, with $U^AU_A = c^2$, and of the projector

$$h_{AB} = \frac{1}{c^2}U_AU_B - g_{AB} \quad (2.20)$$

we may decompose the particle flux vector A^A, the energy-momentum tensor A^{BA}, and the entropy flux vector h^A as follows:

$$\begin{aligned} A^A &= nmU^A, \\ A^{BA} &= t^{\langle BA\rangle} + (p(n,e) + \pi)h^{BA} + \frac{1}{c^2}(U^Bq^A + U^Aq^B) + \frac{e}{c^2}U^AU^B, \\ h^A &= hU^A + \Phi^A. \end{aligned} \quad (2.21)$$

Here m is the atomic rest mass. These equations are nothing but identities, because they define the new quantities

$$n, \; U^A, \; t^{\langle BA\rangle}, \; \pi, \; q^A, \; e, \; h, \; \Phi^A \quad (2.22)$$

in terms of A^A and A^{BA} as follows:

$$\begin{aligned} n &= \frac{1}{mc^2}A^AU_A, && \text{number density} \\ t^{\langle BA\rangle} &= (h^B_Mh^A_N - \frac{1}{3}h^{BA}h_{MN})A^{MN}, && \text{stress deviator} \\ p(n,e) + \pi &= \frac{1}{3}h_{MN}A^{MN}, && \text{pressure} \\ q^A &= -h^A_MU_NA^{MN}, && \text{heat flux} \\ e &= \frac{1}{c^2}U_MU_NA^{MN}, && \text{internal energy density} \\ h &= \frac{1}{c^2}h^AU_A, && \text{entropy density} \\ \Phi^A &= -h^A_Mh^M. && \text{entropy flux} \end{aligned} \quad (2.23)$$

The names that we have attached to these definitions in (2.23) are motivated by the observation that in the Lorentz frame, that is locally and momentarily at rest in the fluid, the non-vanishing components of $n, t^{\langle BA\rangle}, p(n,e)+\pi, q^A, e$, and h, Φ^A have the indicated physical meaning as number density, stress deviator, etc.

The pressure is decomposed into an equilibrium part p, which depends on n and e, and a non-equilibrium part π, which we call the dynamic pressure. Stress-deviator, heat flux, and dynamic pressure are the nine quantities that vanish in equilibrium.

Compared to the original variables A^A, A^{BA} the quantities $n, U^A, t^{\langle AB\rangle}, \pi, q^A, e$ are more suggestive of physical meaning. We therefore shall use them in the sequel.

2.5 Fugacity and Absolute Temperature: Step iv

We eliminate A^A and A^{BA} between the equations (2.15), (2.16), and $(2.21)_{1,2}$ to obtain

$$\begin{aligned} nmU^A &= \left(\frac{\partial\Gamma_0}{\partial\Lambda}+\frac{\partial^2\Gamma_1}{\partial\Lambda\partial G_0}G_1\right)\Lambda^A+\frac{\partial\Gamma_1}{\partial\Lambda}\Lambda^{AB}\Lambda_B, \\ t^{\langle BA\rangle} &+ (p+\pi)h^{BA}+\frac{1}{c^2}(U^Bq^A+U^Aq^B)+\frac{1}{c^2}eU^BU^A \\ &= \left(\Gamma_0+\frac{\partial\Gamma_1}{\partial G_0}G_1\right)g^{BA}+\Gamma_1\Lambda^{BA}+2\left(\frac{\partial\Gamma_0}{\partial G_0}+\frac{\partial^2\Gamma_1}{\partial G_0^2}G_1\right)\Lambda^B\Lambda^A \\ &\quad +2\frac{\partial\Gamma_1}{\partial G_0}\left(\Lambda^B\Lambda^{AC}\Lambda_C+\Lambda^A\Lambda^{BC}\Lambda_C\right). \end{aligned} \tag{2.24}$$

This is a system of 14 equations from which, in principle, we may calculate the Lagrange multipliers Λ, Λ_B, and Λ_{BC} in terms of $n, U^A, t^{\langle AB\rangle}, \pi, q^A$, and e. To do so explicitly is, however, quite impossible in general, because the system of equations is strongly non-linear. We can successfully perform that calculation only in a *linear* theory, i.e., in equilibrium and close to it.

For that purpose we recall that $\Lambda^{AB}|_E=0$ holds and that $t^{\langle BC\rangle}, \pi$, and q^A all vanish in equilibrium. Thus we may hope to determine the equilibrium values of Λ and Λ_A in terms of n, U^A, and e from the equilibrium parts of (2.24), viz.

$$nmU^A=\left(\frac{\partial\Gamma_0}{\partial\Lambda}\Lambda^A\right)\Big|_E, \qquad ph^{AB}+\frac{1}{c^2}eU^AU^B=\left(\Gamma_0 g^{AB}+2\frac{\partial\Gamma_0}{\partial G_0}\Lambda^A\Lambda^B\right)\Big|_E. \tag{2.25}$$

Even this is impossible, however, unless $\Gamma_0=\Gamma_0(\Lambda|_E, G_0|_E)$ is specified as a function of $\lambda|_E$ and $G_0|_E$.

The best way to proceed is to identify $\Lambda|_E$ and $G_0|_E$ with the fugacity α and the absolute temperature T—two suggestive and measurable quantities—and to determine $\Gamma_0(\alpha,T)$ in terms of the thermal equation of state $p=p(n,e)$. To achieve this we first write the equilibrium version of (2.19) with $(2.21)_3$. We obtain

$$h|_E U^A=-\left(\Lambda\frac{\partial\Gamma_0}{\partial\Lambda}+2\frac{\partial\Gamma_0}{\partial G_0}G_0\right)\Lambda^A|_E. \tag{2.26}$$

The equations (2.25), (2.26) imply

$$nmc=\sqrt{G_0}\frac{\partial\Gamma_0}{\partial\Lambda}\Big|_E, \quad p=-\Gamma_0\Big|_E, \quad p+e=2G_0\frac{\partial\Gamma_0}{\partial G_0}\Big|_E, \quad \frac{h|_E}{nm}=-\left(\Lambda+2\frac{\Lambda_0\frac{\partial\Gamma_0}{\partial G_0}}{\frac{\partial\Gamma_0}{\partial\Lambda}}\right)\Big|_E, \tag{2.27}$$

and from these we derive easily

$$d\left(\frac{h|_E}{nm}\right) = -\frac{\sqrt{G_0|_E}}{c}\left(d\left(\frac{e}{nm}\right) + pd\left(\frac{1}{nm}\right)\right) \tag{2.28}$$

and

$$\Lambda|_E = -\frac{\sqrt{G_0|_E}}{c}\left(\frac{e}{nm} + \frac{c}{\sqrt{G_0|_E}}\frac{h|_E}{nm} + \frac{p}{nm}\right). \tag{2.29}$$

By comparison of (2.28) with the Gibbs equation of thermostatics we thus identify the absolute temperature T by the equation

$$\frac{1}{T} = -\frac{\sqrt{G_0|_E}}{c}. \tag{2.30}$$

Knowing this we see that the right-hand side of (2.29) is the fugacity α with

$$\alpha = -\frac{1}{T}\frac{e - Th|_E + p}{nm}. \tag{2.31}$$

We summarize the foregoing arguments about equilibrium and write

$$\Lambda|_E = -\alpha, \qquad \Lambda_A|_E = -\frac{U_A}{T}, \qquad \Lambda_{AB}|_E = 0, \tag{2.32}$$

so that the equilibrium values of the Lagrange multipliers are now known as simple combinations of α, T, and U_A.

In many ways the fugacity and the absolute temperature T are more useful variables than n and e, because they are easier to measure by virtue of their continuity at diathermal walls and permeable surfaces such as phase boundaries. Moreover, α and T are the natural variables of statistical thermodynamics[1]. If $\Gamma_0|_E$, or by $(2.27)_2$ if p is known as a function of α and T, one may use $(2.27)_{1,3}$ to calculate

$$nm = -\frac{1}{T}\left(\frac{\partial p(\alpha,T)}{\partial \alpha}\right)_T \quad \text{and} \quad e = T^2\left(\frac{\partial}{\partial T}\left(\frac{p(\alpha,T)}{T}\right)\right)_\alpha. \tag{2.33}$$

These two relations may be used to switch back and forth between the sets of variables (n, e) and (α, T) provided, of course, that the thermal equation of state $p = p(\alpha, T)$ is known.

2.6 Linear Relations Between Lagrange Multipliers and $\mathbf{n}, \mathbf{U^A}, \mathbf{t^{\langle AB\rangle}}, \pi, \mathbf{q^A}, \mathbf{e}$: Step v

We shall now use the equations (2.24) to calculate the Lagrange multipliers as linear functions of $t^{\langle AB\rangle}$, q^A, and π. In order to linearize the system (2.24) we introduce

$$\begin{aligned}
\Lambda \quad &= -\alpha + \lambda, \\
\Lambda_A \quad &= -\frac{U_A}{T} + \tau_A + \tau U_A, \\
\Lambda_{AB} &= \sigma_{\langle AB\rangle} + \sigma h_{AB} + \frac{1}{c^2}(U_A\sigma_B + U_B\sigma_A) + \frac{3}{c^2}\sigma U_A U_B
\end{aligned} \tag{2.34}$$

on the right-hand side of (2.24) and ignore all products of the quantities λ, τ, and σ. In (2.34) the terms with $\lambda, \tau, \sigma, \tau_A, \sigma_A$, and $\sigma_{\langle AB\rangle}$ represent the non-equilibrium parts of the Lagrange multipliers

[1]It should be said that the present nomenclature deviates from the common one in thermodynamics in the definition (2.31) of the fugacity. Indeed, our α is the natural logarithm of the common fugacity.

that have been decomposed into space-like and time-like components. Due account has been taken of the constraint that Λ_{AB} is trace-less.

Insertion of (2.34) into (2.24) provides a system of linear equations for $\lambda, \tau, \sigma, \tau^A, \sigma^A$, and $\sigma^{\langle AB\rangle}$. We set $\Gamma_0|_E = -p$, according to $(2.27)_2$, and define the derivatives

$$R' = \frac{\partial R}{\partial \ln T} \quad \text{and} \quad \dot{R} = \frac{\partial R}{\partial \alpha}. \tag{2.35}$$

Thus we obtain

$$\begin{bmatrix} -\ddot{p} & \dot{p}-\dot{p}' & 3\left(-\dot{\Gamma}_1+\frac{1}{2}\dot{\Gamma}_1'\right) \\ -\dot{p} & -p' & \left(\frac{3}{2}\Gamma_1'+\Gamma_1\right) \\ \dot{p}-\dot{p}' & p'-p'' & 3\left(\Gamma_1-\frac{3}{2}\Gamma_1'+\frac{1}{2}\Gamma_1''\right) \end{bmatrix} \begin{bmatrix} \lambda \\ (-T\tau) \\ \sigma \end{bmatrix} = \begin{bmatrix} 0 \\ \pi \\ 0 \end{bmatrix}, \tag{2.36}$$

$$\begin{bmatrix} \dot{p} & -\dot{\Gamma}_1 \\ p' & (\Gamma_1-\Gamma_1') \end{bmatrix} \begin{bmatrix} (-T\tau^A) \\ \sigma^A \end{bmatrix} = \begin{bmatrix} 0 \\ q^A \end{bmatrix}, \tag{2.37}$$

$$\Gamma_1 \sigma^{\langle AB\rangle} = t^{\langle AB\rangle}. \tag{2.38}$$

The index E on Γ_1 has been dropped for simplicity.

The solutions of these equations read

$$\begin{aligned}
\lambda &= -\frac{\begin{bmatrix} \dot{p}-\dot{p}' & 3(-\dot{\Gamma}_1+\frac{1}{2}\dot{\Gamma}_1') \\ p'-p'' & 3(\Gamma_1-\frac{3}{2}\Gamma_1'+\frac{1}{2}\Gamma_1'') \end{bmatrix}}{D_1^\pi}\pi = \lambda_\pi \pi, \\
\tau &= -\frac{1}{T}\frac{\begin{bmatrix} -\ddot{p} & 3(-\dot{\Gamma}_1+\frac{1}{2}\dot{\Gamma}_1') \\ \dot{p}-\dot{p}' & 3(\Gamma_1-\frac{3}{2}\Gamma_1'+\frac{1}{2}\Gamma_1'') \end{bmatrix}}{D_1^\pi}\pi = \tau_\pi \pi, \\
\sigma &= -\frac{\begin{bmatrix} -\ddot{p} & \dot{p}-\dot{p}' \\ \dot{p}-\dot{p}' & p'-p'' \end{bmatrix}}{D_1^\pi}\pi = \sigma_\pi \pi, \\
\tau^A &= -\frac{1}{T}\frac{\dot{\Gamma}_1}{D_3}q^A = \tau_q q^A, \\
\sigma^A &= +\frac{\dot{p}}{D_3}q^A = \sigma_q q^A, \\
\sigma^{\langle BA\rangle} &= \frac{1}{\Gamma_1}t^{\langle BA\rangle}.
\end{aligned} \tag{2.39}$$

D_1^π and D_3 are abbreviations for the determinants of the linear system (2.36) and (2.37) respectively. The coefficients $\lambda_\pi, \tau_\pi, \sigma_\pi, \tau_q, \sigma_q$, whose definitions may be read from (2.39), will serve as useful abbreviations in subsequent formulae.

Combining (2.34) and (2.39) we obtain linear expressions for the Lagrange multipliers Λ, Λ_A, and Λ_{AB} in terms of π, q_A, and $t_{\langle AB\rangle}$, viz.

$$\begin{aligned}
\Lambda &= -\alpha + \lambda_\pi \pi, \\
\Lambda_A &= (-\frac{1}{T} + \tau_\pi \pi)U_A + \tau_q q_A, \\
\Lambda_{AB} &= \frac{1}{\Gamma_1}t_{\langle AB\rangle} + \sigma_\pi \pi h_{AB} + \frac{1}{c^2}\sigma_q(U_A q_B + U_B q_A) + \frac{3}{c^2}\sigma_\pi \pi U_A U_B.
\end{aligned} \tag{2.40}$$

2.7 The Linear Flux Tensor: Step vi

In order to obtain a linear expression for the flux tensor A^{ABC} we insert (2.34) into the representation (2.17) and ignore all products of the quantities λ, τ, and σ. The resulting expression reads

$$
\begin{aligned}
A^{BCA} = & \\
= -\frac{1}{T}\Big\{ & \frac{1}{c^2}\left[-\frac{1}{2}\Gamma_1' + \frac{1}{2}\dot{\Gamma}_1'\lambda + \left(\frac{1}{2}\Gamma_1'' - \frac{1}{2}\Gamma_1'\right)(-T\tau) + 3\left(\frac{1}{4}\Gamma_2'' - \frac{1}{2}\Gamma_2' + 2\Gamma_2\right)\sigma\right] U^A U^B U^C \\
& + \left[\frac{1}{2}\Gamma_1 - \frac{1}{2}\dot{\Gamma}_1\lambda + \left(\frac{1}{2}\Gamma_1 - \frac{1}{2}\Gamma_1'\right)(-T\tau) + \left(-\frac{3}{4}\Gamma_2' + \Gamma_2\right)\sigma\right](g^{AB}U^C + g^{BC}U^A + g^{CA}U^B) \\
& + \frac{1}{2c^2}\Gamma_1' T\left[U^A U^B \tau^C + U^B U^C \tau^A + U^C U^A \tau^B\right] \\
& \qquad + \frac{1}{c^2}\left(\Gamma_2 - \frac{1}{2}\Gamma_2'\right)\left[U^A U^B \sigma^C + U^B U^C \sigma^A + U^C U^A \sigma^B\right] \\
& + \frac{1}{2}\Gamma_1 T\left[g^{AB}\tau^C + g^{BC}\tau^A + g^{CA}\tau^B\right] + \frac{1}{2}\Gamma_2\left[g^{AB}\sigma^C + g^{BC}\sigma^A + g^{CA}\sigma^B\right] \\
& + \frac{1}{2}\Gamma_2\left[\sigma^{\langle AB\rangle}U^C + \sigma^{\langle BC\rangle}U^A + \sigma^{\langle CA\rangle}U^B\right]\Big\}.
\end{aligned}
\tag{2.41}
$$

Hence we calculate A^{ABC} as a linear function of π, q^A, and $t^{\langle AB\rangle}$ by introducing (2.39). After a little calculation and by use of the conditions (2.18) we obtain

$$
\begin{aligned}
A^{ABC} = & (C_1^0 + C_1^\pi \pi)U^A U^B U^C + \frac{c^2}{6}(nm - C_1^0 - C_1^\pi \pi)(g^{AB}U^C + g^{BC}U^A + g^{CA}U^B) \\
& + C_3(g^{AB}q^C + g^{BC}q^A + g^{CA}q^B) - \frac{6}{c^2}C_3(U^A U^B q^C + U^B U^C q^A + U^C U^A q^B) \\
& + C_5(t^{\langle AB\rangle}U^C + t^{\langle BC\rangle}U^A + t^{\langle CA\rangle}U^B),
\end{aligned}
\tag{2.42}
$$

where the coefficients C have been introduced as convenient abbreviations defined as

$$
\begin{aligned}
C_1^0 &= \frac{1}{2c^2}\frac{\Gamma_1'}{T}, \\
C_1^\pi &= -\frac{3}{c^2}\frac{1}{T}\frac{\begin{bmatrix} -\dot{\Gamma}_1 & \Gamma_1 - \Gamma_1' & (-\frac{3}{2}\Gamma_2' + 2\Gamma_2) \\ -\ddot{p} & \dot{p} - \dot{p}' & 3(-\dot{\Gamma}_1 + \frac{1}{2}\dot{\Gamma}_1') \\ \dot{p} - \dot{p}' & +p' - p'' & 3(\Gamma_1 - \frac{3}{2}\Gamma_1' + \frac{1}{2}\Gamma_1'' \end{bmatrix}}{D_1^\pi}, \\
C_3 &= \frac{1}{2}\frac{1}{T}\frac{\begin{bmatrix} \Gamma_1 & \Gamma_2 \\ \dot{p} & -\dot{\Gamma}_1 \end{bmatrix}}{D_3}, \\
C_5 &= -\frac{1}{T}\frac{1}{2}\frac{\Gamma_2}{\Gamma_1}.
\end{aligned}
\tag{2.43}
$$

We conclude from (2.43) that the linear expression for the flux tensor A^{ABC} is determined by the three functions p, Γ_1 and Γ_2 which all depend on the fugacity α and the absolute temperature T. Among these three functions Γ_1 and Γ_2 are closely related to p by the conditions (2.18) which we shall exploit later.

2.8 The Entropy Flux Vector: Step vii

From (2.3) and (2.5) we conclude that the entropy flux vector satisfies the condition

$$
dh^A = -\Lambda dA^A - \Lambda_B dA^{BA} - \Lambda_{BC} dA^{BCA}. \tag{2.44}
$$

We insert (2.40) to eliminate the Lagrange multipliers and $(2.21)_{1,2}$, (2.42) to eliminate A^A, A^{BA}, and A^{BCA}. Thus on the right-hand side of (2.44) there appear terms with $d\pi, dq^E$, and $dt^{\langle EF\rangle}$ whose coefficients must be interpreted as derivatives of h^A with respect to π, q^E, and $t^{\langle EF\rangle}$. We get

$$\begin{aligned}
\frac{\partial h^A}{\partial \pi} &= -2c^2 C_1^\pi \sigma_\pi \pi U^A + \left(\tau_q + \frac{c^2}{3} C_1^\pi \sigma_q\right) q^A, \\
\frac{\partial h^A}{\partial q^E} &= \left(\frac{1}{T} - (\tau_\pi - 20C_3\sigma_\pi)\pi\right) h_E^A + (\tau_q - 10C_3\sigma_q)\frac{1}{c^2} U^A q_E + 2C_3 \frac{1}{\Gamma_1} t_{\langle E}^{A\rangle}, \\
\frac{\partial h^A}{\partial t^{\langle EF\rangle}} &= \frac{1}{2}(\tau_q - 2C_5\sigma_q)\left(q_E h_F^A + q_F h_E^A - \frac{2}{3} q^A h_{EF}\right) - C_5 \frac{1}{\Gamma_1} t_{\langle EF\rangle} U^A.
\end{aligned} \tag{2.45}$$

Integration gives after a short calculation

$$h^A = (h|_{\mathrm{E}} + A_1^{\pi^2}\pi^2 + A_1^q q_{\mathrm{E}} q^{\mathrm{E}} + A_1^t t^{\langle EF\rangle} t_{\langle EF\rangle}) U^A + (A_2^0 + A_2^\pi \pi) q^A + A_3^0 t^{\langle AE\rangle} q_E, \tag{2.46}$$

where the coefficients A have been introduced as abbreviations for the following expressions:

$$\begin{aligned}
A_1^{\pi^2} &= -c^2 C_1^\pi \sigma_\pi, \\
A_1^q &= -\frac{1}{c^2}(\tau_q - 10C_5\sigma_q), \\
A_1^t &= -\frac{1}{2} C_5 \frac{1}{\Gamma_1}, \\
A_2^0 &= \frac{1}{T}, \\
A_2^\pi &= -(\tau_\pi - 20C_3\sigma_\pi) = \tau_q + \frac{c^2}{3} C_1^\pi \sigma_q, \\
A_3^0 &= -(\tau_q + 2C_5\sigma_q) = -C_3 \frac{2}{\Gamma_1}.
\end{aligned} \tag{2.47}$$

The definition of $\tau_\pi, \sigma_\pi, \tau_q, \sigma_q$ in terms of the functions p, Γ_1 and the definition of C_1^π, C_3, C_5 in terms of the functions p, Γ_1, Γ_2 may be read from (2.39) and (2.43). We conclude that the entropy flux vector is determined by the functions p, Γ_1, Γ_2 to within third-order terms in π, q^E, and $t^{\langle EF\rangle}$.

The second equations in $(2.47)_{5,6}$ result from the integrability conditions implied by (2.45). These equations are, of course, identically satisfied by the explicit forms (2.39) and (2.43) for $\tau_\pi, \tau_q, \sigma_\pi, \sigma_q$ and C_1^π, C_3, C_5.

2.9 Residual Inequality: Step viii

We recall from (2.7) the residual inequality for the entropy production

$$\Sigma = -\Lambda_{BC} I^{BC} \geq 0 \tag{2.48}$$

and remember that the flux production I^{BC} is a constitutive quantity subject to the principle of relativity, so we must have

$$\frac{\partial x^{*A}}{\partial x^C}\frac{\partial x^{*B}}{\partial x^D} \hat{I}^{CD}(n, U^P, t^{\langle PQ\rangle}, \pi, q^P, e) = \hat{I}^{AB}\left(n, \frac{\partial x^{*P}}{\partial x^R} U^R, \frac{\partial x^{*P}}{\partial x^R}\frac{\partial x^{*Q}}{\partial x^S} t^{\langle RS\rangle}, \pi, \frac{\partial x^{*P}}{\partial x^R} q^R, e\right). \tag{2.49}$$

Thus $\hat{I}^{AB}$ is an isotropic function. From the representation theorems for isotropic functions (see [22], [39]), we obtain the following fairly explicit form for I^{BC}, if we restrict the attention to linear terms in π, q^C, and $t^{\langle CD\rangle}$:

$$I^{BC} = B_1^\pi \pi g^{BC} - \frac{4}{c^2} B_1^\pi \pi U^B U^C + B_3 t^{\langle BC\rangle} + B_4(q^B U^C + q^C U^B). \tag{2.50}$$

In writing (2.50) we have also used the fact that I^{BC} must vanish in equilibrium and that by (1.5) I^{BC} must be trace-less. The coefficients B may depend on n and e.

We insert Λ_{BC} from (2.40) and I^{BC} from (2.50) into the residual inequality and calculate its left-hand side to within third-order terms, because these would not be reliable in the present linear theory. We obtain after a little calculation

$$\Sigma = 12B_1^{\pi}\sigma_{\pi}\pi^2 - 2B_4\sigma_q q^C q_C - B_3\frac{1}{\Gamma_1}t^{\langle BC\rangle}t_{\langle BC\rangle} \geq 0, \tag{2.51}$$

and it follows that B_1^{π}, B_4, and B_3 must satisfy the inequalities

$$B_1^{\pi}\sigma_{\pi} \geq 0, \qquad B_4\sigma_q \geq 0, \qquad B_3\frac{1}{\Gamma_1} \leq 0. \tag{2.52}$$

2.10 Causality and Convexity: Step ix

For a compact statement of the requirement of causality and convexity it is appropriate to introduce a synthetic form of the balance equations (1.2). We let

$$\begin{bmatrix} A_\alpha^A \\ I_\alpha \\ \Lambda_\alpha \end{bmatrix} \quad \text{stand for} \quad \begin{bmatrix} \{A^A, A_B^A, A_{\langle BC\rangle}^A\} \\ \{0,\ 0,\ I_{BC}\} \\ \{\Lambda,\ \Lambda_B, \Lambda_{BC}\} \end{bmatrix} \tag{2.53}$$

and rewrite the field equations (1.2) and the restrictions (2.6) from the entropy principle in the form

$$A_{\alpha,A}^A = I_\alpha \quad \text{and} \quad A_\alpha^A = \frac{\partial h'^A}{\partial \Lambda_\alpha}. \tag{2.54}$$

The Greek indices range over the values $1, 2, \ldots, 14$. The combination of the two equations (2.54) gives

$$\frac{\partial^2 h'^A}{\partial\Lambda_\alpha \partial\Lambda_\beta}\Lambda_{\beta,A} = I_\alpha. \tag{2.55}$$

This is a symmetric system of equations, because the matrix $\partial^2 h'^A/\partial\Lambda_\alpha\partial\Lambda_\beta$ is symmetric in α and β. The system is said to be symmetric *hyperbolic*, if the temporal part of this matrix is positive definite (see Chapter 6). The temporal part is obtained by multiplication of $\partial^2 h'^A/\partial\Lambda_\alpha\partial\Lambda_\beta$ by the four-velocity U_A of the gas. Thus the requirement of causality reads

$$\frac{\partial^2 h'^A}{\partial\Lambda_\alpha \partial\Lambda_\beta}U_A\delta\Lambda_\alpha\delta\Lambda_\beta > 0 \quad \text{or} \quad \frac{\partial A_\alpha^A}{\partial\Lambda_\beta}U_A\delta\Lambda_\alpha\delta\Lambda_\beta > 0, \tag{2.56}$$

and it must hold for arbitrary choices of $\delta\Lambda_\gamma$.

In [40] Ruggeri proved that for a non-degenerate gas this symmetrization condition is equivalent to the convexity of $-h^A U_A$—the entropy density in the rest frame—with respect to the densities $A^A U_A, A^{AB}U_A, A^{ABC}U_A$.

With $(2.54)_2$ we may rewrite (2.56) in the form

$$U_A\delta A_\alpha^A \delta\Lambda_\alpha > 0 \quad \text{or explicitly with (2.53)}$$
$$U_A\delta A^A\delta\Lambda + U_A\delta A^{AB}\delta\Lambda_B + U_A\delta A^{ABC}\delta\Lambda_{BC} > 0. \tag{2.57}$$

This requirement can best be exploited by insertion of the Lagrange multipliers (2.40) and of the representations $(2.21)_{1,2}$ for A^A, A^{BA} and (2.42) for A^{ABC}. There results a quadratic form in

$\delta\alpha, \delta T, \delta\pi, \delta U^C, \delta q^C$, and $\delta t^{\langle CD\rangle}$ that must be positive. The matrix of this quadratic form contains non-equilibrium terms that must be ignored since they are not reliable in the present theory where the $\mathbf{\Lambda}$'s and A^{ABC} are only known linearly. The matrix must be positive definite and a little calculation shows that this requirement implies

$$
\begin{gathered}
C_5\frac{1}{\Gamma_1} > 0, \\
\begin{bmatrix} \frac{1}{T}p' & -\frac{1}{2}\left[-\frac{1}{T} - \frac{1}{2}\frac{1}{T}\left(-\frac{3}{2}\Gamma_1' - m^2c^2\dot{p}\right) + p'\tau_q\right] \\ \odot & -(\tau_q - 10C_3\sigma_q) \end{bmatrix} \text{— positive definite} \\
\begin{bmatrix} \frac{1}{T}\ddot{p} & -\frac{1}{2T}(\dot{p} - \dot{p}') & \frac{1}{2}\left[\tau_\pi(\dot{p} - \dot{p}') - \frac{1}{T}\sigma_\pi(\dot{\Gamma}_1' + m^2c^2\ddot{p})\right] \\ \odot & -\frac{1}{2T}\left(-\frac{1}{2}p'' + \frac{1}{2}p'\right) & \frac{1}{2}\left[-\frac{1}{T}\sigma_\pi\left(-\frac{1}{2}\Gamma_1' + \frac{1}{2}\Gamma_1'' - \frac{1}{2}m^2c^2(\dot{p} - \dot{p}')\right)\right] \\ \odot & \odot & 2c^2\sigma_\pi C_1^\pi \end{bmatrix} \text{— positive definite}
\end{gathered}
\tag{2.58}
$$

Only the upper part of the matrices is written here; the symbols $\odot$ are supposed to indicate that the matrices are symmetric. In particular, it follows from (2.58) and (2.47) that we have

$$A_1^{\pi^2} < 0, \qquad A_1^q > 0, \qquad A_1^t < 0, \tag{2.59}$$

so the equilibrium entropy $\frac{1}{c^2}h^A U_A$ is bigger than the non-equilibrium entropy.

2.11 Summary of Results: Step x

It has proved convenient to introduce the four-velocity U^A with $U_A U^A = c^2$ and the projector $h_{AB} = \frac{1}{c^2}U_A U_B - g_{AB}$. By use of these quantities, the particle flux vector, the energy-momentum tensor, and the entropy flux vector may be decomposed into time-like and space-like components. We have

$$
\begin{aligned}
A^A &= nmU^A, \\
A^{AB} &= t^{\langle AB\rangle} + (p(n,e) + \pi)h^{AB} + \frac{1}{c^2}(U^A q^B + U^B q^A) + \frac{e}{c^2}U^A U^B, \\
h^A &= hU^A + \Phi^A,
\end{aligned}
\tag{2.60}
$$

and the components have suggestive meaning as follows:

n	number density,	
$t^{\langle AB\rangle}$	stress deviator,	
$p+\pi$	pressure,	
q^A	heat flux,	(2.61)
e	energy density,	
h	entropy density,	
Φ^A	(nonconvective) entropy flux.	

The stress deviator, the heat flux, and the dynamic pressure π vanish in equilibrium.

The constitutive theory has provided expressions for the flux tensor A^{ABC} and the flux production I^{AB}, which are linear in π, q^A, and $t^{\langle AB\rangle}$ and have the forms

$$
\begin{aligned}
A^{ABC} &= (C_1^0 + C_1^\pi \pi) U^A U^B U^C + \frac{c^2}{6}(nm - C_1^0 - C_1^\pi \pi)(g^{AB}U^C + g^{BC}U^A + g^{CA}U^B) \\
&\quad + C_3(g^{AB}q^C + g^{BC}q^A + g^{CA}q^B) - \frac{6}{c^2} C_3 (U^A U^B q^C + U^B U^C q^A + U^C U^A q^B) \\
&\quad + C_5 (t^{\langle AB \rangle} U^C + t^{\langle BC \rangle} U^A + t^{\langle CA \rangle} U^B), \qquad (2.62) \\
I^{AB} &= B_1^\pi \pi g^{AB} - \frac{4}{c^2} B_1^\pi \pi U^A U^B + B_3 t^{\langle AB \rangle} + B_4 (q^A U^B + q^B U^A).
\end{aligned}
$$

Also the entropy flux vector h^A has emerged to within third-order terms in π, q^A, and $t^{\langle AB \rangle}$, viz.

$$
h^A = \left(h|_{\mathrm{E}} + A_1^{\pi^2} \pi^2 + A_1^q q_E q^E + A_1^t t_{\langle EF \rangle} t^{\langle EF \rangle} \right) U^A + (A_2^0 + A_2^\pi \pi) q^A + A_3^0 t^{\langle AE \rangle} q_{\mathrm{E}}. \qquad (2.63)
$$

$h|_{\mathrm{E}}$ is the equilibrium entropy density that satisfies the Gibbs equation

$$
d\left(\frac{h|_{\mathrm{E}}}{nm}\right) = \frac{1}{T}\left(d\left(\frac{e}{nm}\right) + p d\left(\frac{1}{nm}\right)\right), \qquad (2.64)
$$

where T is the absolute temperature.

The coefficients C in the flux tensor (2.62) and the coefficients A in the entropy flux vector (2.63) are related to the three functions p, Γ_1, and Γ_2 of the two variables

$$
\text{fugacity:} \quad \alpha = -\frac{1}{T}\frac{e - Th_{\mathrm{E}} + p}{nm} \qquad \text{and} \qquad \text{absolute temperature:} \quad T
$$

in the following manner:

$$
C_1^0 = \frac{1}{2c^2}\frac{\Gamma_1'}{T}, \qquad (2.65)_1
$$

$$
C_1^\pi = -\frac{3}{c^2}\frac{1}{T}\frac{\begin{bmatrix} -\ddot{p} & \dot{p} - \dot{p}' & \dot{\Gamma}_1 \\ \dot{p} - \dot{p}' & p' - p'' & \Gamma_1' - \Gamma_1 \\ \dot{\Gamma}_1 & \Gamma_1' - \Gamma_1 & \frac{5}{3}\Gamma_2 \end{bmatrix}}{\begin{bmatrix} -\ddot{p} & \dot{p} - \dot{p}' & \dot{\Gamma}_1 \\ \dot{p} - \dot{p}' & p' - p'' & \Gamma_1' - \Gamma_1 \\ -\dot{p} & -p' & \frac{5}{3}\Gamma_1 \end{bmatrix}}, \qquad (2.65)_2
$$

$$
C_3 = -\frac{1}{2}\frac{1}{T}\frac{\begin{bmatrix} \dot{p} & -\dot{\Gamma}_1 \\ \Gamma_1 & \Gamma_2 \end{bmatrix}}{\begin{bmatrix} \dot{p} & -\dot{\Gamma}_1 \\ p' & \Gamma_1 - \Gamma_1' \end{bmatrix}}, \qquad (2.65)_3
$$

$$
C_5 = -\frac{1}{2}\frac{1}{T}\frac{\Gamma_2}{\Gamma_1}; \qquad (2.65)_4
$$

$$
A_1^{\pi^2} = -\frac{c^2}{6} C_1^\pi \frac{\begin{bmatrix} -\ddot{p} & \dot{p} - \dot{p}' \\ \dot{p} - \dot{p}' & p' - p'' \end{bmatrix}}{\begin{bmatrix} -\ddot{p} & \dot{p} - \dot{p}' & \dot{\Gamma}_1 \\ \dot{p} - \dot{p}' & p' - p'' & \Gamma_1' - \Gamma_1 \\ -\dot{p} & -p' & \frac{5}{3}\Gamma_1 \end{bmatrix}}, \qquad (2.66)_1
$$

$$
A_1^q = -\frac{1}{2c^2}\frac{-\frac{1}{T}\dot{\Gamma}_1 - 10 C_3 \dot{p}}{\begin{bmatrix} \dot{p} & -\dot{\Gamma}_1 \\ p' & \Gamma_1 - \Gamma_1' \end{bmatrix}}, \qquad (2.66)_2
$$

$$
A_1^t = -\frac{1}{2} C_5 \frac{1}{\Gamma_1}, \qquad (2.66)_3
$$

$$
A_2^0 = \frac{1}{T}, \qquad (2.66)_4
$$

$$A_2^\pi = -\frac{\frac{1}{T}\begin{bmatrix} -\ddot{p} & \dot{\Gamma}_1 \\ \dot{p}-\dot{p}' & \Gamma_1'-\Gamma_1 \end{bmatrix} - \frac{10}{3}C_3\begin{bmatrix} -\ddot{p} & \dot{p}-\dot{p}' \\ \dot{p}-\dot{p}' & p'-p'' \end{bmatrix}}{\begin{bmatrix} -\ddot{p} & \dot{p}-\dot{p}' & \dot{\Gamma}_1 \\ \dot{p}-\dot{p}' & p'-p'' & \Gamma_1'-\Gamma_1 \\ -\dot{p} & -p' & \frac{5}{3}\Gamma_1 \end{bmatrix}} = -\frac{\frac{1}{T}\dot{\Gamma}_1 - \frac{c^2}{3}C_1^\pi \dot{p}}{\begin{bmatrix} \dot{p} & -\dot{\Gamma}_1 \\ p' & \Gamma_1-\Gamma_1' \end{bmatrix}}, \tag{2.66$_5$}$$

$$A_3^0 = \frac{\frac{1}{T}\dot{\Gamma}_1 - 2C_5\dot{p}}{\begin{bmatrix} \dot{p} & -\dot{\Gamma}_1 \\ p' & \Gamma_1-\Gamma_1' \end{bmatrix}} = -C_3\frac{2}{\Gamma_1}. \tag{2.66$_6$}$$

Here dots and primes denote the derivatives with respect to α and $\ln T$, respectively.

Fugacity and absolute temperature are the natural variables of statistical thermodynamics as we shall see later. Also they are easier to measure, at least in principle, than the original variables n and e. If we wish, however, to switch between the two sets of variables (α, T) and (n, e), we may do so done by use of the formulae

$$nm = -\frac{1}{T}\dot{p} \quad \text{and} \quad e = p' - p, \tag{2.67}$$

provided that the thermal equation of state $p = p(\alpha, T)$ is known.

The results (2.65) and (2.66) seem to indicate that the three functions $p(\alpha,T), \Gamma_1(\alpha,T)$, and $\Gamma_2(\alpha,T)$ are needed for the determination of the flux tensor A^{ABC} and I^{AB}. This is not so, however, because the functions Γ_1 and Γ_2 are dependent on $p = p(\alpha,T)$. Indeed we have [see (2.18)]

$$-\frac{1}{2}\Gamma_1' + 3\Gamma_1 = c^2\dot{p} \quad \text{and} \quad -\frac{1}{2}\Gamma_2' + 4\Gamma_2 = -c^2\dot{\Gamma}_1 \tag{2.68}$$

so that Γ_1 and Γ_2 can be related to p by integration. We obtain

$$\begin{aligned} \Gamma_1 &= T^6 \left[-2c^2\int\frac{\dot{p}}{T^7}dT + A_1(\alpha)\right], \\ \Gamma_2 &= T^8\left\{2c^2\int\frac{1}{T^3}\left[-2c^2\int\frac{\ddot{p}}{T^7}dT + \dot{A}_1(\alpha)\right]dT + A_2(\alpha)\right\}, \end{aligned} \tag{2.69}$$

and we conclude that Γ_1 and Γ_2 may be determined from $p(\alpha,T)$, each to within a function of the single variable α, the fugacity.

It follows from (2.65), (2.66) that all coefficients of the flux tensor A^{ABC} and of the entropy flux vector h^A can be determined to within two functions of α, *if only* the thermal equation of state $p = p(\alpha, T)$ is known.

The entropy production Σ in (2.51) contains three terms, one each due to the dynamic pressure, the heat flux, and the stress deviator. Thus there are three dissipative "mechanisms" in the gas. The conditions for a non-negative entropy production are listed in (2.52); more explicitly they read

$$B_1^\pi\frac{\begin{bmatrix} -\ddot{p} & \dot{p}-\dot{p}' \\ \dot{p}-\dot{p}' & p'-p'' \end{bmatrix}}{\begin{bmatrix} -\ddot{p} & \dot{p}-\dot{p}' & \dot{\Gamma}_1 \\ \dot{p}-\dot{p}' & p'-p'' & \Gamma_1'-\Gamma_1 \\ -\dot{p} & -p' & \frac{5}{3}\Gamma_1 \end{bmatrix}} \geq 0, \qquad B_4\frac{\dot{p}}{\begin{bmatrix} \dot{p} & -\dot{\Gamma}_1 \\ p' & \Gamma_1-\Gamma_1' \end{bmatrix}} \geq 0, \qquad B_3\frac{1}{\Gamma_1} \leq 0. \tag{2.70}$$

These inequalities provide restrictions on the coefficients B in the flux production.

In the next section we shall show that the coefficients B may be related to the transport coefficients of ordinary thermodynamics, namely, the heat conductivity, the viscosity, and the bulk viscosity. These are measurable quantities, at least in principle.

The requirement of causality and convexity furnishes restrictions on the coefficients C and A in the flux tensor and in the entropy flux vector [see (2.58), (2.59)]. In effect these conditions restrict the form of the thermal equation of state $p = p(\alpha, T)$ and of the functions $A_1(\alpha), A_2(\alpha)$. Most of the restrictions are difficult to interpret, and we only list a few that can be interpreted, viz.

$$\begin{bmatrix} \frac{1}{T}\ddot{p} & -\frac{1}{2T}(\dot{p}-\dot{p}') \\ -\frac{1}{2T}(\dot{p}-\dot{p}') & \frac{1}{4T}(p''-p') \end{bmatrix} \text{ — pos. def.,} \qquad A_1^{\pi^2} < 0,\ A_1^q > 0,\ A_1^t < 0. \tag{2.71}$$

Inspection of (2.63) shows that the inequalities $(2.71)_{2,3,4}$ are necessary for the entropy density $h = \frac{1}{c^2}h^A U_A$ to have a maximum in equilibrium. The positive-definiteness of the matrix $(2.71)_1$ ensures that the compressibility $\frac{1}{n}\left(\frac{\partial n}{\partial p}\right)_T$ and the specific heat $\left(\frac{\partial e}{\partial T}\right)_n$ are positive; these are two well-known *thermodynamic stability conditions*, that is, requirements on the constitutive functions that guarantee the entropy of a body in homogeneous equilibrium is bigger than in a non-homogeneous equilibrium (e.g., see [15, p. 22 ff]).

3 Identification of Viscosities and Heat Conductivity

3.1 Extended Thermodynamics and Ordinary Thermodynamics

The purpose of this section is to identify the coefficients B_1^π, B_3, B_4 in the flux production by relating them to the bulk viscosity, the shear viscosity, and the heat conductivity. These are the transport coefficients of ordinary thermodynamics which can in principle be measured. In order to find that relation we make a brief detour and show how extended thermodynamics relates to ordinary thermodynamics.

First of all we recall the objective of *extended* thermodynamics, which is the determination of the 14 fields

$$A^A, A^{BA} \quad \text{or} \quad n, U^A, e, \pi, q^A, t^{\langle AB \rangle}. \tag{3.1}$$

The field equations are based upon the equations of balance (1.2), which by $(2.60)_{1,2}$ and (2.62) may be written as

$$\begin{aligned} &(nmU^A)_{,A} = 0, \\ &\left(t^{\langle BA \rangle} + (p+\pi)h^{BA} + \frac{1}{c^2}(q^B U^A + q^A U^B) + \frac{e}{c^2}U^B U^A\right)_{,A} = 0, \\ &A^{BCA}{}_{,A} = B_1^\pi \pi g^{BC} - \frac{4}{c^2}B_1^\pi \pi U^B U^C + B_3 t^{\langle BC \rangle} + B_4(q^B U^C + q^C U^B), \end{aligned} \tag{3.2}$$

where A^{BCA} must be inserted from (2.62).

The objective of *ordinary* thermodynamics is the determination of the five fields

$$n, U^A, e. \tag{3.3}$$

The necessary field equations are based upon the balance equations

$$(nmU^A)_{,A} = 0, \tag*{$(3.4)_1$}$$

$$\left(t^{\langle BA \rangle} + (p+\pi)h^{BA} + \frac{1}{c^2}(q^B U^A + q^A U^B) + \frac{e}{c^2}U^B U^A\right)_{,A} = 0, \tag*{$(3.4)_2$}$$

and constitutive relations must be found for $\pi, t^{\langle BA \rangle}$, and q^A, which relate these quantities to the fields (3.3). For viscous, heat-conducting fluids the constitutive quantities $\pi, t^{\langle BA \rangle}$, and q^A in one event x^D are related to the values of the fields n, U^A, e in that event and to the values of n, U^A, e in the immediate neighborhood of x^D.

3.2 Transition from Extended to Ordinary Thermodynamics

The transition from extended thermodynamics to ordinary thermodynamics proceeds by an approximation: On the left-hand sides of the balance equations (3.2) we replace all quantities by their equilibrium values. This implies that A^{BCA} in $(3.2)_3$ is replaced by the expression

$$A^{BCA}|_E = C_1^0 U^A U^B U^C + \frac{c^2}{2}(nm - C_1^0)(g^{AB}U^C + g^{BC}U^A + g^{CA}U^B),$$

$$A^{BCA}|_E = \frac{1}{2c^2}\frac{\Gamma_1'}{T}U^A U^B U^C - \frac{1}{2}\frac{\Gamma_1}{T}(g^{AB}U^C + g^{BC}U^A + g^{CA}U^B), \tag{3.5}$$

which is derived from (2.62) with $(2.65)_1$. We decompose the energy–momentum balance and the flux balance into time–like and space–like parts and, for the balance equations of particle number, energy momentum, and fluxes, obtain

$$\begin{aligned}
0 &= -\ddot{p} \frac{d\alpha}{d\tau} - (\dot{p}' - \dot{p})\frac{d\ln T}{d\tau} - \dot{p}\, U^A{}_{,A},\\
0 &= (\dot{p}' - \dot{p})\frac{d\alpha}{d\tau} + (p'' - p')\frac{d\ln T}{d\tau} + p'\, U^A{}_{,A},\\
0 &= \dot{p}\, h_M^A \alpha_{,A} + p'\, h_M^A (\ln T)_{,A} - \frac{p'}{c^2}\frac{dU_M}{d\tau},\\
-3B_1^\pi \pi &= \frac{\dot{\Gamma}_1' - 3\dot{\Gamma}_1}{2T}\frac{d\alpha}{d\tau} + \frac{\Gamma_1'' - 4\Gamma_1' + 3\Gamma_1}{2T}\frac{d\ln T}{d\tau} + \frac{\Gamma_1' - \Gamma_1}{2T} U^A{}_{,A},\\
-B_4 q_M &= -\frac{\dot{\Gamma}_1}{2T} h_M^A \alpha_{,A} - \frac{\Gamma_1' - \Gamma_1}{2T} h_M^A (\ln T)_{,A} + \frac{1}{c^2}\frac{\Gamma_1' - \Gamma_1}{2T}\frac{dU_M}{d\tau},\\
B_3 t_{\langle MN\rangle} &= -\frac{\Gamma_1}{2T} h_M^B h_N^C U_{\langle B,C\rangle}.
\end{aligned} \tag{3.6}$$

The equations (2.67) have been used here to replace $\dot{n}, \dot{e}, n'$, and e' by derivatives of p. The symbol $\frac{dR}{d\tau}$ stands for the material time derivative $U^A R_{,A}$.

Elimination of the derivatives $\frac{d\alpha}{dT}, \frac{d\ln T}{d\tau}$ between $(3.6)_{1,2,4}$ and of the derivatives $h_M^A \alpha_{,A}$ between $(3.6)_{3,5}$ allows us to determine π, q_M, and $t_{\langle MN\rangle}$ from $(3.6)_{4,5,6}$ as follows:

$$\pi = -\frac{1}{2T}\frac{1}{B_1^\pi}\frac{\begin{bmatrix} -\ddot{p} & \dot{p} - \dot{p}' & \dot{\Gamma}_1 \\ \dot{p} - \dot{p}' & p' - p'' & \Gamma_1' - \Gamma_1 \\ -\dot{p} & -p' & \frac{5}{3}\Gamma_1 \end{bmatrix}}{\begin{bmatrix} -\ddot{p} & \dot{p} - \dot{p}' \\ \dot{p} - \dot{p}' & p' - p'' \end{bmatrix}}[U^A{}_{,A}], \tag{3.7$_1$}$$

$$q_M = -\frac{1}{2T}\frac{1}{B_4}\begin{bmatrix} \dot{p} & -\dot{\Gamma}_1 \\ +p' & \Gamma_1 - \Gamma_1' \end{bmatrix}\left[h_M^A\left((\ln T)_{,A} - \frac{1}{c^2}\frac{dU_A}{d\tau}\right)\right], \tag{3.7$_2$}$$

$$t_{\langle MN\rangle} = -\frac{1}{T}\frac{1}{B_3}\Gamma_1\left[h_M^B h_N^C U_{\langle B,C\rangle}\right]. \tag{3.7$_3$}$$

The equations $(3.7)_{2,3}$ are the relativistic analogues to the phenomenological equations of Fourier and Navier-Stokes. Equation $(3.7)_1$ gives a linear relation between the dynamic pressure and the divergence of velocity. In analogy to the nomenclature of ordinary irreversible thermodynamics we may call the quantities

$$\pi, q_M, t_{\langle MN\rangle} \quad \text{and} \quad [U^A{}_{,A}]\left[h_M^A T_{,A} - \frac{T}{c^2}\frac{dU_M}{d\tau}\right], [h_M^B h_N^C U_{B,C}] \tag{3.8}$$

thermodynamic forces and fluxes, respectively. The factors of proportionality between these forces and fluxes are called transport coefficients. There are three of them, viz. the bulk viscosity ν, the heat conductivity κ, and the shear viscosity μ. We have

$$\nu + \frac{2}{3}\mu = +\frac{1}{2T}\frac{1}{B_1^\pi}\frac{\begin{bmatrix} -\ddot{p} & \dot{p}-\dot{p}' & \dot{\Gamma}_1 \\ \dot{p}-\dot{p}' & p'-p'' & \Gamma_1'-\Gamma_1 \\ -\dot{p} & -p' & \frac{5}{3}\Gamma_1 \end{bmatrix}}{\begin{bmatrix} -\ddot{p} & \dot{p}-\dot{p}' \\ \dot{p}-\dot{p}' & p'-p'' \end{bmatrix}} \geq 0,$$

$$\kappa = +\frac{1}{2T^2}\frac{1}{B_4}\frac{\begin{bmatrix} \dot{p} & -\dot{\Gamma}_1 \\ p' & \Gamma_1-\Gamma_1' \end{bmatrix}}{\dot{p}} \geq 0, \tag{3.9}$$

$$\mu = -\frac{1}{2T}\frac{1}{B_3}\Gamma_1 \geq 0.$$

By comparison of (3.9) with the inequalities (2.70) we conclude that the requirement of a non-negative entropy production determines the sign of the transport coefficients ν, κ, and μ as indicated in (3.9).

Since in principle viscosity, heat conductivity and bulk viscosity can be measured, we may use (3.9) to identify the coefficients B_1^π, B_3, and B_4.

The equations (3.7) may be regarded as first iterates for π, q_M, and $t_{\langle MN\rangle}$ in an iteration scheme. If these first iterates were inserted on the left-hand sides of (3.2), we could use the equations to calculate second iterates for $\pi, q_M, t_{\langle MN\rangle}$. Such a scheme is akin to the Maxwellian iteration in the kinetic theory of gases that has already been explained in this book (see Paragraph 2.3.2). Purely formal as this iterative scheme may be it achieves some plausibility by producing Fourier's law and the Navier–Stokes equations as first iterates. Also the corresponding scheme in the kinetic theory provides progressively higher powers of the mean free path in higher iterates.

Note that by $(3.7)_2$ the thermodynamic force driving the heat flux is not just equal to the temperature gradient. Rather there is an additional acceleration term, which was first derived by Eckart [3]. This term has the interesting consequence that a gas in equilibrium within a gravitational field has a non-uniform temperature. A discussion of this effect and related ones may be found in [15, p. 427 ff.] or elsewhere in the literature.

4 Specific Results for Relativistic and Degenerate Gases

4.1 Equilibrium Distribution Function

We recall that the coefficients C and A in A^{ABC} and h^A have been related to the thermal equation of state $p = p(\alpha, T)$ [see $(2,62)_1$, (2.65) and (2.63), (2.66)]. Specific expressions for those coefficients require the knowledge of a specific expression for the thermal equation of state. In principle that equation may be determined experimentally, but in practice it is difficult to come up with an analytic expression in this manner, particularly for a relativistic gas. There, however, is an alternative. Indeed, we may calculate the function $p(\alpha, T)$ from statistical thermodynamics of equilibrium.

Statistical thermodynamics defines the particle flux vector A^A, the energy-momentum tensor A^{AB}, and the flux tensor A^{ABC} as moments of the distribution function $F(x^A, p^q)$; in particular,

$$\begin{aligned} A^A &= mc \int p^A \quad F dP, \\ A^{AB} &= c \int p^A p^B \quad F dP, \\ A^{ABC} &= \tfrac{1}{m} c \int p^A p^B p^C \, F dP. \end{aligned} \tag{4.1}$$

Here p^A is the four-momentum of a particle and $dP = \sqrt{-g} p_0 dp^1 dp^2 dp^3$ is the invariant element of momentum space. For more details see Section 6, where we give a brief account of the relativistic kinetic theory of gases.

In particular, in equilibrium the equations (4.1) read with $(2.60)_{1,2}$, $(2.62)_1$ and (2.67), (2.68)

$$\begin{aligned} nmU^A &= mc \int p^A \quad F_{\mathrm{E}} dP, \\ ph^{AB} + \frac{e}{c^2} U^A U^B &= c \int p^A p^B \quad F_{\mathrm{E}} dP, \\ \frac{1}{2c^2} \frac{\Gamma_1'}{T} U^A U^B U^C - \frac{1}{2} \frac{\Gamma_1}{T} (g^{AB} U^C + g^{BC} U^A + g^{CA} U^B) &= \frac{1}{m} c \int p^A p^B p^C F_{\mathrm{E}} dP, \end{aligned} \tag{4.2}$$

where F_{E} is the equilibrium distribution. From (4.2) we obtain

$$\begin{aligned} n &= \frac{1}{c} U_A \int p^A \quad F_{\mathrm{E}} dP, \\ e &= \frac{1}{c} U_A U_B \int p^A p^B \quad F_{\mathrm{E}} dP, \\ p &= \frac{c}{3} h_{AB} \int p^A p^B \quad F_{\mathrm{E}} dP, \\ \frac{\Gamma_1}{T} &= \frac{2}{3mc} h_{AB} U_C \int p^A p^B p^C \, F_{\mathrm{E}} dP. \end{aligned} \tag{4.3}$$

The general form of F_{E} is the Jüttner distribution (see [31], [32]), namely,

$$F_{\mathrm{E}} = \frac{y}{\exp\left(\frac{m}{k}\alpha + \frac{U_A p^A}{kT}\right) \mp 1}. \tag{4.4}$$

Here k is the Boltzmann constant and y is equal to w/h^3, where h is Planck's constant and w is equal to $2s+1$ for particles with spin $s(h/2\pi)$. Thus $1/y$ is the smallest element of phase space that can accommodate a particle. The $\mp$ signs refer to Bosons and Fermions.

The equilibrium distribution function (4.4) assumes different forms for different degrees of degeneracy of the gas and for different degrees to which relativistic effects become important. An appreciation of the whole spectrum of gas properties is obtained by an inspection of Table 1, which exhibits the equilibrium distribution functions—in a rest Lorentz frame—appropriate to various limits. Jüttners's formula is the master formula; it stands in the frame in the center of the table. In the upper left corner we have the classical Maxwell distribution, and in the lower right corner we have the Planck distribution apropriate for a gas of mass-less Bosons. γ in the table is defined as mc^2/kT, and we shall refer to it as the *relativistic coldness*. If γ is much larger than 1, we have a non-relativistic gas; if it is much smaller than 1, we have an ultrarelativistic gas.

We proceed to give specific expressions for the constitutive coefficients in particular cases.

4.2 The Degenerate Relativistic Gas

First of all we consider the generic case that results from insertion of the Jüttner distribution into (4.3). We obtain

	non-relativistic	relativistic	ultrarelativistic
non-degenerate	$y \exp\left\{-\left(\frac{\alpha}{k}+\gamma\right)\right\} \exp\left\{-\frac{p^2}{2mkT}\right\}$	$y \exp\left\{-\frac{\alpha}{k}\right\} \exp\left\{-\gamma\sqrt{1+\frac{p^2}{m^2c^2}}\right\}$	$y \exp\left\{-\frac{\alpha}{k}\right\} \exp\left\{-\frac{cp}{kT}\right\}$
degenerate	$\frac{y}{\exp\left\{\left(\frac{\alpha}{k}+\gamma\right)\right\} \exp\left\{\frac{p^2}{2mkT}\right\} \mp 1}$	$\frac{y}{\exp\left\{\frac{\alpha}{k}+\gamma\sqrt{1+\frac{p^2}{m^2c^2}}\right\} \mp 1}$	$\frac{y}{\exp\left\{\frac{\alpha}{k}+\frac{cp}{kT}\right\} \mp 1}$
strongly degenerate FERMI	$\left[\begin{array}{ll} y & 0 \le p \le \sqrt{-2mkT\left(\frac{\alpha}{k}+\gamma\right)} \\ 0 & p > \sqrt{-2mkT\left(\frac{\alpha}{k}+\gamma\right)} \end{array}\right.$	$\left[\begin{array}{ll} y & 0 \le p \le mc\sqrt{\left(\frac{\alpha/k}{\gamma}\right)^2 - 1} \\ 0 & p > mc\sqrt{\left(\frac{\alpha/k}{\gamma}\right)^2 - 1} \end{array}\right.$	$\left[\begin{array}{ll} y & 0 \le p \le -\alpha\frac{T}{c} \\ 0 & p > -\alpha\frac{T}{c} \end{array}\right.$
strongly degenerate BOSE	$\frac{y}{\exp\left\{\frac{p^2}{2mkT}\right\}-1} \quad p \neq 0$	$\frac{y}{\exp\left\{\gamma\left(\sqrt{1+\frac{p^2}{m^2c^2}}-1\right)\right\}-1} \quad p \neq 0$	$\frac{y}{\exp\left\{\frac{cp}{kT}\right\}-1} \quad p \neq 0$

Table 1: Characterization of Limiting Cases. Equilibrium Distribution Function in Rest Lorentz Frame.

$$
\begin{aligned}
n &= 4\pi y m^3 c^3 \frac{1}{4} \quad I_{3,1}, \\
e &= 4\pi y m^4 c^5 \frac{1}{8} \quad I_{4,0}, \\
p &= 4\pi y m^4 c^5 \frac{1}{24}(I_{4,2} - 3I_{2,0}), \\
\frac{\Gamma_1}{T} &= 4\pi y m^4 c^5 \frac{1}{24}(I_{5,3} - 2I_{3,1}).
\end{aligned}
\tag{4.5}
$$

The right-hand sides of (4.5) have been calculated in the rest Lorentz frame, where U_A equals $(c, 0, 0, 0)$ and the identity $p_A p^A = m^2c^2$ has been used. The expressions $I_{m,n}$ are defined by

$$
I_{m,n} = I_m - I_n \quad \text{with } I_m(\alpha, \gamma) = \int_0^\infty \frac{\cosh m\,\varrho}{\exp\left\{\frac{m}{k}\alpha + \gamma \cosh \varrho\right\} \mp 1} d\varrho, \tag{4.6}
$$

where the integrand results from the one in (4.3) by the substitution $\sinh \varrho = p/mc$, and p is the magnitude of the momentum. The $I_{m,n}$ are functions

$$
\text{of the } \textit{fugacity:} \quad \alpha \qquad \text{and} \qquad \text{the } \textit{relativistic coldness:} \quad \gamma = \frac{mc^2}{kT}. \tag{4.7}
$$

Note that it is impossible to invert the equations (4.5) analytically so as to obtain $\alpha = \alpha(n, e)$ and $T = T(n, e)$. This can only be done numerically.

For subsequent calculations with the functions I_m it is often useful to know two recurrence relations that can easily be derived from (4.6). We have

$$
I'_{n,n-2} = nI_n + (n-2)I_{n-2} \quad \text{and} \quad \dot{I}_{n,n-2} = -\frac{m}{k}\frac{2}{\gamma}(n-1)I_{n-1}, \tag{4.8}
$$

and by use of these relations we can prove the following:

$$
\gamma^5 I_{3,1} = \frac{1}{12}\left[\gamma^5(I_{5,3} - 2I_{3,1})\right]' \quad \text{and} \quad \gamma^6 I_{4,2} = \frac{1}{40}\left[\gamma^6(3I_{6,4} - 5I_{4,2})\right]'. \tag{4.9}
$$

From all we have said before, it is clear that we only need $p = p(\alpha, T)$, i.e., equation $(4.5)_3$, from statistical mechanics, because $n(\alpha, T), e(\alpha, T), \Gamma_1(\alpha, T)$, and $\Gamma_2(\alpha, T)$ follow from $p(\alpha, T)$ by thermodynamic arguments [see (2.67) and (2.69)]; but, of course, the statistical equations $(4.5)_{1,2,4}$ must be compatible with the thermodynamic equations (2.67) and (2.69). This is the case for (2.67) without any further requirement, while insertion of $p(\alpha, T)$ from $(4.5)_3$ into $(2.69)_1$ gives

$$
\frac{\Gamma_1}{T} = 4\pi y m^4 c^5 \frac{1}{24}(I_{5,3} - 2I_{3,1}) + T^5 A_1(\alpha) \tag{4.10}
$$

when $(4.9)_1$ is used. Comparison with $(4.5)_4$ shows that we must have

$$
A_1(\alpha) = 0 \tag{4.11}
$$

for compatibility of thermodynamics and statistical mechanics.

Apart from $p(\alpha, T)$ and $\Gamma_1(\alpha, T)$ we need Γ_2, if we wish to determine the coefficients C and A in (2.65), (2.66). Γ_2 may be calculated from $(2.69)_2$ by insertion of $p(\alpha, T)$. The integration makes use of the identity $(4.9)_2$ and we obtain

$$
\frac{\Gamma_2}{T} = -\frac{1}{60} 4\pi y m^5 c^7 \frac{1}{k\gamma}\left(3I_{6,4} - 5I_{4,2} + \frac{1}{\gamma^6}A(\alpha)\right),
$$

where $A(\alpha)$ is an arbitrary function of α closely related to the function $A_2(\alpha)$ in $(2.69)_2$.

We introduce $p(\alpha, T), \Gamma_1(\alpha, T)$ from $(4.5)_{3,4}$ and $\Gamma_2(\alpha, T)$ into (2.65), (2.66) and get

$$C_1^0 = 4\pi y m^4 c^3 \frac{1}{8} I_{5,3}, \tag{4.12$_1$}$$

$$C_1^\pi = -\frac{3}{2}\frac{1}{c^2}\frac{\begin{bmatrix} I_2 & 3I_3+I_1 & 2I_4 \\ 3I_3+I_1 & 8(I_4+I_2) & 5I_5+3I_3 \\ 2I_4 & 5I_5+3I_3 & 3I_6+I_2+\frac{A}{\gamma^6} \end{bmatrix}}{\begin{bmatrix} I_1 & 2I_2 & 3I_3-I_1 \\ 2I_2 & 3I_3+I_1 & 4I_4 \\ 3I_3-I_1 & 4I_4 & 5I_5-3I_3+2I_1 \end{bmatrix}}, \tag{4.12$_2$}$$

$$C_3 = -\frac{4}{5}\frac{\begin{bmatrix} 3I_{3,1} & 2I_{4,2} \\ 5I_{5,3}+2I_{3,1} & 3I_{6,2}+\frac{A}{\gamma^6} \end{bmatrix}}{\begin{bmatrix} 12I_{3,1} & 8I_{4,2} \\ 8I_{4,2} & 5I_{5,3}+2I_{3,1} \end{bmatrix}}, \tag{4.12$_3$}$$

$$C_5 = \frac{1}{5}\frac{3I_{6,4}-5I_{4,2}+A/\gamma^6}{I_{5,3}-2I_{3,1}}; \tag{4.12$_4$}$$

$$A_1^{\pi^2} = 3\frac{1}{T}\frac{1}{4\pi y}\frac{1}{m^4c^3}C_1^\pi \frac{\begin{bmatrix} 4I_2 & 3I_3+I_1 \\ 3I_3+I_1 & 2(I_4+I_2) \end{bmatrix}}{\begin{bmatrix} I_1 & 2I_2 & 3I_3-I_1 \\ 2I_2 & 3I_3+I_1 & 4I_4 \\ 3I_3-I_1 & 4I_2 & 5I_5-3I_3+2I_1 \end{bmatrix}}, \tag{4.13$_1$}$$

$$A_1^q = -192\frac{1}{T}\frac{1}{4\pi y}\frac{1}{m^4c^7}\frac{I_{4,2}+\frac{15}{2}C_3 I_{3,1}}{\begin{bmatrix} 12I_{3,1} & 8I_{4,2} \\ 8I_{4,2} & 5I_{5,3}+2I_{3,1} \end{bmatrix}}, \tag{4.13$_2$}$$

$$A_1^t = -\frac{12}{5}\frac{1}{T}\frac{1}{4\pi y}\frac{1}{m^4c^5}\frac{3I_{6,4}-5I_{4,2}+A/\gamma^6}{(I_{5,3}-2I_{3,1})^2}, \tag{4.13$_3$}$$

$$A_2^0 = \frac{1}{T}, \tag{4.13$_4$}$$

$$\begin{aligned} A_2^\pi &= -3\frac{1}{T}\frac{1}{4\pi y}\frac{1}{m^4c^5}\frac{\begin{bmatrix} 4I_2 & 3I_3+I_1 \\ 8I_4 & 5I_5+3I_3 \end{bmatrix} + 20C_3\begin{bmatrix} 4I_2 & 3I_3+I_1 \\ 3I_3+I_1 & 2(I_4+I_2) \end{bmatrix}}{\begin{bmatrix} I_1 & 2I_2 & 3I_3-I_1 \\ 2I_2 & 3I_3+I_1 & 4I_2 \\ 3I_3-I_1 & 4I_4 & 5I_5-3I_3+2I_1 \end{bmatrix}} \\ &= 128\frac{1}{T}\frac{1}{4\pi y}\frac{1}{m^4c^5}\frac{3I_{4,2}-\frac{3}{4}c^2C_1^\pi I_{3,1}}{\begin{bmatrix} 12I_{3,1} & 8I_{4,2} \\ 8I_{4,2} & 5I_{5,3}+2I_{3,1} \end{bmatrix}}, \end{aligned} \tag{4.13$_5$}$$

$$A_3^0 = -192\frac{1}{T}\frac{1}{4\pi y}\frac{1}{m^4c^5}\frac{3I_{4,2}-\frac{9}{2}C_5 I_{3,1}}{\begin{bmatrix} 12I_{3,1} & 8I_{4,2} \\ 8I_{4,1} & 5I_{5,3}+2I_{3,1} \end{bmatrix}} = -48\frac{1}{T}\frac{1}{4\pi y}\frac{1}{m^4c^5}C_3\frac{1}{I_{5,3}-2I_{3,1}}. \tag{4.13$_6$}$$

Insertion of $p(\alpha, T)$ and $\Gamma_1(\alpha, T)$ into the inequalities (2.70) gives

$$B_1^\pi \frac{\begin{bmatrix} 4I_2 & 3I_3+I_1 \\ 3I_3+I_1 & 2(I_4+I_2) \end{bmatrix}}{\begin{bmatrix} I_1 & 2I_2 & 3I_3-I_1 \\ 2I_2 & 3I_3+I_1 & 4I_4 \\ 3I_3-I_1 & 4I_4 & 5I_5-3I_3+2I_1 \end{bmatrix}} \le 0, \; B_4\frac{I_{3,1}}{\begin{bmatrix} 12I_{3,1} & 8I_{4,2} \\ 8I_{4,2} & 5I_{5,3}+2I_{3,1} \end{bmatrix}} \le 0, \; B_3\frac{1}{I_{5,3}-2I_{3,1}} \le 0. \tag{4.14}$$

It can easily be shown that $I_{5,3} - 2I_{3,1}$ is positive-valued, so that $(4.14)_3$ implies $B_3 \leq 0$. In the inequalities $(4.14)_{1,2}$ the complexity of the factors of B_1^π and B_4 makes it difficult to evaluate the sign of those coefficients.

With the knowledge $p(\alpha, T)$ and $\Gamma_1(\alpha, T)$ by $(4.5)_{3,4}$ we can also make the expressions for the transport coefficients (3.9) more explicit. We obtain

$$\nu + \frac{2}{3}\mu = -\frac{1}{36}\frac{1}{B_1^\pi}(4\pi y)m^4c^5 \frac{\begin{bmatrix} I_1 & 2I_2 & 3I_3 - I_1 \\ 2I_2 & 3I_3 + I_1 & 4I_4 \\ 3I_3 - I_1 & 4I_4 & 5I_5 - 3I_3 + 2I_1 \end{bmatrix}}{\begin{bmatrix} 4I_2 & 3I_3 + I_1 \\ 3I_3 + I_1 & 2(I_4 + I_2) \end{bmatrix}}, \tag{4.15$_1$}$$

$$\kappa = -\frac{1}{9 \cdot 64}\frac{1}{T}\frac{1}{B_4}(4\pi y)m^4c^5 \frac{\begin{bmatrix} 12I_{3,1} & 8I_{4,2} \\ 8I_{4,2} & 5I_{5,3} + 2I_{3,1} \end{bmatrix}}{I_{3,1}}, \tag{4.15$_2$}$$

$$\mu = -\frac{1}{48}\frac{1}{B_3}(4\pi y)m^4c^5(I_5 - 3I_3 + 2I_1). \tag{4.15$_3$}$$

Recall that by (4.14) all three coefficients are positive.

The relations (4.12) through (4.15) clearly exhibit the main result of the theory so far: All coefficients of the flux tensor A^{ABC} and of the entropy flux vector h^A are determined to within a single function $A(\alpha)$. The coefficients of the flux production I^{AB} are related to the bulk viscosity ν, the heat conductivity κ, and the shear viscosity μ.

We shall now proceed to investigate limiting cases, in which the state functions and the constitutive coefficients assume a more specific form.

4.3 Non-Degenerate Relativistic Gas

The non-degenerate gas is characterized by large values of the fugacity

$$\frac{m}{k}\alpha \gg 1 \tag{4.16}$$

so that the term ± 1 in the denominator of (4.6) may be neglected. In the rest Lorentz frame we may write

$$I_n(\alpha, \gamma) \approx e^{-\frac{m}{k}\alpha} \int_0^\infty \cosh(n\varrho)e^{-\gamma\cosh\varrho}d\varrho. \tag{4.17}$$

The integrals in (4.17) define Bessel functions of the second kind, which we shall denote by K_n,

$$K_n(\gamma) = \int_0^\infty \cosh(n\varrho)e^{-\gamma\cosh\varrho}d\varrho. \tag{4.18}$$

The K's are the natural special functions of non-degenerate relativistic gases. They satisfy the recurrence relations

$$K_{n+1}(\gamma) - K_{n-1}(\gamma) = \frac{2n}{\gamma}K_n(\gamma), \tag{4.19}$$

$$\frac{d}{d\gamma}[\gamma^n K_n(\gamma)] = -\gamma^n K_{n-1}(\gamma), \qquad \frac{d}{d\gamma}[\gamma^{-n} K_n(\gamma)] = -\gamma^{-n} K_{n+1}(\gamma).$$

With this knowledge it is a trivial though cumbersome task to write the formulae (4.5) and (4.12) through (4.15) in a form appropriate to a non-degenerate relativistic gas. From (4.17), (4.18), and

(4.19) we obtain

$$
\begin{aligned}
n &= e^{-\frac{m}{k}\alpha} 4\pi y m^3 c^3 \frac{K_2}{\gamma}, \\
e &= nmc^2 \left(G - \frac{1}{\gamma} \right), \\
p &= nmc^2 \frac{1}{\gamma} = nkT, \\
\frac{\Gamma_1}{T} &= nmc^2 \frac{2}{\gamma} G,
\end{aligned}
\tag{4.20}
$$

where G stands for K_3/K_2.

Note that the pressure is related to the particle density and the temperature in the same manner as in a classical ideal gas. This coincidence was first remarked on by Jüttner [41].

The coefficients C and A in the flux tensor A^{ABC} and the entropy flux vector h^A read[2]

$$
C_1^0 = nm \left(1 + \frac{6}{\gamma} G \right), \tag{4.21$_1$}
$$

$$
C_1^\pi = -\frac{6}{c^2} \frac{\left(2 - \frac{5}{\gamma^2}\right) + \left(\frac{19}{\gamma} - \frac{30}{\gamma^3}\right) G - \left(2 - \frac{45}{\gamma^2}\right) G^2 - \frac{9}{\gamma} G^3 + \left(1 - \frac{1}{\gamma^2} + \frac{5}{\gamma} G - G^2\right) \frac{1}{48} m^3 c^3 \frac{A(\alpha)}{n\gamma^5}}{\frac{3}{\gamma} - \left(2 - \frac{20}{\gamma^2}\right) G - \frac{13}{\gamma} G^2 + 2G^3}, \tag{4.21$_2$}
$$

$$
C_3 = -\frac{1}{\gamma} \frac{1 + \frac{6}{\gamma} G - G^2 + \frac{1}{240} m^3 c^3 \frac{A(\alpha)}{n\gamma^5}}{1 - \frac{5}{\gamma} G - G^2}, \tag{4.21$_3$}
$$

$$
C_5 = \left(\frac{6}{\gamma} + \frac{1}{G} \right) + \frac{1}{G} \frac{1}{240} m^3 c^3 \frac{A(\alpha)}{n\gamma^5}; \tag{4.21$_4$}
$$

$$
A_1^{\pi^2} = \frac{1}{4} \frac{k}{m^3 c^2} \frac{\gamma^2}{n} \frac{1 - \frac{1}{\gamma^2} + \frac{5}{\gamma} G - G^2}{\frac{3}{\gamma} - \left(2 - \frac{20}{\gamma}\right) G - \frac{13}{\gamma} G^2 + 2G^3} \cdot C_1^\pi, \tag{4.22$_1$}
$$

$$
A_1^q = \frac{1}{2} \frac{k}{m^2 c^6} \frac{\gamma}{n} \frac{\frac{5}{\gamma} + \left(\frac{30}{\gamma^2} - 1\right) G - \frac{10}{\gamma} G^2 + G^3 + \frac{1}{48} m^3 c^3 \frac{A}{n\gamma^6}}{\left(1 + \frac{5}{\gamma} G - G^2\right)^2}, \tag{4.22$_2$}
$$

$$
A_1^t = -\frac{1}{4} \frac{k}{m^2 c^4} \frac{\gamma^2}{n} \frac{1}{G^2} \left(1 + \frac{6}{\gamma} G + \frac{1}{240} m^3 c^3 \frac{A}{n\gamma^5} \right), \tag{4.22$_3$}
$$

$$
A_2^0 = \frac{1}{T}, \tag{4.22$_4$}
$$

$$
\begin{aligned}
A_2^\pi = {} & \frac{k}{m^2 c^4} \frac{\gamma}{n} \\
& \cdot \frac{-2 + \frac{5}{\gamma^2} - \left(\frac{22}{\gamma} - \frac{30}{\gamma^3}\right) G + \left(4 - \frac{65}{\gamma^2}\right) G^2 + \frac{22}{\gamma} G^3 - 2G^4 + \left(1 - \frac{1}{\gamma^2} + \frac{5}{\gamma} G - G^2\right) \frac{1}{48} m^3 c^3 \frac{A}{n\gamma^5}}{\left(1 + \frac{5}{\gamma} G - G^2\right) \left(-\frac{3}{\gamma} + (2 - \frac{20}{\gamma^2}) G + \frac{13}{\gamma} G^2 - 2G^3\right)},
\end{aligned}
\tag{4.22$_5$}
$$

$$
A_3^0 = \frac{k}{m^2 c^4} \frac{\gamma}{n} \frac{1 + \frac{6}{\gamma} G - G^2 + \frac{1}{240} m^3 c^3 \frac{A}{n\gamma^5}}{G \left(1 - \frac{5}{\gamma} G - G^2\right)}. \tag{4.22$_6$}
$$

In the sequel we shall drop the function $A(\alpha)$ since we suspect that it is zero. That suspicion is based on the kinetic theory of gases. The transport coefficients (4.15) for a non-degenerate gas

---

[2]We are indebted to S. Pennisi who has corrected several errors in these formulae.

assume the forms

$$\begin{aligned}
\nu + \frac{2}{3}\mu &= \frac{1}{3}\frac{nmc^2}{\gamma}\frac{1}{B_1^\pi}\frac{-\frac{3}{\gamma} + (2 - \frac{20}{\gamma^2})G + \frac{13}{\gamma}G^2 - 2G^3}{1 - \frac{1}{\gamma^2} + \frac{5}{\gamma}G - G^2}, \\
\kappa &= -nk \quad \frac{1}{B_4}(\gamma + 5G - \gamma G^2), \\
\mu &= -nkT \ \frac{1}{B_3}G.
\end{aligned} \tag{4.23}$$

Unlike the general case the non-degenerate one has n depending explicitly on α, so that we were able to eliminate the fugacity from the equations $(4.20)_{2,3,4}$ and (4.21), (4.22), and (4.23).

4.4 Degenerate Non-Relativistic Gas

The non-relativistic gas is characterized by large values of the relativistic coldness

$$\gamma \gg 1. \tag{4.24}$$

In that case the integrals (4.6) may be approximated by polynomials in $1/\gamma$ so that we have

$$\begin{aligned}
I_0 &= \frac{\sqrt{2mkT}}{mc}\left(i_0 - \frac{1}{4}\ \frac{1}{\gamma}i_2 + \frac{3}{32}\ \frac{1}{\gamma^2}i_4 + \cdots\right), \\
I_1 &= \frac{\sqrt{2mkT}}{mc}\left(i_0 + \frac{3}{4}\ \frac{1}{\gamma}i_2 - \frac{5}{32}\ \frac{1}{\gamma^2}i_4 + \cdots\right), \\
I_2 &= \frac{\sqrt{2mkT}}{mc}\left(i_0 + \frac{15}{4}\ \frac{1}{\gamma}i_2 + \frac{35}{32}\ \frac{1}{\gamma^2}i_4 + \cdots\right), \\
I_3 &= \frac{\sqrt{2mkT}}{mc}\left(i_0 + \frac{35}{4}\ \frac{1}{\gamma}i_2 + \frac{315}{32}\ \frac{1}{\gamma^2}i_4 + \cdots\right), \\
I_4 &= \frac{\sqrt{2mkT}}{mc}\left(i_0 + \frac{63}{4}\ \frac{1}{\gamma}i_2 + \frac{1155}{32}\ \frac{1}{\gamma^2}i_4 + \cdots\right), \\
I_5 &= \frac{\sqrt{2mkT}}{mc}\left(i_0 + \frac{99}{4}\ \frac{1}{\gamma}i_2 + \frac{3003}{32}\ \frac{1}{\gamma^2}i_4 + \cdots\right), \\
I_6 &= \frac{\sqrt{2mkT}}{mc}\left(i_0 + \frac{143}{4}\ \frac{1}{\gamma}i_2 + \frac{6435}{32}\ \frac{1}{\gamma^2}i_4 + \cdots\right),
\end{aligned} \tag{4.25}$$

where the i_n are defined by

$$i_n\left(\frac{m}{k}\alpha + \gamma\right) = \int_0^\infty \frac{x^n}{\exp\{\frac{m}{k}\alpha + \gamma + x^2\} \mp 1}dx. \tag{4.26}$$

The functions i_n are the special functions familar from non-relativistic statistical mechanics of degenerate gases (see Paragraph 2.4.4); they satisfy the recurrence relation

$$\frac{di_n(a)}{da} = \frac{n-1}{2}i_{n-2}(a). \tag{4.27}$$

With the knowledge of the limiting forms (4.25) of the functions $I_n(\alpha, \gamma)$ it is easy to calculate from (4.5) the equations of state appropriate to a degenerate non-relativistic gas, viz.

$$\begin{aligned}
n &= y8\pi\sqrt{2}(kT)^{3/2}m^{3/2} \left(i_2\left(\frac{m}{k}\alpha + \gamma\right) + \frac{5}{4}i_4\left(\frac{m}{k}\alpha + \gamma\right)\frac{1}{\gamma} + \frac{7}{32}i_6\left(\frac{m}{k}\alpha + \gamma\right)\frac{1}{\gamma^2}\right) \\
e &= nmc^2 + y8\pi\sqrt{2}(kT)^{5/2}m^{3/2}\left(i_4\left(\frac{m}{k}\alpha + \gamma\right) + \frac{5}{4}i_6\left(\frac{m}{k}\alpha + \gamma\right)\frac{1}{\gamma}\right) \\
p &= \frac{2}{3}y8\pi\sqrt{2}(kT)^{5/2}m^{3/2} \left(i_4\left(\frac{m}{k}\alpha + \gamma\right) + \frac{3}{4}i_6\left(\frac{m}{k}\alpha + \gamma\right)\frac{1}{\gamma}\right) \\
\frac{\Gamma_1}{T} &= \frac{4}{3}y8\pi\sqrt{2}(kT)^{5/2}m^{3/2} \left(i_4\left(\frac{m}{k}\alpha + \gamma\right) + \frac{7}{4}i_6\left(\frac{m}{k}\alpha + \gamma\right)\frac{1}{\gamma}\right)
\end{aligned} \tag{4.28}$$

Higher-order terms in $1/\gamma$ are neglected. The coefficients C and A in the flux tensor A^{ABC} and the entropy flux vector h^A in the non-relativistic case read

$$C_1^0 = nm + 4my8\pi\sqrt{2}(kT)^{3/2}m^{3/2}\left(i_4\left(\frac{m}{k}\alpha+\gamma\right)\frac{1}{\gamma}+\frac{7}{4}i_6\left(\frac{m}{k}\alpha+\gamma\right)\frac{1}{\gamma^2}\right)+\mathcal{O}\left(\frac{1}{\gamma^3}\right), \quad (4.29)_1$$

$$C_1^\pi = -6\frac{m}{kT}\frac{1}{\gamma}+\mathcal{O}\left(\frac{1}{\gamma^2}\right), \quad (4.29)_2$$

$$C_3 = -\frac{2}{5}-\frac{27i_2i_8-35i_4i_6}{105i_2i_6-125i_4^2}\frac{1}{\gamma}+\mathcal{O}\left(\frac{1}{\gamma^2}\right), \quad (4.29)_3$$

$$C_5 = 1+\frac{7}{5}\frac{i_6}{i_4}\frac{1}{\gamma}+\mathcal{O}\left(\frac{1}{\gamma^2}\right); \quad (4.29)_4$$

$$A_1^{\pi^2} = -\frac{9}{nkT^2}\gamma^2\frac{5i_0i_2i_4-8i_2^3}{5i_0i_4i_8-49i_0i_6^2-81i_2^2i_8+210i_2i_4i_6-125i_4^3}+\mathcal{O}(\gamma), \quad (4.30)_1$$

$$A_1^q = \frac{9}{2}\frac{m}{nk^2T^3}\frac{i_2^2}{21i_2i_6-25i_4^2}+\mathcal{O}\left(\frac{1}{\gamma}\right), \quad (4.30)_2$$

$$A_1^t = -\frac{3}{8}\frac{1}{nkT^2}\frac{i_2}{i_4}+\mathcal{O}\left(\frac{1}{\gamma}\right), \quad (4.30)_3$$

$$A_2^0 = \frac{1}{T}, \quad (4.30)_4$$

$$A_2^\pi = \frac{18}{nkT^2}\gamma\frac{i_2^2}{21i_2i_6-25i_4^2}+\mathcal{O}(\gamma^0), \quad (4.30)_5$$

$$A_3^0 = \frac{3}{5}\frac{1}{nkT^2}\frac{i_2}{i_4}+\mathcal{O}\left(\frac{1}{\gamma}\right). \quad (4.30)_6$$

The transport coefficients ν, κ, and μ assume the forms

$$\nu+\frac{2}{3}\mu = -\frac{1}{9}\frac{1}{B_1^\pi c^4}\frac{nk^3T^3}{m^2}\frac{45i_0i_4i_8-49i_0i_6^2-81i_2^2i_8+210i_2i_4i_6-125i_4^3}{5i_0i_2i_4-9i_2^3}, \quad (4.31)_1$$

$$\kappa = -\frac{1}{9}\frac{1}{B_4c^2}\frac{nk^2T}{m}\frac{21i_2i_6-25i_4^2}{i_2^2}, \quad (4.31)_2$$

$$\mu = -\frac{2}{3}\frac{1}{B_3}nkT\frac{i_4}{i_2}. \quad (4.31)_3$$

4.5 Non-Degenerate Non-Relativistic Gas

It is an easy matter to calculate the constitutive coefficients in the purely classical case, that is, for the non-degenerate, non-relativistic gas. This can be achieved two ways:

i. Either we start from the equations (4.20) through (4.23) for the non-degenerate relativistic case and let the relativistic coldness γ go to infinity, thus getting the non-relativistic limit. In this case we make use of the asymptotic expansion

$$K_n(\gamma)\underset{\gamma\to\infty}{\longrightarrow}\sqrt{\frac{\pi}{2\gamma}}e^{-\gamma}\left(1+\frac{4n^2-1}{1!8\gamma}+\frac{(4n^2-1)(4n^2-9)}{2!(8\gamma)^2}+\cdots\right) \quad (4.32)$$

for the Bessel functions (4.18).

ii. Or we start from the equations (4.28) through (4.31) and let $\frac{m}{k}\alpha + \gamma$ go to infinity as is appropriate for the non-degenerate limit. In this case we make use of the asymptotic formula

$$i_{2n}(a) \underset{a\to\infty}{\longrightarrow} e^{-a} \int_0^\infty x^{2n} e^{-x^2} dx = e^{-a} \frac{\sqrt{\pi}}{2} \frac{(2n-1)!!}{2^n} \tag{4.33}$$

for the functions (4.26).

In either way, for the equations of state, we obtain

$$\begin{aligned} n &= (4\pi y) \sqrt{2mkT}^3 e^{-(\frac{m}{k}\alpha+\gamma)} \frac{3}{8}\sqrt{\pi}, \\ e &= nmc^2 + \frac{3}{2} nkT, \\ p &= nkT, \\ \frac{\Gamma_1}{T} &= 2nkT. \end{aligned} \tag{4.34}$$

The equations $(4.34)_{2,3}$ represent the classical caloric and thermal equations of state for a monatomic ideal gas. The only difference to those classical constitutive equations is the term nmc^2 in e. This term renders specific the constant of integration that remains unknown in classical thermodynamics.

The coefficients C and A in the flux tensor A^{ABC} and in the entropy flux vector h^A read

$$C_1^0 = nm\left(1 - \frac{6}{\gamma} + \frac{15}{\gamma^2}\right), \tag{4.35$_1$}$$

$$C_1^\pi = -\frac{1}{c^2}\left(6 + \frac{63}{\gamma} + \frac{693}{4}\frac{1}{\gamma^2}\right), \tag{4.35$_2$}$$

$$C_3 = -\frac{1}{\gamma}\left(\frac{2}{5} + \frac{7}{5}\frac{1}{\gamma} + \frac{21}{20}\frac{1}{\gamma^2}\right), \tag{4.35$_3$}$$

$$C_5 = 1 + \frac{7}{2}\frac{1}{\gamma} + \frac{35}{8}\frac{1}{\gamma^2}; \tag{4.35$_4$}$$

$$A_1^{\pi^2} = -\frac{3}{5}\frac{m^2}{nk^3T^4} c^4, \tag{4.36$_1$}$$

$$A_1^q = \frac{1}{5}\frac{m}{nk^2T^3}, \tag{4.36$_2$}$$

$$A_1^t = -\frac{1}{4}\frac{1}{nkT^2}, \tag{4.36$_3$}$$

$$A_2^0 = \frac{1}{T}, \tag{4.36$_4$}$$

$$A_2^\pi = \frac{4}{5}\frac{m}{nk^2T^3} c^2, \tag{4.36$_5$}$$

$$A_3^0 = \frac{2}{5}\frac{1}{nkT^2}. \tag{4.36$_6$}$$

The transport coefficients ν, κ, and μ are given by

$$\nu + \frac{2}{3}\mu = -\frac{5}{3}\frac{1}{B_1^\pi c^4}\frac{nk^3T^3}{m^2}, \tag{4.37$_1$}$$

$$\kappa = -\frac{5}{2}\frac{1}{B_4 c^2}\frac{nk^2T}{m}, \tag{4.37$_2$}$$

$$\mu = -\frac{1}{B_3} nkT. \tag{4.37$_3$}$$

4.6 Strongly Degenerate Relativistic Fermi Gas

For Fermi particles strong degeneration means

$$\frac{m}{k}\alpha + \gamma \ll -1 \tag{4.38}$$

so that $g_c/kT = -((m/k)\alpha + \gamma)$—with g_c being the free enthalpy—tends to infinity as the degeneration increases. In that case the distribution function (4.4) has a nearly constant value, equal to y for

$$0 \leq p < mc\sqrt{\left(1 + \frac{gc}{mc^2}\right)^2 - 1}$$

and practically vanishes for

$$p > mc\sqrt{\left(1 + \frac{gc}{mc^2}\right)^2 - 1}.$$

Therefore the function $F(\varrho) = \left(\exp\{\frac{\alpha}{k} + \gamma\cosh\varrho\} + 1\right)^{-1}$ in the integrand of $I_m(\alpha, \gamma)$ in (4.6) assumes the form shown in Figure 1. The sudden decrease occurs at ϱ_F, where the argument of the exponential vanishes.

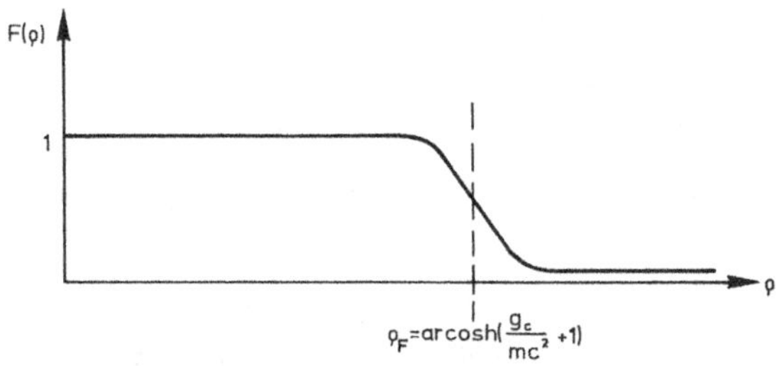

Figure 1. Strong degeneration in Fermions.

The slope of the function at ϱ_F is equal to $-\sqrt{(1 + g_c/mc^2)^2 - 1}$ so that by (4.38) the curve is very steep there. This fact enables us to calculate approximate expressions for the integrals $I_n(\alpha, \gamma)$ as follows. By partial integration we may write these integrals in the form

$$I_n(\alpha, \gamma) = -\frac{1}{n}\int_0^\infty \sinh\varrho \left[-\gamma \sinh\varrho \frac{e^{\left(\frac{m}{k}\alpha + \gamma\cosh\varrho\right)}}{\left(e^{\frac{m}{k}\alpha + \gamma\cosh\varrho} + 1\right)^2}\right] d\varrho, \tag{4.39}$$

where the square bracket encloses the derivative of $F(\varrho)$. Since that derivative has appreciable values only in the neighbourhood of ϱ_F, we obtain approximate values for $I_n(\alpha, \gamma)$ by expanding $\sinh n\varrho$ in terms of $\xi = \gamma(\cosh\varrho - \cosh\varrho_F)$ up to second order

$$\sinh n\varrho = \sinh n\varrho_F \left(1 + \frac{n}{\gamma}\frac{\cosh n\varrho_F}{\sinh n\varrho_F \sinh\varrho_F}\xi + \frac{1}{2}\frac{n}{\gamma^2}\frac{n\sinh n\varrho_F \sinh\varrho_F - \cosh n\varrho_F \cosh\varrho_F}{\sinh n\varrho_F \sinh^3\varrho_F}\xi^2\right). \tag{4.40}$$

Insertion into (4.39) provides integrals of the type

$$\int_{\frac{\alpha}{k}+\gamma}^{\infty} \xi^n \frac{e^\xi}{(e^\xi + 1)^2} d\xi \approx \int_{-\infty}^{\infty} \xi^n \frac{e^\xi}{(e^\xi + 1)^2} d\xi = \begin{cases} 0, & \text{n odd,} \\ 2n(n-1)!(1 - 2^{1-n})\zeta_n, & \text{n even,} \end{cases} \tag{4.41}$$

where the first equation is approximately valid by (4.38) and the second one is taken from a mathematical handbook. We obtain

$$I_n(\alpha,\gamma) = \frac{\sinh n\varrho_F}{n}\left(1+\frac{\pi^2}{6}\frac{n}{\gamma^2}\frac{n\sinh n\varrho_F \sinh\varrho_F - \cosh n\varrho_F\cosh\varrho_F}{\sinh n\varrho_F \sinh^3\varrho_F}\right). \tag{4.42}$$

ζ_n in (4.41) stands for $\sum_{r=1}^{\infty}(1/r^n)$. Note that with increasing degeneration g_c and, hence ϱ_F, becomes bigger and the second term in (4.42) becomes negligible.

Inspection of the results (4.5) and (4.12) through (4.15) shows that we need the function I_n for $n = 0,\dots,6$. For those few values of n it is easy to write $I_n(\alpha,\gamma)$ as functions of the variables

$$\gamma \quad \text{and} \quad x = \sinh\varrho_F = \sqrt{\left(1+\frac{g_c}{mc^2}\right)^2 - 1}. \tag{4.43}$$

With a little calculation we obtain

$$\begin{aligned}
I_0(\alpha,\gamma) &= ar\sinh x - \frac{\pi^2}{6}\frac{1}{\gamma^2}\frac{\sqrt{1+x^2}}{x^3},\\
I_1(\alpha,\gamma) &= x - \frac{\pi^2}{6\gamma^2}\frac{1}{x^3},\\
I_2(\alpha,\gamma) &= \sqrt{1+x^2}\left(x - \frac{\pi^2}{6\gamma^2}\left(\frac{1}{x^3}-\frac{2}{x}\right)\right),\\
I_3(\alpha,\gamma) &= \left(x+\frac{4}{5}x^3\right) - \frac{\pi^2}{6\gamma^2}\left(\frac{1}{x^3}-\frac{4}{x}-8x\right),\\
I_4(\alpha,\gamma) &= \sqrt{1+x^2}\left(x+2x^3 - \frac{\pi^2}{6\gamma^2}\left(\frac{1}{x^3}-\frac{8}{x}-24x\right)\right),\\
I_5(\alpha,\gamma) &= \frac{16}{5}x^5+4x^3+x - \frac{\pi^2}{6\gamma^2}\left(\frac{1}{x^3}-\frac{12}{x}-72x-64x^3\right),\\
I_6(\alpha,\gamma) &= \sqrt{1+x^2}\left(\frac{16}{3}x^5+\frac{16}{3}x^3+x - \frac{\pi^2}{6\gamma^2}\left(\frac{1}{x^3}-\frac{18}{x}-144x-160x^3\right)\right).
\end{aligned} \tag{4.44}$$

This knowledge enables us to write the state functions (4.5) in a form appropriate to a strongly degenerate relativistic Fermi gas.

$$\begin{aligned}
n &= 4\pi y m^3 c^3 \frac{1}{3}\; x^3 \left[1+\frac{\pi^2}{6\gamma^2}\frac{6x^2+3}{x^4}\right],\\
e &= 4\pi y m^4 c^5 \frac{1}{24}\, g(x) \left[1+\frac{\pi^2}{6\gamma^2}\frac{24(3x^2+1)\sqrt{1+x^2}}{g(x)x}\right],\\
p &= \frac{1}{3}\,4\pi y m^4 c^5 \frac{1}{8}\; f(x) \left[1+\frac{\pi^2}{6\gamma^2}\frac{24x\sqrt{1+x^2}}{f(x)}\right],\\
\frac{\Gamma_1}{T} &= \frac{2}{3}\,4\pi y m^5 c^5 \frac{1}{5}\; x^5 \left[1+\frac{\pi^2}{6\gamma^2}\frac{4x^6+3x^4}{\frac{1}{5}x^8}\right].
\end{aligned} \tag{4.45}$$

Following Chandrasekhar [37] we have introduced $g(x)$ and $f(x)$ as abbreviations defined by

$$g(x) = 3x(2x^2+1)\sqrt{1+x^2} - 3\text{arsinh}x \quad \text{and} \quad f(x) = x(2x^2-3)\sqrt{1+x^2} + 3\text{arsinh}x. \tag{4.46}$$

Note that in the case of complete degeneration the second term in the square brackets of (4.45) may be neglected,

$$n \to 4\pi y m^3 c^3 \frac{1}{3} x^3, \quad e \to 4\pi y m^4 c^5 \frac{1}{24} g(x), \quad p \to 4\pi y m^4 c^5 \frac{1}{8} f(x), \quad \frac{\Gamma_1}{T} \to 4\pi y m^5 c^5 \frac{2}{15} x^5. \tag{4.47}$$

The temperature is then no longer a variable, because it is effectively equal to zero at full degeneration.

With the functions (4.44) known it is an easy though cumbersome task to calculate the relations (4.12) through (4.15) in a form appropriate to a strongly degenerate relativistic Fermi gas. We leave that to the reader.

4.7 A Remark on the Strongly Degenerate Relativistic Bose Gas

For Bose particles strong degeneration means

$$0 < \frac{m}{k}\alpha + \gamma \ll 1. \tag{4.48}$$

In that case the integrals $I_m(\alpha, \gamma)$ in (4.6) may be approximately represented in terms of the Bessel functions (4.32) as

$$I_m(\alpha,\gamma) = e^{-(m/k)\alpha} \int_0^\infty \cosh m\varrho\, e^{-\gamma\cosh\varrho} \frac{1}{1 - e^{-((m/k)\alpha + \gamma\cosh\varrho)}} d\varrho,$$

$$I_m(\alpha,\gamma) \approx e^{-(m/k)\alpha} \int_0^\infty \cosh m\varrho\, e^{-\gamma\cosh\varrho} \sum_{r=0}^{\infty} (-1)^r e^{-r((m/k)\alpha + \gamma\cosh\varrho)} d\varrho,$$

$$I_m(\alpha,\gamma) \approx \sum_{r=0}^{\infty} (-1)^r e^{-(1+r)(m/k)\alpha} K_m\left((1+r)\gamma\right). \tag{4.49}$$

The sum, however, must contain many terms, if the approximation is to be good. Therefore no simple formulae can be had for the constitutive coefficients in this case.

4.8 Equilibrium Properties of an Ultrarelativistic Gas

The ultrarelativistic gas is characterized by a very small relativistic coldness

$$\gamma \ll 1. \tag{4.50}$$

This condition is realized in a gas at very high temperatures. The distribution function (4.4) in the local rest frame assumes the form

$$F_{\mathrm{E}} = \frac{y}{e^{(m/k)\alpha + cp/kT} \mp 1}, \tag{4.51}$$

and, instead of (4.5), from (4.3) we obtain

$$\begin{aligned} n &= 4\pi y \left(\frac{kT}{c}\right)^3 j_2(\alpha), \\ e &= 4\pi y\, c \left(\frac{kT}{c}\right)^4 j_3(\alpha), \\ p &= \frac{1}{3} 4\pi y\, c \left(\frac{kT}{c}\right)^4 j_3(\alpha), \\ \frac{\Gamma_1}{T} &= \frac{2}{3} 4\pi y \left(\frac{kT}{c}\right)^5 j_4(\alpha). \end{aligned} \tag{4.52}$$

The functions $j_n(\alpha)$ are defined by

$$j_n(\alpha) = \int_0^\infty \frac{x^n}{e^{(m/k)\alpha + x} \mp 1} dx, \tag{4.53}$$

and they satisfy the recurrence formula

$$\frac{dj_n(\alpha)}{d\alpha} = -\frac{m}{k} n j_{n-1}(\alpha). \tag{4.54}$$

5 Two Applications

5.1 The Limit Mass of White Dwarfs

We assume that a star as a white dwarf is fully ionized and that the free electrons are the only component that is thermodynamically active. Due to the high density the electrons form a fully degenerate Fermi gas, and the particle number, i.e., electron number and the pressure, are

$$\begin{aligned} n &= 4\pi y m^3 c^3 \frac{1}{3} x^3 = A x^3, \\ p &= \frac{1}{24} 4\pi y m^4 c^5 f(x) = B f(x) \end{aligned} \tag{5.1}$$

according to (4.45). $f(x)$ is given by $(4.46)_2$, and for a later purpose we note that

$$\frac{df}{dx} = \frac{8x^4}{\sqrt{1+x^2}} \tag{5.2}$$

holds. x, or n, is the only relevant variable here, since the star is considered to be at rest. The temperature plays no role in a fully degenerate gas, and stress deviator and heat flux are ignored. Thus this consideration is not really an example for relativistic extended thermodynamics but rather for ordinary relativistic thermostatics. It does, however, serve to show the relevance of relativistic effects in thermodynamics and that is why it is included. The arguments presented here are due to Chandrasekhar (see [37]).

While the ions of hydrogen, helium, and traces of heavier elements are unimportant thermodynamically, they do of course create the gravitational field, and therefore we must find the relation between n, the number density of electrons, and ϱ, the mass density of the star. If X, Y, and Z denote the mass fractions of hydrogen, helium, and heavier elements with a mean molecular weight of M_z (say) and order number $\frac{1}{2}M_z$, we have

$$n = \varrho\frac{X}{m_{\mathrm{H}}} + 2\varrho\frac{Y}{4m_{\mathrm{H}}} + \frac{1}{2}M_z\varrho\frac{Z}{M_z m_{\mathrm{H}}}.$$

Here m_{H} is the mass of the hydrogen nucleus. We ignore X in comparison with Y and Z, since the star will have very little hydrogen left, and obtain with $Y + Z = 1$ and $(5.1)_1$

$$\varrho = 2m_{\mathrm{H}} n = 2m_{\mathrm{H}} A x^3. \tag{5.3}$$

The radial component of the momentum balance is the only relevant equation for the single variable x or n or p. It reads

$$\begin{aligned} \frac{dp(r)}{dr} &= -\varrho(r) G \frac{M(r)}{r^2}, \\ \frac{dp(r)}{dr} &= -\frac{\varrho(r)}{r^2} 4\pi G \int_0^r \varrho(r') r'^2 dr', \end{aligned} \tag{5.4}$$

where G is the gravitation constant. By use of (5.3) and of the state equations we convert this equation into an equation that determines x as a function of r. Several substitutions are made so as to give the equation its canonical form. First we differentiate, introduce $(5.1)_1$ and (5.3), and use (5.2). Thus we obtain

$$\frac{1}{r^2}\frac{d}{dr}\left(\frac{r^2}{\varrho}\frac{dp}{dr}\right) = -4\pi G\varrho,$$

$$\frac{1}{r^2}\frac{d}{dr}\left(\frac{r^2}{x^3}\frac{df(x)}{dr}\right) = -4\pi G(2m_{\mathrm{H}})^2\frac{A^2}{B}x^3,$$

$$\frac{1}{r^2}\frac{d}{dr}\left(r^2\frac{d\sqrt{1+x^2}}{dr}\right) = -\frac{2\pi G m_{\mathrm{H}}^2 A^2}{B}x^3.$$

We eliminate the square root on the left-hand side by introducing the new dependent variable $y = \sqrt{1+x^2}$

$$\frac{1}{r^2}\frac{d}{dr}\left(r^2\frac{dy}{dr}\right) = -\underbrace{\frac{2\pi G m_{\mathrm{H}}^2 A^2}{B}}_{1/l^2}(y^2-1)^{3/2}. \tag{5.5}$$

We define the length l as indicated in (5.5) and use it to introduce a dimensionless radius. It is also useful for the solution of the differential equation to normalize y by its (unknown) central value y_0 and introduce

$$\Phi = \frac{y}{y_0} \quad \text{and} \quad \eta = y_0\frac{r}{l}. \tag{5.6}$$

In this manner the momentum balance reads

$$\frac{1}{\eta^2}\frac{d}{d\eta}\left(\eta^2\frac{d\Phi}{d\eta}\right) = -\left(\Phi^2 - \frac{1}{y_0^2}\right)^{3/2}. \tag{5.7}$$

It has to be solved for the boundary conditions

$$\Phi(0) = 1 \quad \text{and} \quad \frac{\partial\Phi}{\partial\eta}(0) = 0. \tag{5.8}$$

The latter condition is required by smoothness in the center.

Numerically found solutions for different values of y_0, or of the central density n_0, are shown in Figure 2(a). For the outer radius R of the star we must have $n_R = 0$; hence $x_R = 0$ and $y_R = 1$. Therefore $\Phi(\eta_R) = \Phi(y_0(R/l)) = 1/y_0$ and hence the central value y_0 determines the radius R or vice versa. We obtain

$$y_0 = y_0(R) \tag{5.9}$$

albeit only numerically. The function is shown in Figure 2(b). But one observation is clear from Figure 2; namely, if the central number density, and hence y_0, go to infinity, $\Phi = 0$ occurs at a value $\eta \approx 6.9$ and therefore R must be zero. This then represents the ultimate white dwarf, and we shall now proceed to investigate its mass.

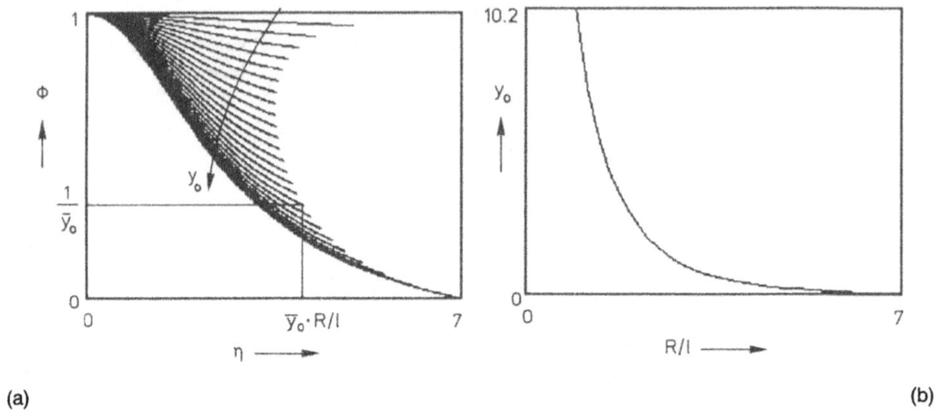

Figure 2. (a) $\Phi(\eta, y_0)$ for different values of y_0 with an indication of how to find $y_0 = y_0(R)$. (b) y_0 as a function of R.

The mass of the star is also a function of R. Indeed we have

$$\begin{aligned} M(r) &= 4\pi \int_0^r \varrho(r')r'^2 dr' \quad \text{with } \varrho = 2m_{\mathrm{H}} A x^3 \\ &\qquad\qquad\qquad\qquad = 2m_{\mathrm{H}} A \sqrt{y^2-1}^{\,3} \\ &\qquad\qquad\qquad\qquad = 2m_{\mathrm{H}} A y_0^3 \left(\Phi^2 - 1/y_0^2\right)^{3/2}, \\ M(\eta) &= 8\pi m_{\mathrm{H}} A l^3 \int_0^\eta \left(\Phi^2 - \frac{1}{y_0^2}\right)^{3/2} \eta'^2 d\eta' \quad \text{with (5.7)}, \\ M(\eta) &= -2\sqrt{\frac{2}{\pi}} \frac{B^{3/2}}{m_{\mathrm{H}}^2 G^{3/2} A^2} \int_0^\eta \frac{d}{d\eta'}\left(\eta'^2 \frac{d\Phi}{d\eta'}\right) d\eta', \\ M(\eta) &= -2\sqrt{\frac{2}{\pi}} \frac{B^{3/2}}{m_{\mathrm{H}}^2 G^{3/2} A^2} \eta^2 \frac{d\Phi}{d\eta}. \end{aligned}$$

The total mass M is therefore given by

$$M = -2\sqrt{\frac{2}{\pi}} \frac{B^{3/2}}{m_{\mathrm{H}}^2 G^{3/2} A^2} \left(\eta^2 \frac{d\Phi}{d\eta}\right) \Big|_{\eta = y_0(R/l)}. \tag{5.10}$$

Since y_0 is known as a function of R and $d\Phi/d\eta$ is known for $\eta = y_0(R/l)$ (see Figure 2), we may calculate

$$M = M(R). \tag{5.11}$$

This function has been drawn in Figure 3. We see that M is a decreasing function of R and that for $R = 0$—the ultimate dwarf—it assumes the value

$$M_{\min} \approx 2\sqrt{\frac{2}{\pi}} \frac{B^{3/2}}{m_{\mathrm{H}}^2 G^{3/2} A^2} 2 \approx 1,4 M_\odot, \tag{5.12}$$

where $M_\odot = 1.97\, 10^{30}$kg is the mass of the sun.

It follows that the mass of a white dwarf cannot be bigger than roughly one and a half times the mass of the sun. If any star's mass *is* bigger, its final stage cannot be the white dwarf, that is, the electrons cannot withstand the gravitational contraction.

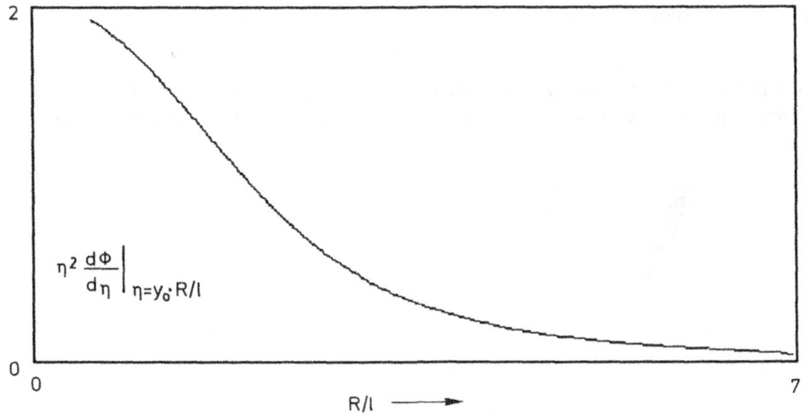

Figure 3. The mass of a white dwarf as a function of its radius.

The discussion of this example would be incomplete, if no mention were made of the non-relativistic limit. Of course, the value of the relativistic coldness cannot serve to distinguish the relativistic case from the non-relativistic one in the case of full degeneration because γ has dropped out of the formulae, but there is the value of x. By (4.43) we have

$$x = \sqrt{\left(1 + \frac{g_c}{mc^2}\right)^2 - 1},$$

where g_c is the free enthalpy. Obviously for a non-relativistic gas we have $g_c \ll mc^2$ and x is small compared to 1. In this case $f(x) \underset{x \to \infty}{\longrightarrow} \frac{8}{5} x^5$ and from (4.45) we obtain $p \sim n^{5/3}$, which indeed conforms to the pressure-density relation (2.1.14) appropriate to the non-relativistic strongly degenerate Fermi gas.

Instead of (5.7) we now arrive at the differential equation

$$\frac{1}{\xi^2}\frac{d}{d\xi}\left(\xi^2 \frac{d\Psi}{d\xi}\right) = -\frac{2}{x_0}\Psi^{3/2} \quad \text{with } \Psi = \left(\frac{x}{x_0}\right)^2, \; \xi = x_0 \frac{r}{l}, \tag{5.13}$$

which must be solved for the boundary conditions

$$\Psi(0) = 1, \qquad \frac{d\Psi}{d\xi}(0) = 0. \tag{5.14}$$

Figure 4 shows plots of the solution for different values of x_0. R can be calculated from the point ξ where Ψ becomes zero and we obtain

$$x_0 = x_0(R). \tag{5.15}$$

The mass of the star is calculated as before and we obtain

$$M = -4\pi m_{\rm H} A l^3 x_0 \left(\xi^2 \frac{d\Psi}{d\xi}\right)|_{\xi=x_0(R/l)}. \tag{5.16}$$

The slope of Ψ is known at $\xi = x_0(R/l)$ from Figure 4, and we obtain

$$M = M(R)$$

as shown in Figure 5. We conclude that M grows to infinity as R tends to zero, so that there is no restriction on the mass of the white dwarf, if the non-relativistic equation of state is assumed to be valid.

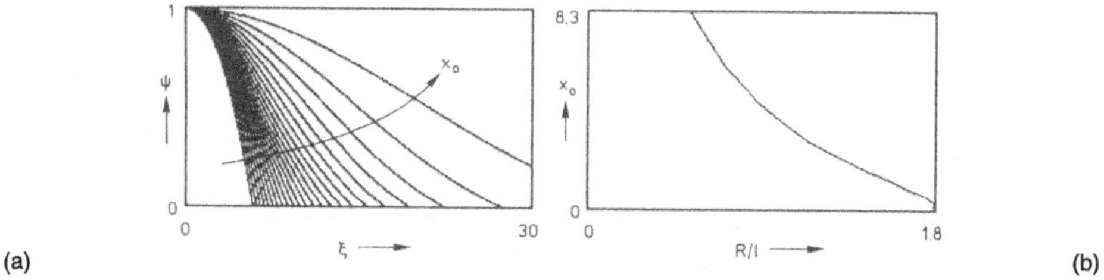

Figure 4. (a) $\Psi(\xi; x_0)$ for different values of x_0.
(b) x_0 as a function of R.

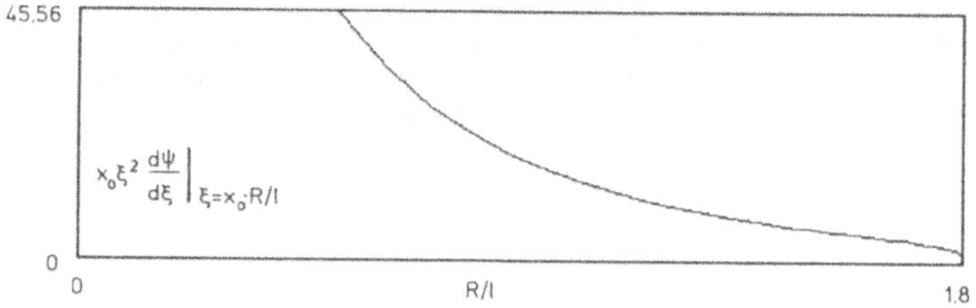

Figure 5. The mass as a function of R for a non-relativistic state function.

5.2 Thermo Acceleration Waves

We consider a particular class of solutions of the system (1.2) of balance laws of relativistic extended thermodynamics, namely, weak discontinuity waves, or, in the language of continuum mechanics, acceleration waves. These are smooth fields on either side of a moving surface Σ, but on Σ, while continuous, they have jumps in the normal derivatives. Σ is called the wave front.

Let the representation of Σ in space-time be

$$\varphi(x^A) = 0 \quad \text{so that} \quad \begin{aligned} &\frac{\partial \varphi}{\partial t} + V n^a \frac{\partial \varphi}{\partial x^a} = 0, \\ &\frac{\partial \varphi}{\partial t} + V|\text{grad}\varphi| = 0 \end{aligned} \tag{5.17}$$

hold, where V is the normal speed of the surface and n^a its unit normal vector. If ψ is a general quantity with the jump $\delta\psi$ in its normal derivative, for the jumps of $\frac{\partial \psi}{\partial t}$ and $\frac{\partial \psi}{\partial x^a}$ we have

$$\left[\frac{\partial \psi}{\partial t}\right] = -\delta\psi V \quad \text{and} \quad \left[\frac{\partial \psi}{\partial x^a}\right] = \delta\psi n_a. \tag{5.18}$$

We subtract the balance equations (1.2) in front of the wave and behind it and obtain by use of $(5.17)_2$, (5.18)

$$\varphi_{,A}\,\delta A^A = 0, \qquad \varphi_{,A}\,\delta A^{AB} = 0, \qquad \varphi_{,A}\,\delta A^{ABC} = 0. \tag{5.19}$$

These equations will be referred to as the discontinuity equations.

We decompose $\varphi_{,A}$ into space–like and time–like components relative to the proper time direction U_A of the fluid; in particular,

$$\varphi_{,A} = |\text{grad}\,\varphi|\left(n_A - \frac{\lambda}{c}U_A\right) \quad \text{with} \quad n_A n^A = -1, \; n_A U^A = 0, \tag{5.20}$$

where $|\text{grad}\,\varphi| \equiv \sqrt{h^{AB}\varphi_{,A}\,\varphi_{,B}}$. Thus, by (5.17) λ is the normal speed of the surface relative to the fluid and divided by c; n_a is the normal vector to Σ in the rest frame of the fluid.

A more explicit form of the discontinuity equations (5.19) is obtained by insertion of (5.20) and of the decompositions $(2.60)_{1,2}$ and (2.62), viz.

$$\begin{aligned}
&\left(n_A - \frac{\lambda}{c}U_A\right)\delta\left\{nmU^A\right\} = 0,\\
&\left(n_A - \frac{\lambda}{c}U_A\right)\delta\left\{t^{\langle AB\rangle} + \left(p(n,e) - \frac{C_1^0}{C_1^\pi} + \frac{1}{C_1^\pi}\chi\right)h^{AB} + \frac{1}{c^2}\left(U^Aq^B + U^Bq^A\right) + \frac{e}{c^2}U^AU^B\right\} = 0,\\
&\left(n_A - \frac{\lambda}{c}U_A\right)\delta\left\{\chi U^AU^BU^C + \frac{c^2}{6}(nm-\chi)\left(g^{AB}U^C + g^{BC}U^A + g^{CA}U^B\right)\right.\\
&\qquad + C_3\left(g^{AB}q^C + g^{BC}q^A + g^{CA}q^B\right)\\
&\qquad - \frac{6}{c^2}C_3\left(U^AU^Bq^C + U^BU^Cq^A + U^CU^Aq^B\right)\\
&\qquad \left. + C_5\left(t^{\langle AB\rangle}U^C + t^{\langle BC\rangle}U^A + t^{\langle CA\rangle}U^B\right)\right\} = 0.
\end{aligned} \tag{5.21}$$

Here we have introduced $\chi = C_1^0 + C_1^\pi\pi$ as an abbreviation, so $\pi = (\chi - C_1^0)/C_1^\pi$ holds. In this paragraph χ will replace the variable π, because it is easier notationally.

We shall investigate the case that the wave propagates into a region where equilibrium prevails so that $t^{\langle AB\rangle}, q^A$, and π are all zero. The discontinuity equations are further simplified by projecting the vector and tensor equation on the orthonormal set of base vectors

$$U_A, n_A, \overset{k}{e}_A \qquad (k = 1, 2), \tag{5.22}$$

which obey the conditions

$$\begin{aligned}
U^AU_A = c^2, \qquad & U^An_A = 0, & U^A\overset{k}{e}_A &= 0,\\
& n^An_A = -1, & n^A\overset{l}{e}_A &= 0,\\
& & \overset{l}{e}{}^A\overset{k}{e}_A &= \delta^{lk}.
\end{aligned} \tag{5.23}$$

For the jumps of the normal derivatives of the 14 variables $n, e, \chi, U^A, q^A, t^{\langle AB\rangle}$ we introduce the following relations or definitions:

$$\begin{aligned}
&\delta n,\\
&\delta e,\\
&\delta\chi;
\end{aligned} \tag{5.24$_1$}$$

$$\delta U^A: \quad \begin{aligned} U_A \delta U^A &= 0, \\ n_A \delta U^A &= \delta U, \\ \overset{k}{e}_A \, \delta U^A &= \delta U^k; \end{aligned} \tag{5.24$_2$}$$

$$\delta q^A: \quad \begin{aligned} U_A \delta q^A &= 0, \\ n_A \delta q^A &= \delta q, \\ \overset{k}{e}_A \, \delta q^A &= \delta q^k; \end{aligned} \tag{5.24$_3$}$$

$$\delta t^{\langle AB \rangle}: \quad \begin{aligned} & U_A n_B \, \delta t^{\langle AB \rangle} = 0, \quad U_A n_B \, \delta t^{\langle AB \rangle} = 0, && U_A \overset{k}{e}_B \, \delta t^{\langle AB \rangle} = 0, \\ & \qquad\qquad\qquad\quad n_A n_B \delta t^{\langle AB \rangle} = \delta t, && n_A \overset{k}{e}_B \, \delta t^{\langle AB \rangle} = \delta t^k, \\ & && \overset{k}{e}_A \overset{l}{e}_B \, \delta t^{\langle AB \rangle} = \delta t^{kl}. \end{aligned} \tag{5.24$_4$}$$

There are 14 independent jumps among $\delta n, \delta e, \delta \chi, \delta U, \delta U^k, \delta q, \delta q^k, \delta t, \delta t^k, \delta t^{kl}$. Since $t^{\langle AB \rangle}$ contains only 5 independent components, only two of the jumps δt^{kl} are independent.

With all this the discontinuity equations form a homogeneous algebraic system of equations, which is listed on the next page as equation (5.25).

Non-trivial solutions result when the determinant vanishes, and there are three possibilities for that.

First of all

$$\lambda = 0. \tag{5.26}$$

In this case we may have $\delta t^{jk} \neq 0$. This represents what is called a contact wave. It does not move with respect to the body, and it follows from (5.25) that we have

$$\begin{aligned} \delta \chi &= \frac{C_1^\pi}{C_1^\pi - \frac{6}{c^2} C_5} \left[\left(m + \frac{6}{c^2} C_5(\quad)_n \right) \delta n + \frac{6}{c^2} C_5(\quad)_e \delta e \right], \\ \delta U &= 0, \\ \delta q &= 0, \\ \delta q^k &= -\frac{1}{6} c^2 \frac{nm - C_1^0}{C_3} \delta U^k, \\ \delta t &= \frac{c^2}{6C_5} (m \delta n - \delta \chi), \\ \delta t^k &= 0. \end{aligned} \tag{5.27}$$

Thus six jumps may be assigned on the contact wave, viz. $\delta n, \delta e, \delta U^k$, and the two independent components of δt^{kl}; therefore, we say that this wave has *multiplicity* 6.

Now we assume $\lambda \neq 0$. In this case $\delta t^{jk} = 0$ holds, and the system (5.25) uncouples into two systems for

$$\delta U^k, \delta q^k, \delta t^k \quad \text{and} \quad \delta u, \delta e, \delta \chi, \delta U, \delta q, \delta t, \tag{5.28}$$

respectively. We look at the former. It reads

$$\begin{bmatrix} \frac{\lambda}{c}(e+p) & \frac{\lambda}{c} & -1 \\ \lambda c^2 (nm + 5C_1^0) & -30\lambda C_3 & -6cC_5 \\ c^2(nm - C_1^0) & 6C_3 & 6\lambda c C_5 \end{bmatrix} \begin{bmatrix} \delta U^k \\ \delta q^k \\ \delta t^k \end{bmatrix} = 0. \tag{5.29}$$

$$
\begin{bmatrix}
-\lambda c & 0 & 0 & n & 0 & 0 & 0 & 0 & 0 & 0 \\
()_n & ()_e & \frac{1}{C_1^{\pi}} & -\frac{1}{c}\lambda(p+e) & 0 & -\frac{1}{c}\lambda & 0 & 1 & 0 & 0 \\
0 & -\lambda c & 0 & -\frac{1}{c}\lambda(p+e) & 0 & 1 & 0 & 0 & 0 & 0 \\
0 & 0 & 0 & 0 & -\frac{1}{c}\lambda(p+e) & 0 & -\frac{1}{c}\lambda & 0 & 1 & 0 \\
-\frac{m}{2}\lambda c^5 & 0 & -\frac{1}{2}\lambda c^5 & \frac{c^4}{6}(mn+5C_1^0) & 0 & -5C_3c^2 & 0 & C_5c^2 & 0 & 0 \\
-\frac{m}{6}c^4 & 0 & \frac{1}{6}c^4 & -\frac{c^3}{6}\lambda(mn+5C_1^0) & 0 & 0 & 0 & 1 & 0 & 0 \\
0 & 0 & 0 & 0 & -\frac{c^3}{6}\lambda(mn+5C_1^0) & 0 & 5c\lambda C_3 & 0 & C_5c^2 & 0 \\
\frac{m}{6}\lambda c^4 & 0 & -\frac{1}{6}\lambda c^3 & -\frac{c^2}{2}(mn-C_1^0) & 0 & -3C_3 & 0 & -c\lambda C_5 & 0 & 0 \\
0 & 0 & 0 & 0 & -\frac{c^2}{6}(mn-C_1^0) & 0 & -C_3 & 0 & -c\lambda C_5 & 0 \\
-\frac{m}{6}\lambda c^3\delta^{lk} & 0 & \frac{1}{6}\lambda c^3\delta^{lk} & \frac{c^2}{6}(mn-C_1^0)\delta^{lk} & 0 & C_3\delta^{lk} & 0 & 0 & 0 & -c\lambda C_5
\end{bmatrix}
\begin{bmatrix}
\delta n \\ \delta e \\ \delta\chi \\ \delta U \\ \delta U^k \\ \delta q \\ \delta q^k \\ \delta t \\ \delta t^k \\ \delta t^{kl}
\end{bmatrix}
= 0
$$

$$
(\ )_n = \frac{\partial p}{\partial n} - \frac{1}{C_1^{\pi}}\frac{\partial C_1^0}{\partial n}; \qquad (\ )_e = \frac{\partial p}{\partial e} - \frac{1}{C_1^{\pi}}\frac{\partial C_1^0}{\partial e}. \qquad (5.25)
$$

For non-trivial solutions the determinant must vanish, which requires that the speeds be given by

$$\lambda = \pm\sqrt{\frac{\{6(e+p)C_3 + c^2 C_1^0\}\, C_5 + (6C_3 + C_5)nmc^2}{\{nmc^2 + 5\left[6(e+p)C_3 + c^2 C_1^0\right]\}\, C_5}}. \tag{5.30}$$

Of course, now the second set of quantities in (5.28) must vanish along with δt^{jk}. Therefore the wave has multiplicity 2, since the two values δU^k (say) may be assigned on it. The waves are called transversal, since δU^k is perpendicular to the normal on the wave front.

Finally we consider the case

$$\delta t^{jk} = 0, \qquad \delta U^k = 0, \qquad \delta q^k = 0, \qquad \delta t^k = 0. \tag{5.31}$$

In this case the system (5.25) reduces to 6 equations for

$$\delta n,\ \delta e,\ \delta \chi,\ \delta U,\ \delta q,\ \delta t, \tag{5.32}$$

which are all coupled. By a short calculation we eliminate $\delta U, \delta q, \delta t$ and obtain a linear algebraic system for $\delta u, \delta n$, and $\delta \chi$ alone, with the coefficient matrix

$$\begin{bmatrix} 15C_5(\quad)_n - 2mc^2 & 15C_5\left[(\quad)_e - \lambda^2\right] & 15C_5\dfrac{1}{C_1^\pi} + 2c^2 \\ -\{(5\lambda^2 - 3)\left[c^2 C_1^0 + 6C_3(e+p)\right] + (\lambda^2+3)nmc^2\} & 6nC_3(5\lambda^2 - 3) & 0 \\ -\{c^2\left(5C_1^0 - 2nm\right) + 30C_3(e+p)\} & 30C_3 n & 3c^2 n \end{bmatrix} = 0. \tag{5.33}$$

The requirement that the system have non-trivial solutions leads to a biquadratic equation for λ. Accordingly there are 4 speeds of which two are different in value. Their explicit forms are exceedingly long and, therefore, are not written here. The corresponding waves are longitudinal waves, because δU is the component of δU^A in the direction of n_A, the normal to the wave front. The multiplicity of this wave is 1, since only one jump may be assigned, e.g., δn; all others may then be calculated.

Since we know the coefficients C_1^0 through C_5, which determine the wave speeds, we can calculate them explicitly for a relativistic gas.

Indeed, we may use (4.21)—appropriate to a non-degenerate relativistic gas—to obtain the λ's. Setting $A(\alpha) = 0$, we obtain

$$\text{for the transversal wave:} \qquad \lambda = \pm\sqrt{\frac{G\left[\gamma^2(G^2-1) + G(6G - 5\gamma)\right]}{\left[\gamma^2(G^2-1) + 5G(G-\gamma)\right](6G+\gamma)}}. \tag{5.34}$$

Once again, the longitudinal wave speed, which is the solution of the biquadratic equation that results from setting the determinant of (5.33) equal to zero, is impossible to write here. Seccia and Strumia [38] published an abbreviated expression.

In the non-relativistic limit we obtain from (4.35) for the transversal wave

$$\lambda_{\text{trans}} = \pm\frac{1}{c}\sqrt{\frac{7}{5}\frac{k}{m}T} \tag{5.35}$$

and for the two longitudinal waves

$$\begin{aligned} \lambda^1_{\text{long}} &= \pm\frac{1}{c}\sqrt{1,35\frac{kT}{m}}, \\ \lambda^2_{\text{long}} &= \pm\frac{1}{c}\sqrt{5.18\frac{kT}{m}}. \end{aligned} \tag{5.36}$$

These expressions agree with the values calculated by Boillat [43] and with earlier results reported in [44], [45], [46] based on the kinetic theory. Also, of course λ_{trans} and λ^2_{long} agree with the speeds (1.3.38) calculated in Paragraph 1.3.7 within even the early theory of extended thermodynamics.

Seccia and Strumia also provide the speeds for the ultrarelativistic case in which $\gamma \ll 1$ holds. They obtain

$$\begin{aligned} \lambda_{\text{trans}} &= \pm\sqrt{\frac{1}{5}}, \\ \lambda^1_{\text{long}} &= \pm\sqrt{\frac{1}{3}}, \\ \lambda^2_{\text{long}} &= \pm\sqrt{\frac{3}{5}}. \end{aligned} \tag{5.37}$$

We cannot confirm these results in this book, because we have not given the values of the coefficients C_0^1 through C_5 when we treated the ultrarelativistic case in Paragraph 4.8.

6 The Relativistic Kinetic Theory for Non-Degenerate Gases

6.1 Boltzmann–Chernikov Equation

In previous chapters we have on several occasions pointed to the close analogy between the results of extended thermodynamics and of the kinetic theory of gases in non-relativistic gases. The same analogy can be exhibited for relativistic gases. The relativistic formulation of the Boltzmann equation has been given by Chernikov in a series of remarkable papers, see [47], [48], [49]. Here we describe only the bare bones of that theory.

In the relativistic kinetic theory a certain complexity is introduced by the fact that

$$d\mathbf{P} = \frac{\sqrt{-g}}{p_0}dp^1dp^2dp^3 \tag{6.1}$$

is the scalar element of momentum space. Thus

$$F(x^A, p^a)\frac{\sqrt{-g}}{p_0}dp^1dp^2dp^3 \tag{6.2}$$

may be called the number density of the atoms with momentum between p^a and $p^a + dp^a$[3]. The phase density F must obey the *Boltzmann–Chernikov equation*

$$p^A \frac{\partial F}{\partial x^A} - \Gamma^d_{AB} p^A p^B \frac{\partial F}{\partial p^d} = \int (F(p'_C)F(q'_C) - F(p_C)F(q_C)h\langle pq\rangle \, de \, d\mathbf{Q}. \tag{6.3}$$

The left-hand side may simply be written as the derivative of $F(x^A, p^a)$ with respect to the proper time of the atoms with four-momentum p^A. The Christoffel symbols $\mathbf{\Gamma}$ enter by use of the equation of motion of such atoms. The right-hand side is the collision integral and must be compared with the corresponding integral in the classical Boltzmann equation (1.3.2) in Chapter 1. p_C, q_C and p'_C, q'_C are the four-momenta of two colliding atoms before and after the collision. $\langle pq\rangle$ is the scalar relative speed of the atoms, and h is the scattering factor for scattering into the solid angle element de. Conservation of energy and momentum requires

$$p_C + q_C = p'_C + q'_C. \tag{6.4}$$

6.2 Equations of Transfer

Multiplication of the Boltzmann–Chernikov equation by a tensor $\psi^{A_1A_2\cdots A_n}$ and integration over $d\mathbf{P}$ leads to the generic equation of transfer

$$\begin{aligned} \left(\int p^A \psi^{A_1\cdots A_n} F d\mathbf{P}\right)_{,A} - \int p^A \psi^{A_1\cdots A_n}{}_{,A} F dP + \int \Gamma^d_{AB} p^A p^B \frac{\partial \psi^{A_1\cdots A_n}}{\partial p^d} F d\mathbf{P} \\ = -\frac{1}{2} \int \left[\psi^{A_1\cdots A_n}; 1\right] (F(p'_C)F(q'_C) - F(p_C)F(q_C))\, h\langle pq\rangle \, de \, d\mathbf{Q} \, d\mathbf{P}, \end{aligned} \tag{6.5}$$

where [ ] stands for

$$\left[\psi^{A_1\cdots A_n}; 1\right] = \frac{1}{2}(\psi^{A_1\cdots A_n}(x^D, p^{d'}) + \psi^{A_1\cdots A_n}(x^D, q^{d'}) - \psi^{A_1\cdots A_n}(x^D, p^d) - \psi^{A_1\cdots A_n}(x^D, q^d)). \tag{6.6}$$

The nth moment of the phase density is defined by

$$A^{A_1\cdots A_n}(x^D) = m^{2-n} c \int p^{A_1} \cdots p^{A_n} F dP. \tag{6.7}$$

If $\psi^{A_1\cdots A_n}$ in (6.5) is chosen as $p^{A_1}\cdots p^{A_n}$, it turns out that the general equation of transfer assumes the form

$$A^{AA_1\cdots A_n}{}_{,A} = -\frac{1}{2} \int m^{2-n} c \left[p^{A_1}\cdots p^{A_n}; 1\right] (F(p'_C) + F(q'_C) - F(p_C) - F(q_C))\, h\langle pq\rangle \, de \, d\mathbf{Q} \, d\mathbf{P}, \tag{6.8}$$

which is an equation of balance for moments of the phase density.

6.3 Equations of Balance for Particle Number, Energy-Momentum, Fluxes, and Entropy

For the special choices $n = 1, 2$, and 3 we obtain from (6.7)

$$\begin{aligned} A^A &= mc \int p^A F d\mathbf{P}, \\ A^{AB} &= \int p^A p^B F d\mathbf{P}, \\ A^{ABC} &= \tfrac{1}{m} c \int p^A p^B p^C F d\mathbf{P}, \end{aligned} \tag{6.9}$$

[3]Actually the number of atoms between x^a, p^a and $x^a + dx^a, p^a + dp^a$ is obtained by mulitplying (6.2) by $\sqrt{-g}dx^1dx^2dx^3$, but these fine points can best be studied in Chernikov's papers.

which are the first moments of the phase density. Because of (6.4) the corresponding balance equations read

$$\begin{aligned} A^A{}_{,A} &= 0, \\ A^{BA}{}_{,A} &= 0, \\ A^{BCA}{}_{,A} &= I^{BC}. \end{aligned} \tag{6.10}$$

They represent the conservation laws of rest mass, energy and momentum, and the balance of fluxes, that is, the 15 equations that we have chosen as the basis of relativistic extended thermodynamics (see Section 1). Since $p^A p_A = c^2$, we have from (6.7) and (6.8)

$$A^B{}_B{}^A = c^2 A^A \quad \text{and} \quad I^B_B = 0, \tag{6.11}$$

two relations that we had to postulate in the phenomenological theory in order to reduce the number of independent equations in (6.10) to fourteen.

If we insert the function

$$\psi = -kc \ln F \tag{6.12}$$

into the generic equation of transfer (6.5) we obtain

$$\left(-kc \int p^A \ln F F d\mathbf{P}\right)_{,A} = \frac{kc}{4} \int \ln \frac{F(p'_C)F(q'_C)}{F(p_C)F(q_C)} \left(F(p'_C)F(q'_C) - F(p_C)F(q_C)\right) h\langle pq\rangle \, de \, d\mathbf{Q} \, d\mathbf{P}. \tag{6.13}$$

The right-hand side is non-negative, and therefore we come to the conclusion that

$$h^A = -kc \int p^A \ln F F d\mathbf{P} \tag{6.14}$$

represents the entropy flux vector in the kinetic theory. The right-hand side of (6.13) is the entropy production.

6.4 Maxwell–Jüttner Distribution: Equilibrium Properties

Equilibrium is characterized by a vanishing entropy production. It follows by inspection of (6.13) that $\ln F_{\mathrm{E}}$ must be a summational invariant in the collisions that occur in equilibrium. Since there are only five independent summational invariants, viz. the rest mass and the components of the four-momentum p^A, $\ln F_{\mathrm{E}}$ must be a linear combination of those and we obtain

$$F_{\mathrm{E}} = A e^{-\lambda_B p^B}. \tag{6.15}$$

The coefficients A and λ_B may be calculated in terms of n, T, and U^A as follows: We insert F_{E} into (6.9) and make use of the decompositions $(2.60)_{1,2}$ to obtain

$$\begin{aligned} nmU^A &= mc \int p^A \quad A e^{-\lambda_B p^B} d\mathbf{P}, \\ p(n,e)h^{AB} + \frac{e}{c^2} U^A U^B &= \quad c \int p^A p^B \, A e^{-\lambda_B p^B} d\mathbf{P}. \end{aligned} \tag{6.16}$$

We introduce $\gamma = mc\sqrt{\lambda_A \lambda^A}$ as an abbreviation and obtain

$$\lambda^A = \frac{\gamma}{mc^2} U^A, \qquad A = \frac{n}{4\pi m^3 c^3} \frac{\gamma}{K_2(\gamma)}, \qquad \gamma = \frac{nmc^2}{p(n,e)}, \tag{6.17}$$

where $K_2(\gamma)$ is one of the Bessel functions introduced in (4.18). This set of equations identifies λ^A and A in terms of n, U^A, and $p(n, e)$. If we wish to relate λ^A and A to n, T, and U^A we have to calculate e from (6.16) and the entropy density

$$h|_{\mathrm{E}} = \frac{1}{c^2} h^A|_{\mathrm{E}} U_A = -\frac{k}{c} U_A \int p^A \ln F_{\mathrm{E}} F_{\mathrm{E}} \, d\mathbf{P}. \tag{6.18}$$

We obtain with $G = K_3/K_2$

$$e = nmc^2 \left(G - \frac{1}{\gamma} \right), h|_{\mathrm{E}} = -kn \ln A + \frac{k\gamma}{mc^2} e. \tag{6.19}$$

From $(6.19)_2$ we derive easily

$$d\left(\frac{h|_{\mathrm{E}}}{n}\right) = \frac{k\gamma}{mc^2} \left[d\left(\frac{e}{n}\right) - \frac{1}{n^2} dp(n, e) \right], \tag{6.20}$$

whence we conclude by reference to the Gibbs equation that γ is the relativistic coldness

$$\gamma = \frac{mc^2}{kT}. \tag{6.21}$$

Thus the equilibrium phase density is given by

$$F_{\mathrm{E}} = \frac{n}{4\pi m^3 c^3} \frac{\gamma}{K_2(\gamma)} e^{-U_A p^A/kT}. \tag{6.22}$$

This is indeed the Maxwell–Jüttner distribution for a non-degenerate gas [see (4.4) and $(4.20)_1$]. Previously, in Paragraph 4.2, we made use of that phase density to derive the equations of state.

6.5 Possible Thermodynamic Fields in Equilibrium

Inspection of the Boltzmann–Chernikov equation (6.3) shows that its right-hand side vanishes for the Maxwell–Jüttner distribution (6.15) irrespective of how A and λ_B might depend on x^B. If the equation is to be satisfied, its left-hand side must also vanish, and this requirement imposes restrictions on the possible fields $A(x^B)$ and $\lambda_A(x^B)$. In fact, insertion of (6.15) into (6.3) leads to the condition

$$p^A \frac{\partial A}{\partial x^A} - A \lambda_{B,A} p^A p^B = 0. \tag{6.23}$$

Since this equation must hold for all values of p^a, we conclude that it implies

$$\frac{\partial A}{\partial x^A} = 0 \quad \text{and} \quad \lambda_{B,A} + \lambda_{A,B} = 0. \tag{6.24}$$

It follows that A must be independent of x^B and that $\lambda_{B,A}$ must be antisymmetric in equilibrium. We recall that $\lambda_A = \frac{1}{kT} U_A$ and $A \sim n(\gamma/K_2(\gamma))$ and conclude from (6.24) that

$$U^A \frac{\partial T}{\partial x^A} = 0 \quad \text{and} \quad \frac{\partial \ln T}{\partial x^A} = \frac{1}{c^2} U^B U_{A,B}. \tag{6.25}$$

This means that the temperature of a volume element moving with the gas must be constant in time while the temperature gradient need not vanish. Both results differ from the non-relativistic case where, according to Paragraph 1.3.3, the temperature could depend on time but had to be independent of $\mathbf{x}$. The result $(6.25)_2$ is nicely consistent with equation (3.7) that predicts a heat flux, if $\partial T/\partial x^A$ is not balanced by the expression $T/c^2 U^B U_{A,B}$.

The equations (6.24) permit a wide variety of motions, which we do not discuss in detail (e.g., see [44] and [15]). In a gas at rest in a stationary gravitational field (6.25) implies

$$T(x^a) = \frac{c}{\sqrt{g_{00}(x^a)}} \tag{6.26}$$

so that an isothermal atmosphere does not represent a possible equilibrium condition. This observation is also confirmed by $(6.24)_1$, because, since A must be independent of x^a, T must be non-uniform to offset the barometric distribution of the particle number density n.

7 The Non-Relativistic Limit of Relativistic Thermodynamics

7.1 The Problem

There is more to the problem of finding the non-relativistic limit of relativistic extended thermodynamics than the calculations of Paragraph 4.5 of the limiting values (4.34) and (4.35) through (4.37) of the state functions and of the coefficients C and A. Indeed, that calculation still leaves us with 14 field equations, while in the non-relativistic theory the natural number of field equations seemed to be 13.

Also in the kinetic theory the non-relativistic theory required moments of the distribution function of up to fourth order, while in the relativistic theory we used only first- and second-order moments.

Superficially viewed these discrepancies seem to pose serious problems, but in fact they can be smoothly resolved. The easiest way to do so is to make use of the kinetic theory and compare the formulae of Section 6 of this chapter to those of Section 1.3 of Chapter 1. This topic was investigated by Dreyer and Weiss [50], whose analysis is condensed here into this short section.

7.2 Variables and Constitutive Quantities

We recall that in the relativistic theory as well as in the non-relativistic one the variables and constitutive quantities are moments of the phase density, but moments are differently defined in the two theories. We recall their forms

relativistic	non-relativistic
$A^{A_1 A_2 \cdots A_n} = m^{2-n} c \int p^{A_1} p^{A_2} \cdots p^{A_N} F d\mathbf{P}$	$F^{a_1 a_2 \cdots a_n} = m \int c^{a_1} c^{a_2} \cdots c^{a_n} f d\mathbf{c}$

(7.1)

where c^a are velocity components and p^A are components of the four-momentum. By (6.1) we have $d\mathbf{P} = \frac{1}{p_0} dp^1 dp^2 dp^3$ in a Lorentz frame while $d\mathbf{c}$ is equal to $dc^1 dc^2 dc^3$. The four momentum p^A satisfies the constraint $p_A p^A = m^2 c^2$, so we have

$$p_0 = p^0 = mc\sqrt{1 + \frac{p^2}{m^2 c^2}} = mc\left(1 + \frac{1}{2}\frac{p^2}{m^2 c^2} - \frac{1}{8}\frac{p^4}{m^4 c^4} + \mathcal{O}\left(\frac{1}{c^6}\right)\right). \tag{7.2}$$

Momentum and velocity and the two phase densities are related by

$$p^a = mc^a + \mathcal{O}\left(\frac{1}{c^2}\right) \quad \text{and} \quad F = \frac{1}{m^3} f + \mathcal{O}\left(\frac{1}{c^2}\right). \tag{7.3}$$

It is then easy to show that the relativistic and the non-relativistic moments are related as follows:

$$
\begin{aligned}
A^0 &= cF + \mathcal{O}\left(\frac{1}{c}\right), & A^c &= F^c + \mathcal{O}\left(\frac{1}{c^2}\right), \\
A^{a0} &= cF^a + \mathcal{O}\left(\frac{1}{c}\right), & A^{ac} &= F^{ac} + \mathcal{O}\left(\frac{1}{c^2}\right), \\
A^{00} &= \underline{cA^0} + \frac{1}{2}F^{aa} + \mathcal{O}\left(\frac{1}{c^2}\right), & A^{0c} &= \underline{cA^c} + \frac{1}{2c}F^{cbb} + \mathcal{O}\left(\frac{1}{c^3}\right), \\
A^{ab0} &= cF^{ab} + \mathcal{O}\left(\frac{1}{c}\right), & A^{abc} &= F^{abc} + \mathcal{O}\left(\frac{1}{c^2}\right), \\
A^{a00} &= \underline{cA^{a0}} + \frac{1}{2}F^{abb} + \mathcal{O}\left(\frac{1}{c^2}\right), & A^{aoc} &= \underline{cA^{ac}} + \frac{1}{2c}F^{abbc} + \mathcal{O}\left(\frac{1}{c^3}\right), \\
A^{000} &= \underline{-c^2A^0 + 2cA^{00}} + \frac{1}{4c}F^{aabb} + \mathcal{O}\left(\frac{1}{c^3}\right) & A^{00c} &= \underline{-c^2A^c + 2cA^{0c}} + \frac{1}{4c^2}F^{aabbc} + \mathcal{O}\left(\frac{1}{c^4}\right).
\end{aligned}
\tag{7.4}
$$

Insertion of the expansions (7.4) into the balance equations (1.2) leads to some cancellation, since the underlined terms drop out because of conservation of particle number, momentum, and energy. What remains can be written as follows up to terms of $\mathcal{O}(1/c^2)$:

$$
\begin{aligned}
&\frac{1}{c}\frac{\partial A^0}{\partial t} + \frac{\partial A^c}{\partial x^c} = 0 && \Rightarrow \text{conservation of mass:} && \frac{\partial F}{\partial t} + \frac{\partial F^c}{\partial x^c} = 0, \\
&\frac{1}{c}\frac{\partial A^{a0}}{\partial t} + \frac{\partial A^{ac}}{\partial x^c} = 0 && \Rightarrow \text{conservation of momentum:} && \frac{\partial F^a}{\partial t} + \frac{\partial F^{ac}}{\partial x^c} = 0, \\
&\frac{1}{c}\frac{\partial A^{00}}{\partial t} + \frac{\partial A^{0c}}{\partial x^c} = 0 && \Rightarrow \text{conservation of energy:} && \frac{\partial \frac{1}{2}F^{aa}}{\partial t} + \frac{\partial \frac{1}{2}F^{aac}}{\partial x^c} = 0, \\
&\frac{1}{c}\frac{\partial A^{\langle ab\rangle 0}}{\partial t} + \frac{\partial A^{\langle ab\rangle c}}{\partial x^c} = I^{\langle ab\rangle} && \Rightarrow \text{balance of momentum flux:} && \frac{\partial F^{\langle ab\rangle}}{\partial t} + \frac{\partial F^{\langle abc\rangle}}{\partial x^c} = I^{\langle ab\rangle}, \\
&\frac{1}{c}\frac{\partial A^{a00}}{\partial t} + \frac{\partial A^{a0c}}{\partial x^c} = I^{a0} && \Rightarrow \text{balance of energy flux:} && \frac{\partial \frac{1}{2}F^{abb}}{\partial t} + \frac{\partial \frac{1}{2}F^{abbc}}{\partial x^c} = cI^{a0}, \\
&\frac{1}{c}\frac{\partial A^{000}}{\partial t} + \frac{\partial A^{00c}}{\partial x^c} = I^{00} && \Rightarrow \text{balance of 4th moment:} && \frac{\partial F^{aabb}}{\partial t} + \frac{\partial F^{aabbc}}{\partial x^c} = 4c^2 I^{00}.
\end{aligned}
\tag{7.5}
$$

The equations on the right-hand side of (7.5) are the 13 balance laws of non-relativistic extended thermodynamics *plus* a 14th equation for the fourth moment F^{aabb}. We note that these 14 non-relativistic equations contain moments of up to fifth order while the relativistic equations have only first- and second-order moments. This phenomenon is due to the occurrence of p_0 in the denominator of the element $d\mathbf{P}$ of momentum space and to the expansion (7.2).

From these arguments we conclude that the relativistic theory seems to dictate a 14-moment theory for non-relativistic extended thermodynamics as the most natural choice. Kremer [51] has been motivated by this consideration to formulate, within the non-relativistic regime, an extended thermodynamics of 14 moments.

In order to summarize we compare variables and constitutive quantities in the relativistic and non-relativistic case and write

relativistic variables	non-relativistic variables	
A^A, A^{AB} or $n,\ U^A,\ e,\ t^{\langle AB\rangle},\ q^A,\ \pi$	$F, F^a, F^{aa}, F^{\langle ab\rangle} F^{abb}, F^{aabb}$	(7.6)
relativistic constitutive quantities	non-relativistic constitutive quantities	
$A^{ABC},\ I^{AB}$	$F^{\langle abc\rangle}, F^{abbc}, F^{aabbc}, I^{\langle ab\rangle}, cI^{a0}, 4c^2 I^{00}$	(7.7)

7.3 The Dynamic Pressure

The physical significance of most of the non-relativistic variables is clear from their definition (7.1). Thus F and F^a are mass and momentum density, $\frac{1}{2}F^{aa}$ is the energy density, $F^{\langle ab\rangle}$ is the deviator of the momentum flux, and $\frac{1}{2}F^{abb}$ is the energy flux; these quantities all have a suggestive physical meaning and a clear conceptual connection to the corresponding relativistic variables n through q^A. There is no suggestive meaning, however, for the fourth momentum F^{aabb}, nor is there a clear connection of this variable to the dynamic pressure π. This connection will have to be found, and we proceed to do this.

The easiest way is to write down A_a^{a0} and A^{000} in the rest frame in two ways: First by use of the representation (2.62) and then, approximately, by use of $(7.4)_{10,11}$. Equating those formulae we obtain

$$
\begin{aligned}
-\frac{c^3}{2}(nm - C_1^0 - C_1^\pi \pi) &= 3c(p(n,e) + \pi) + \frac{1}{2c}\varrho^{aacc} + \mathcal{O}\left(\frac{1}{c^3}\right), \\
\frac{c^3}{2}(nm + C_1^0 + C_1^\pi \pi) &= -c^3 nm + 2ce \ + \frac{1}{4c}\varrho^{aacc} + \mathcal{O}\left(\frac{1}{c^3}\right),
\end{aligned}
\tag{7.8}
$$

where ϱ^{aacc} is the non-relativistic *central* fourth moment of the phase density. We substract the two equations (7.8) and obtain

$$
p(n,e) + \pi = \frac{2}{3}(e - nmc^2) - \frac{1}{12c^2}\varrho^{aacc} + \mathcal{O}\left(\frac{1}{c^4}\right). \tag{7.9}
$$

The non-equilibrium part of this equation reads

$$
\pi = -\frac{1}{12c^2}\varrho^{aacc}|_{\mathrm{NE}} + \mathcal{O}\left(\frac{1}{c^4}\right). \tag{7.10}
$$

We conclude that the occurrence of the dynamic pressure π in an ideal gas is a relativistic phenomenon. Therefore we were justified to ignore it in Chapters 1 and 2. The *new* result in (7.10) is that π does not appear as a pressure at all but rather as a fourth moment of the phase density.

Chapter 5

Thermodynamics of Light and Sound

Light in a cavity and sound in a solid may be considered as gases of quasi-particles, the photons, and phonons respectively. It is then possible to treat them thermodynamically. The first successful attempt in that direction was Boltzmann's derivation of what we now call the Stefan–Boltzmann law. That law relates the energy density of cavity radiation in equilibrium with the wall to the temperature of the wall.

Radiation hydrodynamics or radiation thermodynamics is important for the theory of stellar structure, because the radiation pressure in stars equilibrates the gravitational pull. Eddington, a pioneer of stellar physics, was able to derive the general form of the stress tensor of radiation, which he wrote as

$$-P_{ij} = -e\left(\frac{1-\chi}{2}\delta_{ij} + \frac{3\chi-1}{2}\frac{p_i p_j}{p^2}\right).$$

Here e and p_i are the densities of energy and momentum. χ is called the Eddington factor; it is a function of e and p^2.

The form of that function is discussed significantly in the literature. In 1984 Levermore [52] compared several propositions for the Eddington factors. More recently Anile, Pennisi, and Sammartino [53] approached the problem by use of extended thermodynamics. They showed that χ must have the form

$$\chi = \frac{5}{3} - \frac{2}{3}\sqrt{4 - 3\frac{c^2p^2}{e^2}}.$$

This form was also confirmed by Kremer and Müller [54] by a different approach.

This chapter presents a thermodynamic theory of light and sound. It demonstrates that extended thermodynamics permits the explicit calculation of the main part of the equations of balance of energy for photons and phonons. Wave speeds are calculated and the limiting cases of near-equilibrium and free streaming are discussed.

The chapter ends with a discussion of shock waves in a gas of photons and phonons and includes results by Dreyer and Seelecke [55].

1 Phonons and Photons

1.1 Cavity Radiation and Sound in Single Crystals

Light and sound are waves, electromagnetic and elastic, respectively, but in some ways, which physicists understand, they may be considered as a gas of particles, the photons and phonons. Thus a wave of frequency ω and wave number $\mathbf{k}$ corresponds to a particle of energy $\hbar\omega$ and momentum $\hbar\mathbf{k}$. The frequency and the wave number are not independent, of course. For light we have

$$\omega = ck, \tag{1.1}$$

where c is the speed of light. For sound the *dispersion relation* $\omega = \omega(\mathbf{k})$ is more complex, but it reduces to (1.1)—with c as the speed of sound—for longitudinal sound waves of small frequencies. This is the only case we shall consider here so that equation (1.1) holds for both photons and phonons. We shall refer to these particles as pho□ons.

To fix the ideas about the systems under consideration we have drawn Figure 1. On the left-hand side we show a cylinder closed off by a piston and filled with radiation, i.e., a gas of photons. In equilibrium the temperature of the wall of this cavity is T everywhere and the photons exert a pressure p on the piston. Kirchhoff's experiments, of more than 100 years ago, have shown that the energy density of the radiation is a function only of temperature

$$e = e(T). \tag{1.2}$$

On the right-hand side of Figure 1 we see a single crystal specimen of some solid "filled with sound," a gas of phonons. In equilibrium the temperature is uniformly equal to T and specific heat measurements show that the energy density is only a function of T, at least if we can assume incompressiblity of the body.

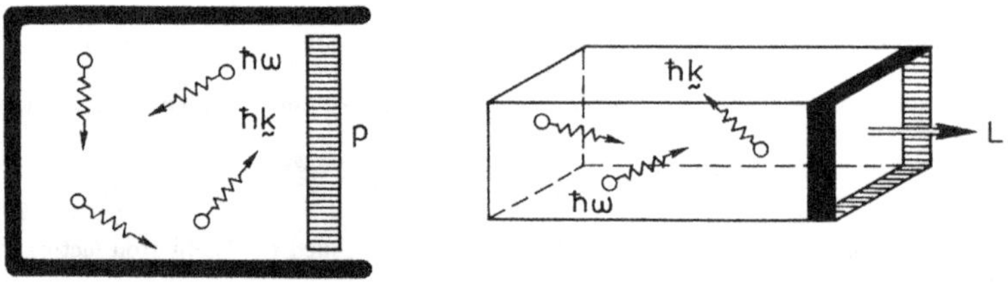

Figure 1. A photon gas and a phonon gas.

1.2 Thermodynamic Processes

The objective of thermodynamics of pho□ons is the determination of 8 fields, viz.

$$\begin{array}{llll} \text{pho□on number density} & n, & \text{pho□on flux} & j_k, \\ \text{energy density} & e, & \text{energy flux} & Q_k. \end{array} \tag{1.3}$$

These quantities are objective scalars and vectors respectively. The necessary field equations are based upon the balance equations for

$$\text{pho□on number}\quad \frac{\partial n}{\partial t}+\frac{\partial j_k}{\partial x_k}=P_n,\qquad \text{pho□on flux}\quad \frac{\partial j_i}{\partial t}+\frac{\partial J_{ik}}{\partial x_k}=P_{j_i},$$
$$\text{energy}\quad \frac{\partial e}{\partial t}+\frac{\partial Q_k}{\partial x_k}=P_e,\qquad \text{momentum}\quad \frac{\partial p_i}{\partial t}+\frac{\partial P_{ik}}{\partial x_k}=P_{p_i}. \tag{1.4}$$

We make some specific assumptions relating the quantities in (1.4). We assume

$$J_{ik}, P_{ik}\text{- symmetric tensors},\quad J_{ii}=n\,c^2,\quad P_{ii}=e,\quad Q_i=p_ic^2. \tag{1.5}$$

These assumptions seem arbitrary at this stage, but they are really well motivated as we shall see in the next paragraph. First, however, we complete the discussion of the structure of the theory.

The system of equations (1.4) is closed by considering the flux tensors J_{ik}, P_{ik} and the productions P_n through P_{p_i} as constitutive quantities. As usual in extended thermodynamics the generic form of the constitutive equations reads

$$C=\hat{C}(n,j_i,e,p_i) \tag{1.6}$$

so that the value of a constitutive quantity at a point and time depends only on the values of the fields (1.3) at that point and time.

If the constitutive functions $\hat{C}$ were known, we could eliminate the constitutive quantities from the balance equations and obtain a specific set of field equations. Every solution of these equations is called a *thermodynamic process.*

1.3 Motivations

There are two sources of motivation for the assumptions (1.5), the first valid only for photons.

The Maxwell equations and the Maxwell–Lorentz nether relations

$$D_i=\varepsilon_0E_i,\qquad H_i=\frac{1}{\mu_0}B_i \tag{1.7}$$

imply equations of balance for energy and momentum of the electromagnetic field. In terms of the electric field $\mathbf{E}$, the dielectric displacement $\mathbf{D}$, the magnetic field $\mathbf{H}$, and the magnetic flux density $\mathbf{B}$, we have (e.g., see [15, p. 319]).

$$\overset{\text{e.m}}{e}=\frac{1}{2}(\mathbf{ED}+\mathbf{BH})\ ,\qquad \overset{\text{e.m}}{p_k}=(\mathbf{D}\times\mathbf{B})_k\ ,$$
$$\overset{\text{e.m}}{Q_k}=(\mathbf{E}\times\mathbf{H})_k,\qquad \overset{\text{e.m}}{P_{ik}}=\frac{1}{2}(\mathbf{ED}+\mathbf{BH})\delta_{ik}-E_iD_k-B_iH_k. \tag{1.8}$$

Inspection shows that $\overset{\text{e.m}}{P_{ik}}$ is symmetric and obviously we have $\overset{\text{e.m}}{P_{ii}}=\overset{\text{e.m}}{e}$. Also, by (1.7) we see that $\overset{\text{e.m}}{Q_k}=\overset{\text{e.m}}{p_k}\,c^2$ holds, since $c^2=1/\varepsilon_0\mu_0$. Thus the Maxwell theory of electrodynamics supports some of the assumptions (1.5).

The second source of motivation for those assumptions may be found in the kinetic theory of pho□ons. In that theory one introduces a phase density $f(\mathbf{x},\mathbf{k},t)$ such that $f(\mathbf{x},\mathbf{k},t)d\mathbf{x}d\mathbf{k}$ is the number of pho□ons with $\mathbf{k}$ at $\mathbf{x}, t$. This phase density obeys a balance law, the *Boltzmann–Peierls equation*, namely,

$$\frac{\partial f}{\partial t} + \frac{\partial w_i f}{\partial x_i} = P \quad \text{with } w_i = \frac{\partial \omega}{\partial k_i} \text{ or with (1.1) } w_i = c\frac{k_i}{k}. \tag{1.9}$$

P is the production density of pho□ons in the phase space spanned by $\mathbf{k}$ and $\mathbf{x}$.

Integration over $\mathbf{k}$ after multiplication by 1, $\frac{\partial \omega}{\partial k_i} = c\frac{k_i}{k}$, $\hbar\omega = \hbar ck$, and $\hbar k_i$ leads to the following four balance laws:

$$\begin{aligned}
\frac{\partial}{\partial t}\int f\,d\mathbf{k} + \frac{\partial}{\partial x_l}\int c\frac{k_l}{k} f\,d\mathbf{k} &= \int P\,d\mathbf{k},\\
\frac{\partial}{\partial t}\int c\frac{k_i}{k} f\,d\mathbf{k} + \frac{\partial}{\partial x_l}\int c^2\frac{k_j k_l}{k^2} f\,d\mathbf{k} &= \int c\frac{k_i}{k} P\,d\mathbf{k},\\
\frac{\partial}{\partial t}\int \hbar ck f\,d\mathbf{k} + \frac{\partial}{\partial x_l}\int \hbar c^2 k_l\, f\,d\mathbf{k} &= \int \hbar\omega P\,d\mathbf{k},\\
\frac{\partial}{\partial t}\int \hbar k_i f\,d\mathbf{k} + \frac{\partial}{\partial x_l}\int \hbar c\frac{k_i k_l}{k} f\,d\mathbf{k} &= \int \hbar k_i P\,d\mathbf{k}.
\end{aligned} \tag{1.10}$$

Here $\int f\,d\mathbf{k}$ is obviously the number density of pho□ons while $\int \hbar\omega f\,d\mathbf{k}$ and $\int \hbar k_i f\,d\mathbf{k}$ are the densities of energy and momentum. The equations (1.10), therefore, must be interpreted as the balance equations for pho□on number, pho□on flux, pho□on energy, and pho□on momentum. Comparison with (1.4) provides the following interpretations:

$$\begin{aligned}
n &= \int f\,d\mathbf{k}, & j_i &= \int c\frac{k_i}{k} f\,d\mathbf{k}, & P_n &= \int P\,d\mathbf{k},\\
& & J_{ik} &= \int c^2\frac{k_i k_k}{k^2} f\,d\mathbf{k}, & P_{j_i} &= \int c\frac{k_i}{k} P\,d\mathbf{k},\\
e &= \int \hbar ck f\,d\mathbf{k}, & Q_i &= \int \hbar c^2 k_i\, f\,d\mathbf{k}, & P_e &= \int \hbar\omega\, P\,d\mathbf{k},\\
p_i &= \int \hbar k_i f\,d\mathbf{k}, & P_{ik} &= \int \hbar c\frac{k_i k_k}{k} f\,d\mathbf{k}, & P_{p_i} &= \int \hbar k_i P\,d\mathbf{k}.
\end{aligned} \tag{1.11}$$

Inspection shows that the assumptions (1.5) are confirmed by the equations (1.11). Also, since ω and k_i are an objective scalar and vector respectively—at least in a non-relativistic theory—we conclude that the quantities n through P_{p_i} in (1.3) are all objective.

1.4 Principles of the Constitutive Theory: The Entropy Inequality

In order to restrict the generality of the constitutive functions we exploit the three restrictive principles of the constitutive theory of extended thermodynamics, viz.

— the principle of relativity
— the entropy principle, and
— the requirement of convexity and causality.

The principle of relativity requires that the field equations be invariant under changes of frame. In the present case where the equations (1.4) are themselves invariant, this implies that the constitutive functions $\hat{C}$ in (1.6) are isotropic functions.

The formulation of the entropy principle requires prior discussion of a subtlety that occurs in a gas of pho□ons: There is no isolated pho□on gas; rather there is always a background of material particles and that background furnished the temperature of the system.

In the case of phonons the background is the crystalline body of temperature T in which the phonons exist and with which they interact, if only weakly. Also photons interact with a background of walls or traces of dust or gas of temperature T. This is why the pho□ons exhibit production terms in their balance equations for the number, the energy, and the momentum [see (1.4)].

The entropy inequality must hold for the system as a whole. Thus we have

$$\frac{\partial h^{\mathrm{tot}}}{\partial t} + \frac{\partial h_i^{\mathrm{tot}}}{\partial x_i} \geq 0 \quad \text{where } h^{\mathrm{tot}} = h + h^B, \; h_i^{\mathrm{tot}} = h_i + h_i^B. \tag{1.12}$$

Here h and h_i are the entropy density and entropy flux of the pho□ons, while h^B and h_i^B are the corresponding quantities of the background. The background absorbs the energy production $-P_e$ at its temperature T so that it experiences a *radiative* entropy supply $-P_e/T$. The balance of entropy of the background therefore reads

$$\frac{\partial h^B}{\partial t} + \frac{\partial h_i^B}{\partial x_i} + \frac{P_e}{T} = 0, \tag{1.13}$$

provided that it does not itself contribute to the entropy production. Combining (1.12) and (1.13) we obtain for the balance of entropy of pho□ons

$$\frac{\partial h}{\partial t} + \frac{\partial h_i}{\partial x_i} - \frac{P_e}{T} \geq 0. \tag{1.14}$$

This inequality must hold for all thermodynamic processes. h and h_i are constitutive quantities of the generic form (1.6).

The requirement of convexity and causality states that the matrix of second derivatives of h with respect to its variables n, j_i, e, p_i must be negative definite.

1.5 Exploitation of the Entropy Principle

The key to the exploitation of the entropy inequality is the statement that (1.14) must hold for all thermodynamic processes rather than for all fields. We may eliminate that constraint by the use of Lagrange multipliers (see [18]). Indeed the larger inequality

$$\begin{aligned} \frac{\partial h}{\partial t} + \frac{\partial h_e}{\partial x_e} - \frac{1}{T} P_e - \lambda \left(\frac{\partial n}{\partial t} + \frac{\partial j_e}{\partial x_e} - P_n \right) - \lambda_i \left(\frac{\partial j_i}{\partial t} + \frac{\partial J_{ie}}{\partial x_e} - P_{j_i} \right) & \\ - \Lambda \left(\frac{\partial e}{\partial t} + \frac{\partial Q_e}{\partial x_e} - P_e \right) - \Lambda_i \left(\frac{\partial p_i}{\partial t} + \frac{\partial P_{ie}}{\partial x_e} - P_{p_i} \right) \geq 0 & \end{aligned} \tag{1.15}$$

must hold for *all* fields n, j_i, e, p_i. The Lagrange multipliers $\lambda, \lambda_i, \Lambda, \Lambda_i$ may be functions of n, j_i, e, and p_i; they must be isotropic functions because of the principle of relativity.

We insert the constitutive relations (1.6) and employ the chain rule to write the inequality (1.15) in a more explicit form, viz.

$$
\begin{aligned}
&\underline{\left(\frac{\partial h}{\partial n}-\lambda\right)}\frac{\partial n}{\partial t}+\underline{\left(\frac{\partial h}{\partial j_i}-\lambda_i\right)}\frac{\partial j_i}{\partial t}+\underline{\left(\frac{\partial h_e}{\partial n}-\lambda_i\frac{\partial J_{ie}}{\partial n}-\Lambda_i\frac{\partial P_{ie}}{\partial n}\right)}\frac{\partial n}{\partial x_e}\\
&\qquad\qquad\qquad+\underline{\left(\frac{\partial h_e}{\partial j_n}-\quad\lambda\delta_{en}-\lambda_i\frac{\partial J_{ie}}{\partial j_n}-\Lambda_i\frac{\partial P_{ie}}{\partial j_n}\right)}\frac{\partial j_n}{\partial x_e}\\
&\underline{\left(\frac{\partial h}{\partial e}-\Lambda\right)}\frac{\partial e}{\partial t}+\underline{\left(\frac{\partial h}{\partial p_i}-\Lambda_i\right)}\frac{\partial p_i}{\partial t}+\underline{\left(\frac{\partial h_e}{\partial e}-\lambda_i\frac{\partial J_{ie}}{\partial e}-\Lambda_i\frac{\partial P_{ie}}{\partial e}\right)}\frac{\partial e}{\partial x_e}\\
&\qquad\qquad\qquad+\underline{\left(\frac{\partial h_e}{\partial p_n}-c^2\Lambda\delta_{en}-\lambda_i\frac{\partial J_{ie}}{\partial p_n}-\Lambda_i\frac{\partial P_{ie}}{\partial p_n}\right)}\frac{\partial p_n}{\partial x_e}\\
&\qquad\qquad\qquad+\lambda P_n+\lambda_i P_{j_i}+\left(\Lambda-\frac{1}{T}\right)P_e-\Lambda_i P_{p_i}\geq 0.
\end{aligned}
\tag{1.16}
$$

The left-hand side of this inequality is explicitly linear in the derivatives

$$
\frac{\partial n}{\partial t},\ \frac{\partial j_i}{\partial t},\ \frac{\partial n}{\partial x_e},\ \frac{\partial j_n}{\partial x_e},\ \frac{\partial e}{\partial t},\ \frac{\partial p_i}{\partial t},\ \frac{\partial e}{\partial x_e},\ \frac{\partial p_n}{\partial x_e}. \tag{1.17}
$$

Since these derivatives are arbitrary we could violate the inequality if they were to contribute to its left-hand side; therefore, the underlined quantities in (1.16) must vanish, a requirement that we summarize as

$$
\begin{aligned}
dh &= \lambda dn+\lambda_i dj_i+\Lambda de+\Lambda_i dp_i,\\[1ex]
dh_e &= \lambda dj_e+\lambda_i dJ_{ie}+\Lambda c^2 dp_e+\Lambda_i dP_{ie}.
\end{aligned}
\tag{1.18}
$$

There remains the residual inequality

$$
\lambda P_n+\lambda_i P_{j_i}+\left(\Lambda-\frac{1}{T}\right)P_e+\Lambda_i P_{p_i}\geq 0. \tag{1.19}
$$

In equilibrium all productions P vanish so that the left-hand side of (1.19) has its minimum, namely, zero. Hence follows

$$
\lambda_{\mathrm{E}}=0,\qquad \lambda_i|_{\mathrm{E}}=0,\qquad \Lambda_{\mathrm{E}}=\frac{1}{T},\qquad \Lambda_i|_{\mathrm{E}}=0. \tag{1.20}
$$

The equations $(1.20)_{2,4}$ are trivially satisfied, if the pho□on flux and momentum density vanish in equilibrium as they must. The other two equations require by $(1.18)_1$

$$
\frac{\partial h_{\mathrm{E}}}{\partial n}=0 \quad\text{and}\quad \frac{\partial h_{\mathrm{E}}}{\partial e}=\frac{1}{T},\qquad \text{hence}\qquad T=T(e)\quad\text{or}\quad e=e(T). \tag{1.21}
$$

$\partial h/\partial n$ equals $-g/nT$ in equilibrium thermodynamics, where $g=e-Th+\frac{1}{3}P_{ii}$ is the enthalpy density of the pho□on gas. Equation $(1.21)_2$ implies that g_{E} vanishes, and with $P_{ii}=e$ and $(1.21)_2$ we thus conclude after a trivial calculation

$$
\frac{de}{dT}=4\frac{e}{T}\quad\text{hence}\quad e=\sigma T^4. \tag{1.22}
$$

This is the well-known *law of Stefan-Boltzmann*, found empirically by Stefan and derived, essentially in the above manner, by Boltzmann.

For the exploitation of the equations (1.18) it is convenient to introduce the *potentials*

$$
h'=-h+\lambda n+\lambda_i j_i+\Lambda e+\Lambda_i p_i\quad\text{and}\quad h_i'=-h_i+\lambda j_i+\lambda_e J_{ei}+\Lambda c^2 p_i+\Lambda_e P_{ei} \tag{1.23}
$$

so that (1.18) assumes the form

$$dh' = nd\lambda + j_j d\lambda_j + ed\Lambda + p_i d\Lambda_i,$$
$$dh'_i = j_i d\lambda + J_{ij} d\lambda_j + p_i c^2 d\Lambda + P_{ij} d\Lambda_j. \tag{1.24}$$

It follows that the knowledge of h' and h'_i as functions of the variables $\lambda, \lambda_j, \Lambda, \lambda_j$ suffices to calculate n, j_j, e, p_i, J_{ij}, and P_{ij} as functions of these variables; therefore h' and h'_i are properly called potentials.

The conditions (1.24) have not been exploited yet in full generality nor will this be done here. We shall, however, proceed to illustrate their restrictive character by investigating a special case, the case where e and p_i are the only fields.

2 Energy and Momentum of Light and Sound

2.1 Eddington Factor, Entropy, and Entropy Flux

We restrict the scope of the theory by considering its objective to be the determination of 4 fields, namely,

$$\text{energy density} \quad e, \qquad\qquad \text{momentum density} \quad p_i. \tag{2.1}$$

In this case the relevant balance laws are those for energy and momentum, viz.

$$\frac{\partial e}{\partial t} + \frac{\partial Q_k}{\partial x_k} = P_e \quad \text{and} \quad \frac{\partial p_i}{\partial t} + \frac{\partial P_{ik}}{\partial x_k} = P_{p_i}. \tag{2.2}$$

As before we assume

$$P_{ik} \sim \text{symmetric}, \qquad P_{ii} = e, \qquad Q_i = p_i c^2. \tag{2.3}$$

We close the system (2.2) by the formulation of constitutive equations for P_{ik} and P_e, P_{pi} of the generic form

$$C = \hat{C}(e, p_i). \tag{2.4}$$

The functions $\hat{C}$ must be isotropic functions so that we have the representation

$$P_{ik} = e\left(\frac{1-\chi}{2}\delta_{ik} + \frac{3\chi - 1}{2}\frac{p_i p_k}{p^2}\right) \quad \text{where } \chi = \chi(e, p^2). \tag{2.5}$$

There is only one scalar coefficient in (2.5), because of the constraint $(2.3)_2$. χ is called the Eddington factor, because Eddington first introduced it in connection with the radiative stress in stars.

The entropy principle will restrict the function χ. The restrictions were given in Paragraph 1.5 for the general case. In the present case there remain only those parts of the conditions that refer to the fields e and $\mathbf{p}$; therefore we have

$$\begin{aligned} dh' &= e \quad d\Lambda + p_j \ d\Lambda_j, \\ dh'_i &= p_i c^2 \, d\Lambda + P_{ij} \, d\Lambda_j \end{aligned} \quad \text{where} \quad \begin{aligned} h &= -h' + \Lambda e \quad + \Lambda_j p_j, \\ h_i &= -h'_i + \Lambda c^2 p_i + \Lambda_j P_{ji}, \end{aligned} \tag{2.6}$$

and the residual inequality reads

$$\left(\Lambda - \frac{1}{T}\right) P_e + \Lambda_i P_{p_i} \geq 0. \tag{2.7}$$

Here h' and h'_i are functions of Λ and Λ_i so that we have

$$h' = h'(\Lambda, L) \quad \text{and} \quad h'_i = \varphi(\Lambda, L)\Lambda_i \tag{2.8}$$

where L stands for $\Lambda_i\Lambda_i$. More explicitly the equations (2.6) therefore read

$$\frac{\partial h'}{\partial \Lambda} = e, \qquad 2\frac{\partial h'}{\partial L}\Lambda_i = p_i, \qquad \frac{\partial \varphi}{\partial \Lambda}\Lambda_i = c^2 p_i, \qquad 2\frac{\partial \varphi}{\partial L}\Lambda_i\Lambda_j + \varphi\delta_{ij} = P_{ij}. \tag{2.9}$$

The exploitation proceeds as follows: Elimination of p_i between $(2.9)_{2,3}$ gives

$$\frac{\partial h'}{\partial L} = \frac{1}{2c^2}\frac{\partial \varphi}{\partial \Lambda}, \quad \text{hence by } (2.9)_1 \quad \frac{\partial e}{\partial L} = \frac{1}{2c^2}\frac{\partial^2 \varphi}{\partial \Lambda^2}. \tag{2.10}$$

The trace of $(2.9)_4$ reads with

$$2\frac{\partial \varphi}{\partial L}L + 3\varphi = e, \tag{2.11}$$

and if this is combined with $(2.10)_2$, we obtain a differential equation for φ, viz.

$$4\frac{\partial^2 \varphi}{\partial L^2}L + 10\frac{\partial \varphi}{\partial L} = \frac{1}{c^2}\frac{\partial^2 \varphi}{\partial \Lambda^2}. \tag{2.12}$$

It is possible to find the general solution of this differential equation—hyperbolic in the L direction when $L \neq 0$ and singular for $L = 0$—for a given function $\varphi_0(\Lambda, 0)$. Introducing the characteristic curves

$$\xi = \Lambda + \frac{1}{c}\sqrt{L}, \qquad \eta = \Lambda - \frac{1}{c}\sqrt{L},$$

we obtain

$$\frac{1}{2}\varphi_{\xi\eta} = \frac{\varphi_\eta - \varphi_\xi}{\eta - \xi},$$

the solution of which can be found [56]. Here we shall follow [54] and derive the solution by making a power Ansatz of the form

$$\varphi = \sum_{r=0}^{\infty} \varphi_r(\Lambda)L^r. \tag{2.13}$$

Insertion into the differential equation (2.12) leads to a recurrence relation of the form

$$\varphi_{r+1} = \frac{1}{2c^2}\frac{1}{(r+1)(2r+5)}\frac{\partial^2 \varphi_r}{\partial \Lambda^2}. \tag{2.14}$$

The initiation for the solution of this relation is provided by (2.11) and $(1.20)_3$, (1.22) from which formulae we obtain

$$\varphi_0(\Lambda) = \frac{1}{3}e_{\mathrm{E}}(\Lambda) = \frac{1}{3}\sigma\frac{1}{\Lambda^4}. \tag{2.15}$$

A simple calculation will then show that the solution of the recurrence relation (2.14) reads

$$\varphi_r = \frac{\sigma}{3}\frac{1}{\Lambda^4}(r+1)\left(\frac{1}{c^2\Lambda^2}\right)^r. \tag{2.16}$$

Hence we obtain

$$\varphi = \frac{\sigma}{3}\frac{1}{\Lambda^4}\sum_{r=0}^{\infty}(r+1)\left(\frac{L}{c^2\Lambda^2}\right)^r = \frac{\sigma}{3}\frac{1}{[\Lambda^2 - L/c^2]^2}, \tag{2.17}$$

provided $L < c^2\Lambda^2$. Now that φ is known as an explicit function of Λ and L, we may calculate e and p^2 from (2.11) and $(2.9)_3$, respectively,

$$e = \frac{\sigma}{3}\frac{3\Lambda^2 + L/c^2}{[\Lambda^2 - L/c^2]^3}, \qquad c^2p^2 = \frac{\sigma^2}{9}\frac{16\Lambda^2(L/c^2)}{[\Lambda^2 - L/c^2]^6}. \tag{2.18}$$

We may invert these equations and obtain Λ and L as functions of e and p^2. We obtain with $\surd = \left(1 - \frac{3}{4}\frac{c^2p^2}{e^2}\right)^{1/2}$

$$\Lambda = \frac{\sigma^{1/4}}{2^{1/2}}\frac{1}{e^{1/4}}\frac{(1+\surd)^{5/4}}{\left[1+\surd - \frac{3}{2}\frac{c^2p^2}{e^2}\right]^{3/4}}, \qquad \frac{L}{c^2} = \frac{3^2\sigma^{1/2}}{2^3}\frac{1}{e^{1/2}}\frac{c^2p^2}{e^2}\frac{(1+\surd)^{1/2}}{\left[1+\surd - \frac{3}{2}\frac{c^2p^2}{e^2}\right]^{3/2}}. \tag{2.19}$$

It follows from (2.17) that

$$\varphi = \frac{1}{3}\left(\sqrt{4e^2 - 3c^2p^2} - e\right), \qquad \frac{\partial\varphi}{\partial L}L = e - \frac{1}{2}\sqrt{4e^2 - 3c^2p^2} \tag{2.20}$$

so that by $(2.9)_4$ we have

$$P_{ij} = e\left(\frac{1}{3}\left(\sqrt{4 - 3\frac{c^2p^2}{e^2}} - 1\right)\delta_{ij} + \left(2 - \sqrt{4 - 3\frac{c^2p^2}{e^2}}\right)\frac{p_ip_j}{p^2}\right). \tag{2.21}$$

Comparison with (2.5) shows that the Eddington factor has thus been identified as

$$\chi = \frac{5}{3} - \frac{2}{3}\sqrt{4 - 3\frac{c^2p^2}{e^2}}. \tag{2.22}$$

We shall represent all subsequent results in terms of the Eddington factor, and for that purpose we list the following identities that follow from (2.22) and (2.19):

$$\frac{c^2p^2}{e^2} = \frac{3}{4}(3-\chi)\left(\chi - \frac{1}{3}\right), \quad \Lambda = \frac{1}{2^{3/4}}\frac{1}{3^{1/4}}\frac{\sigma^{1/4}}{e^{1/4}}\frac{(3-\chi)^{1/2}}{(1-\chi)^{3/4}}, \quad \Lambda^2 - \frac{L}{c^2} = \frac{2^{1/2}}{3^{1/2}}\frac{\sigma^{1/2}}{e^{1/2}}\frac{1}{(1-\chi)^{1/2}}. \tag{2.23}$$

We proceed to calculate the entropic quantities—first, the entropy density and its flux. By $(2.10)_1, (2.9)_1$, and $(2.8)_2$ we have

$$h' = -\frac{\sigma}{3}\frac{\Lambda}{[\Lambda^2 - L/c^2]^2}, \quad \text{hence by } (2.6)_1 \quad h = \frac{4}{3}\sigma\frac{\Lambda}{[\Lambda^2 - L/c^2]^2}; \tag{2.24}$$

$$h_i' = \varphi\Lambda_i, \quad \text{hence by } (2.6)_2 \quad h_i = -\frac{4}{3}\sigma\frac{\Lambda_i}{[\Lambda^2 - L/c^2]^2} = \frac{[\Lambda^2 - L/c^2]}{\Lambda}c^2p_i. \tag{2.25}$$

By use of (2.23) we therefore obtain for the entropy density and the entropy flux as functions of e and p

$$h = \frac{4}{3}\sigma^{1/4}e^{3/4}\frac{3^{3/4}}{2^{7/4}}(3-\chi)^{1/4}(1-\chi)^{1/4}, \qquad h_i = \frac{2^{5/4}}{3^{1/4}}\frac{(1-\chi)^{1/4}}{(3-\chi)^{1/2}}\left(\frac{\sigma}{e}\right)^{1/4}c^2p_i. \tag{2.26}$$

The above results were also derived by Larecki [57] on the basis of the kinetic theory of phonons.

2.2 Summary of Results and Limiting Cases

The field equations for the determination of energy and momentum of the pho□ons read

$$\frac{\partial e}{\partial t} + c^2 \frac{\partial p_k}{\partial x_k} = P_e, \qquad \frac{\partial p_i}{\partial t} + \frac{\partial P_{ik}}{\partial x_k} = P_{p_i}. \tag{2.27}$$

The main part of these equations has become fully explicit by the application of the entropy principle. We have obtained

$$P_{ik} = e\left(\frac{1-\chi}{2}\delta_{ik} + \frac{3\chi - 1}{2}\frac{p_i p_k}{p^2}\right) \quad \text{where } \chi = \frac{5}{3} - \frac{2}{3}\sqrt{4 - 3\frac{c^2 p^2}{e^2}} \tag{2.28}$$

is the Eddington factor.

Moreover, we have determined the entropy density and its flux, viz.

$$h = \frac{4}{3}\sigma^{1/4} e^{3/4} \frac{3^{3/4}}{2^{7/4}}(3-\chi)^{1/2}(1-\chi)^{1/4}, \qquad h_i = \frac{2^{5/4}}{3^{1/4}}\frac{(1-\chi)^{1/4}}{(3-\chi)^{1/2}}\left(\frac{\sigma}{e}\right)^{1/4} c^2 p_i. \tag{2.29}$$

We distinguish two limiting cases as follows:

i. The near–equilibrium case $cp \ll e$.
In this case $\chi \to \frac{1}{3}$ and the momentum flux becomes isotropic

$$P_{ik} = \frac{e}{3}\delta_{ik}. \tag{2.30}$$

The entropy density and entropy flux assume the forms

$$h = \frac{4}{3}\sigma^{1/4} e^{3/4}, \qquad h_i = \left(\frac{\sigma}{e}\right)^{1/4} c^2 p_i, \tag{2.31}$$

which by use of the equilibrium relation (1.22) we may write in the form

$$h = \frac{4}{3}\sigma T^3, \qquad h_i = \frac{c^2 p_i}{T}. \tag{2.32}$$

It is remarkable that the entropy flux for pho□ons near equilibrium has the same form as in a particle gas near equilibrium, namely, energy flux divided by absolute temperature.

ii. The free streaming case $cp_i \lesssim en_i$ where n_i is a unit vector in the streaming direction. In this case $\chi \to 1$ and the momentum flux reads

$$P_{ik} = e\, n_i n_k. \tag{2.33}$$

The entropy density is constant and the entropy flux vanishes.

2.3 Waves

It is instructive to investigate linear wave propagation in the two limiting cases that we have discussed.

In the near-equilibrium case, we linearize the equations (2.27) in $\mathbf{p}$ and obtain with (2.30)

$$\frac{\partial e}{\partial t} + c^2 \frac{\partial p_k}{\partial x_k} = 0, \qquad \frac{\partial p_i}{\partial t} + \frac{1}{3}\frac{\partial e}{\partial x_i} = B_0 p_i \tag{2.34}$$

where B_0 is the equilibrium value of the coefficient B that determines the momentum production $P_{p_i} = B_0 p_i$. We eliminate $\mathbf{p}$ from these equations and obtain a telegraph equation of the form

$$\frac{\partial^2 e}{\partial t^2} - B_0 \frac{\partial e}{\partial t} - \frac{c^2}{3} \frac{\partial^2 e}{\partial x_i \partial x_i} = 0. \tag{2.35}$$

This equation describes the propagation of a damped wave in all directions with the speed

$$V = \frac{c}{\sqrt{3}}. \tag{2.36}$$

In the case of phonons this mode of propagation is often called the *second sound.*

In the free streaming case we have $P_e = 0$ and $P_{p_i} = 0$. The two equations (2.27) read with $p_i = (e/c)n_i$ and $P_{ik} = e n_i n_k$ [see (2.33)]

$$\frac{\partial e}{\partial t} + c n_k \frac{\partial e}{\partial x_k} = 0 \quad \text{and} \quad \frac{1}{c} n_i \frac{\partial e}{\partial t} + n_i n_k \frac{\partial e}{\partial x_k} = 0. \tag{2.37}$$

These equations are identical. They describe the propagation of a wave in the direction of n_i with the speed

$$V = c. \tag{2.38}$$

This mode of propagation is called ballistic in the case of phonons, while in the case of light it represents the free streaming of photons in a laser beam (say).

3 Shock Waves

3.1 Rankine–Hugoniot Equations and the Speed of Shock Waves

The equations of balance of energy and momentum in a volume Ω whose surface $\partial\Omega$ moves with the speed s in the normal direction read

$$\begin{aligned} \frac{d}{dt}\int_{\Omega} e \, d\Omega + \int_{\partial\Omega} (-es + Q_k n_k)\, dA &= \int_{\Omega} P_e \, d\Omega, \\ \frac{d}{dt}\int_{\Omega} p_i \, d\Omega + \int_{\partial\Omega} (-p_i s + P_{ik} n_k)\, dA &= \int_{\Omega} P_{p_i} \, d\Omega. \end{aligned} \tag{3.1}$$

Applied to a "pill box" at a shock surface these equations give rise to the *Rankine–Hugoniot equations*

$$\begin{aligned} -s[\![e]\!] + [\![Q_k]\!] n_k &= 0, \\ -s[\![p_i]\!] + [\![P_{ik}]\!] n_k &= 0 \end{aligned} \tag{3.2}$$

where the bracket $[\![\psi]\!]$ denotes the difference of the quantity ψ before and behind the shock, i.e., on the sides + and − respectively. For a systematic and general derivation of the Rankine–Hugoniot equations see Chapter 7, Paragraph 5.2.

By use of (2.3) and (2.5) we obtain more explicit equations, namely,

$$\begin{aligned} -s[e\,] + \qquad\qquad\qquad & c^2[p_k]n_k = 0, \\ -s[p_i] + \left[e\left(\frac{1-\chi}{2}\delta_{ik} + \frac{3\chi-1}{2}\frac{p_ip_k}{p^2}\right)\right] & n_k = 0. \end{aligned} \tag{3.3}$$

We shall assume that the shock travels in the 1-direction into a region that is in thermal equilibrium so that we have

$$n_k = (1,0,0) \quad \text{and} \quad p_i^+ = 0, \qquad \text{hence} \quad \chi^+ = \frac{1}{3}. \tag{3.4}$$

It is then obvious from $(3.3)_2$ that p_i^- has only the 1-component p^-, so the whole problem becomes one-dimensional and there are only two relevant equations in (3.3), particularly,

$$\begin{aligned} & -s(e^+ - e^-) - c^2p^- = 0, \\ & +sp^- + \frac{1}{3}e^+ - e^-\chi^- = 0 \quad \text{where } \chi^- = \frac{5}{3} - \frac{2}{3}\sqrt{4 - 3\frac{c^2p_-^2}{e_-^2}}. \end{aligned} \tag{3.5}$$

These are two equations for the three unknowns s, p^-, e^- with e^+ as a parameter. We define

$$X = \frac{e^-}{e^+}$$

as the strength of the shock and calculate the speed s as a function of X.

Introducing X into the equations (3.5) we obtain

$$\begin{aligned} & -\frac{s}{c}(1-X) - \frac{cp^-}{e^+} = 0, \\ & \frac{s}{c}\frac{cp^-}{e^+} + \frac{1}{3}(1-5X) + \frac{2}{3}X\sqrt{4 - 3\frac{1}{X^2}\frac{c^2p_-^2}{e_+^2}} = 0. \end{aligned} \tag{3.6}$$

We eliminate s and obtain an equation for cp_-/e_+, viz.

$$\frac{cp^-}{e^+} = \pm\sqrt{\frac{1}{3}(X-1)\left[3X + 1 \overset{(+)}{-} 4\sqrt{X}\right]}. \tag{3.7}$$

The solution with the + sign in the radicand exists only for $X > 1$. This solution therefore does not contain the weak shock with $X \approx 1$, and we ignore it in the present analysis. The outer sign of the square root must be chosen so that $s > 0$ holds, i.e., for $X > 1$ we choose +, for $X < 1$ we choose −. Thus we obtain by $(3.6)_1$ for the speed of the shock wave

$$s = \frac{c}{\sqrt{3}}\sqrt{\frac{3X + 1 - 4\sqrt{X}}{X-1}}. \tag{3.8}$$

Note that

$$\left(\frac{cp^-}{e^-}\right)^2 = \frac{1}{3}\frac{X-1}{X^2}\left[3X + 1 - 4\sqrt{X}\right] \tag{3.9}$$

is restricted to the range $(cp^-/e^-)^2 \leq 1$ implying that the shock parameter has the range

$$\frac{1}{4} \leq X \leq \infty.$$

The speeds corresponding to the limiting cases are

$$s \underset{X \to 1/4}{\longrightarrow} \frac{c}{3} \quad \text{and} \quad s \underset{X \to \infty}{\longrightarrow} c,$$

while a weak shock with X approximately equal to 1 has the speed

$$s_{X \approx 1} = \frac{c}{\sqrt{3}}. \tag{3.10}$$

For a phonon gas we may say that the weak shocks with $X \approx 1$ propagate with the speed of the second sound while strong shocks with $X \gg 1$ propagate with the ordinary speed of sound.

3.2 Entropy Criterion for the Admissibility of Shock Waves

For the entropy we have an integral balance of the type

$$\frac{d}{dt}\int_{\Omega} h \, d\Omega + \int_{\partial\Omega} (-hs + h_k n_k) \, dA = \int_{\Omega} \Sigma \, d\Omega, \tag{3.11}$$

which refers to a volume Ω whose surface moves with speed s in the normal direction. This is quite analogous to the balance equations (3.1). A contrast to these equations appears when we consider Ω to be a "pill box" volume at a shock surface, because we shall assume that the shock has a non-negative entropy production. Thus the local form of (3.11) on the shock reads

$$-s[h] + [h_k]n_k = \Sigma_s \geq 0. \tag{3.12}$$

Now the question arises whether all shock parameters in the possible range $\frac{1}{4} \leq X \leq \infty$ conform to this inequality. Those which do not conform are not permissible. Let us check:

We use (2.49) and (3.7), (3.8) to write the condition (3.12) in the explicit form

$$-1 + X^{3/4}\frac{3^{3/4}}{2^{7/4}}(3-\chi_-)^{1/2}(1-\chi_-)^{1/4} + (1-X)3^{3/4}2^{1/4}\frac{(1-\chi_-)^{1/4}}{(3-\chi_-)^{1/2}}\frac{1}{X^{1/4}} \geq 0. \tag{3.13}$$

The left-hand side is a complicated function of X, because by (3.5) and (3.9) we have

$$\chi_- = \frac{5}{3} - \frac{2}{3}\frac{1}{X}\left|X + 2\sqrt{X} - 1\right|, \tag{3.14}$$

which must be inserted in (3.13). Dreyer and Seelecke [52] have plotted the left-hand side of (3.13), see Figure 2, and we conclude that the inequality is not satisfied for $X < 1$, that is, if $e^- < e^+$. This means that shocks will always increase the energy density.

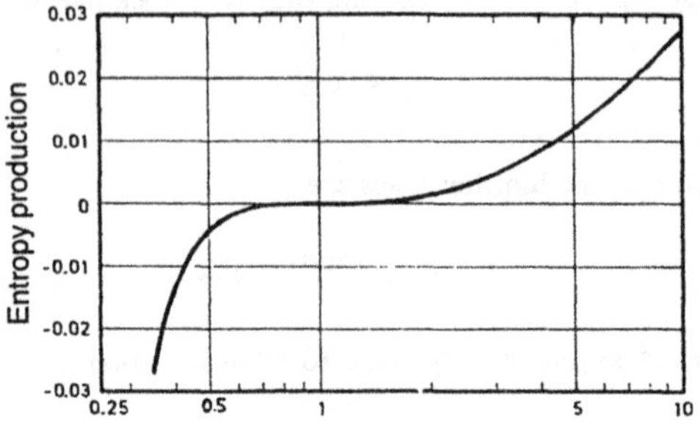

Figure 2. Entropy production versus shock parameter.

Chapter 6

Formal Structure of Extended Thermodynamics

Reflecting on the essentials of extended thermodynamics one realizes that the following ingredients are important:

— basic equations of balance type;
— constitutive equations local in space-time,
hence quasilinear first-order field equations;
— entropy inequality with concave entropy density,
hence symmetric hyperbolic field equations.

Ruggeri [19] first recognized these essential ingredients and combined them and formulated a properly systematic theory. That theory is the subject of this chapter.

The exploitation of the entropy principle introduces Lagrange multipliers and a four-vector potential from which all constitutive relations may be derived. The Lagrange multipliers may be considered as privileged fields in extended thermodynamics, because the four-vector potential depends on them and because the system of field equations exhibits its symmetry most clearly when written in terms of the Lagrange multipliers.

In all thermodynamic theories the velocity belongs to the basic fields and, in fact, the velocity plays a prominent role among the thermodynamic variables, because the constitutive quantities depend on it in a specific manner. This dependence can be determined by exploiting the requirement of Galilean invariance of the field equations. This has been done in a systematic manner by Ruggeri [21].

In the case that the variables are tensors of increasing rank, the dependence of the variables on velocity can be determined explicitly. Such is the case in extended thermodynamics of 13 variables as we saw in Chapter 2. Here we treat the more general case of an arbitrary number of variables, following Ruggeri [21].

1 The Formal Structure of Extended Thermodynamics

1.1 The Axioms of Extended Thermodynamics: Thermodynamic Processes

The objective of thermodynamics is the determination of n fields, which we combine in the n-vector $\mathbf{u}(x^D)$. Here x^D are space-time coordinates of an event with $t = x^0$ and $x^a = (x^1, x^2, x^3)$.

The necessary field equations are based upon n equations of balance of the form

$$\mathbf{F}^A{}_{,A} = \mathbf{f}. \tag{1.1}$$

The comma denotes partial differentiation. The components of $\mathbf{F}^0$ are called densities and the components of $\mathbf{F}^i$ are called fluxes. We refer to $\mathbf{F}^A$ as four-dimensional fluxes. Their divergence in space-time is equal to the production vector $\mathbf{f}$.

In order to provide field equations for the fields $\mathbf{u}$, the balance equations must be supplemented by constitutive equations. Such constitutive equations relate the fluxes $\mathbf{F}^A$ and the productions $\mathbf{f}$ to the fields $\mathbf{u}$ in a materially dependent manner. In extended thermodynamics the constitutive equations are assumed to be of the forms

$$\mathbf{F}^A \equiv \mathbf{F}^A(\mathbf{u}) \quad \text{and} \quad \mathbf{f} \equiv \mathbf{f}(\mathbf{u}) \tag{1.2}$$

so that the fluxes $\mathbf{F}^A$ and the productions $\mathbf{f}$ at one event depend only on the values of the fields $\mathbf{u}$ at that same event. We may say that the constitutive relations are *local* in space-time.

If the constitutive functions $\mathbf{F}^A(\mathbf{u})$ were known explicitly, we could eliminate $\mathbf{F}^A$ and $\mathbf{f}$ between (1.1) and (1.2) and obtain an explicit set of field equations. Every solution of that set of equations is called a *thermodynamic process.*

1.2 Universal Principles of the Constitutive Theory

In reality of course, the constitutive functions are not known and it is the task of the constitutive theory to determine these functions or at least to reduce their generality. The tools of the constitutive theory are certain universal principles which represent expectations whose validity we accept from long experience. The main principles are

i. the entropy principle,

ii. the requirement of convexity and causality, and

iii. the principle of relativity.

The most important part of the entropy principle is the entropy inequality

$$h^A{}_{,A} \geq 0, \tag{1.3}$$

which must hold for all thermodynamic processes. h^A is the four-dimensional entropy flux, which is a constitutive quantity such that in extended thermodynamics we have

$$h^A \equiv h^A(\mathbf{u}). \tag{1.4}$$

The requirement of convexity and causality demands that the field equations be symmetric hyperbolic. This requirement ensures the well-posedness of Cauchy problems for the field equations and it makes all wave speeds finite. Without essential loss of generality we may always choose the fields $\mathbf{u}$ to be the densities $\mathbf{F}^0$. If this is done and if the entropy principle is accepted as described above, the requirement of convexity and causality is equivalent to the statement that [1]

$$h^0(\mathbf{u}) \text{ is a concave function, i.e., } \frac{\partial^2 h^0}{\partial \mathbf{u} \partial \mathbf{u}} \sim \text{negative definite.} \tag{1.5}$$

The principle of relativity states that the field equations (1.1), (1.2) have the same form in all Galilean frames or in all Lorentz frames depending on whether the theory is non-relativistic or relativistic.

1.3 Exploitation of the Entropy Principle

We recall that the entropy inequality must hold for all thermodynamic processes, that is, solutions of the field equations. In a manner of speaking the field equations provide constraints for the fields that must satisfy the entropy inequality. A lemma by Liu [18] proves that it is possible to use Lagrange multipliers $\mathbf{\Lambda}$ to eliminate such constraints. Indeed, the new inequality

$$h^A{}_{,A} - \mathbf{\Lambda}(\mathbf{F}^A{}_{,A} - \mathbf{f}) \geq 0 \tag{1.6}$$

must hold for *all* continuous differentiable fields $\mathbf{u}$. The Lagrange multipliers are functions of $\mathbf{u}$.

The inequality (1.6) may be written in the form

$$\left(\frac{\partial h^A}{\partial \mathbf{u}} - \mathbf{\Lambda}\frac{\partial \mathbf{F}^A}{\partial \mathbf{u}}\right) \mathbf{u}_{,A} + \mathbf{\Lambda}\mathbf{f} \geq 0, \tag{1.7}$$

and, as this must hold for all fields $\mathbf{u}$, it must hold in particular for all $\mathbf{u}_{,A}$ at an event. The factor of $\mathbf{u}_{,A}$ must therefore vanish, lest the inequality be violated for some choice of $\mathbf{u}_{,A}$. Thus we conclude

$$\frac{\partial h^A}{\partial \mathbf{u}} = \mathbf{\Lambda}\frac{\partial \mathbf{F}^A}{\partial \mathbf{u}} \quad \text{and} \quad \mathbf{\Lambda}\mathbf{f} \geq 0. \tag{1.8}$$

The condition $(1.8)_1$ may be written in the short form

$$dh^A = \mathbf{\Lambda} d\mathbf{F}^A, \tag{1.9}$$

which makes it clear that the vector $\mathbf{\Lambda}$ is independent of the choice of variables $\mathbf{u}$. In particular, if we choose $\mathbf{u} = \mathbf{F}^0$, we obtain from (1.8) or (1.9)

$$\frac{\partial h^0}{\partial \mathbf{u}} = \mathbf{\Lambda} \quad \text{or by differentiation with respect to } \mathbf{u} \quad \frac{\partial^2 h^0}{\partial \mathbf{u} \partial \mathbf{u}} = \frac{\partial \mathbf{\Lambda}}{\partial \mathbf{u}}. \tag{1.10}$$

By the requirement of convexity h^0 is a concave function of $\mathbf{u}$; therefore we have

$$\left(\frac{\partial \mathbf{\Lambda}}{\partial \mathbf{u}}\right) = \left(\frac{\partial^2 h^0}{\partial \mathbf{u} \partial \mathbf{u}}\right) \tag{1.11}$$

symmetric and negative definite, and thus we conclude by a theorem of non-linear analysis (see [58]) that the map from $\mathbf{u}$ to $\mathbf{\Lambda}$ is not only locally but also globally invertible. We may thus write the constitutive quantities in the form

$$\mathbf{F}^A \equiv \mathbf{F}^A(\mathbf{\Lambda}), \qquad \mathbf{f} \equiv \mathbf{f}(\mathbf{\Lambda}), \qquad h^A \equiv h^A(\mathbf{\Lambda}). \tag{1.12}$$

We introduce the four-vector potential

$$h'^A = \mathbf{\Lambda}\mathbf{F}^A - h^A \tag{1.13}$$

[1] Often in the literature $-h^A$ is considered as the entropy flux. $h^0(\mathbf{u})$ is then a convex function, and the requirement of convexity received its name from that convention.

and rewrite (1.9) in the form

$$dh'^A = d\mathbf{\Lambda}\mathbf{F}^A. \tag{1.14}$$

Therefore we have

$$\mathbf{F}^A = \frac{\partial h'^A}{\partial \mathbf{\Lambda}} \quad \text{hence by (1.13)} \quad h^A = -h'^A + \mathbf{\Lambda}\frac{\partial h'^A}{\partial \mathbf{\Lambda}}. \tag{1.15}$$

It follows that the constitutive quantities $\mathbf{F}^A$ and h^A may be derived from a single vector, viz. h'^A. This fact justifies the name "vector potential" for h'^A.

There is a residual inequality, namely,

$$\Sigma = \mathbf{\Lambda} \cdot \mathbf{f}(\mathbf{\Lambda}) \geq 0, \tag{1.16}$$

which ensures that the entropy production Σ is non-negative.

The conditions (1.15), (1.16) have been derived as necessary conditions for the validity of the entropy principle (1.3). They can be shown to be sufficient as well. Indeed, if (1.15), (1.16) hold and if we define

$$h^A = -h'^A + \mathbf{\Lambda}\mathbf{F}^A,$$

we obtain

$$\begin{aligned} h^A{}_{,A} &= -h'^A{}_{,A} + \mathbf{\Lambda}_{,A}\,\mathbf{F}^A + \mathbf{\Lambda}\mathbf{F}^A{}_{,A} \\ &= -\left(\frac{\partial h'^A}{\partial \mathbf{\Lambda}} - \mathbf{F}^A\right)\Lambda_{,A} + \mathbf{\Lambda}\mathbf{f} \quad \text{or by } (1.15)_1,\ (1.16) \\ &\geq 0. \end{aligned}$$

Therefore we may summarize the contents of this section in the following theorem:

Theorem 1 *The constitutive relations $\mathbf{F}^A \equiv \mathbf{F}^A(\mathbf{u})$ and $\mathbf{f} \equiv \mathbf{f}(\mathbf{u})$ in the system of balance laws*

$$\mathbf{F}^A{}_{,A} = \mathbf{f}$$

are compatible with the entropy inequality

$$h^A{}_{,A} \geq 0 \quad \text{with } h^A \equiv h^A(\mathbf{u}),$$

if and only if there exists a four-vector potential h'^A and a privileged field $\mathbf{\Lambda}$ such that

$$\mathbf{F}^A = \frac{\partial h'^A}{\partial \mathbf{\Lambda}}, \qquad h^A = -h'^A + \mathbf{\Lambda}\frac{\partial h'^A}{\partial \mathbf{\Lambda}}, \qquad \mathbf{\Lambda}\mathbf{f}(\mathbf{\Lambda}) \geq 0.$$

The relations (1.15), (1.16) exhaust the consequences of the entropy principle. Note that the integrability condition for h'^A implied by $(1.15)_1$ states that the four $n \times n$ matrices

$$\frac{\partial \mathbf{F}^A}{\partial \mathbf{\Lambda}} \quad \text{must be symmetric.} \tag{1.17}$$

Obviously these provide many severe restrictions on the functions $\mathbf{F}^A(\mathbf{\Lambda})$.

So far so good; however, what we want are restrictions on the functions $\mathbf{F}^A(\mathbf{u})$. Therefore, before we obtain "useful" results, we must reverse the switch of variables from $\mathbf{u}$ to $\mathbf{\Lambda}$. This is usually a cumbersome task, which cannot be performed on the level of generality that we are maintaining in this section. In previous sections we performed the switch of variables for specific sets of examples. In most cases it could only be done for constitutive relations that are linear in the non-equilibrium variables.

1.4 Symmetric Hyperbolic Field Equations

We proceed to show that the field equations (1.1), (1.2) or (1.1), $(1.12)_{1,2}$ form a symmetric hyperbolic system, if the entropy principle and the requirement of convexity and causality are satisfied. Indeed, if we combine (1.1) and $(1.12)_1$, we obtain for the field equations

$$\frac{\partial \mathbf{F}^A}{\partial \mathbf{\Lambda}}\mathbf{\Lambda}_{,A} = \mathbf{f} \quad \text{or by } (1.15)_1 \quad \frac{\partial^2 h'^A}{\partial \mathbf{\Lambda}\partial \mathbf{\Lambda}}\mathbf{\Lambda}_{,A} = \mathbf{f}. \tag{1.18}$$

The system of equations therefore has the form

$$\mathbf{A}^A(\mathbf{\Lambda})\mathbf{\Lambda}_{,A} = \mathbf{f}(\mathbf{\Lambda}) \quad \text{with } \mathbf{A}^A = \frac{\partial^2 h'^A}{\partial \mathbf{\Lambda}\partial \mathbf{\Lambda}}. \tag{1.19}$$

$\mathbf{A}^A$ are Hessian matrices, hence, symmetric, and $\mathbf{A}^0$ is negative-definite by the requirement (1.5) of convexity, if $\mathbf{u} = \mathbf{F}^0$ holds.

Indeed from (1.13) with $A = 0$ we have for that choice of $\mathbf{u}$

$$h'^0 = \mathbf{\Lambda}\mathbf{F}^0 - h^0 = \mathbf{\Lambda}\mathbf{u} - h^0$$

so that h'^0 is the Legendre transform of h^0. This implies that h'^0 is a concave function of $\mathbf{\Lambda}$, if h^0 is a concave function of $\mathbf{u}$. The proof is simple. We consider the quadratic form $Q = \delta\mathbf{\Lambda}\delta\mathbf{u}$ and write

i. by $\mathbf{\Lambda} = \frac{\partial h^0}{\partial \mathbf{u}}$ [see $(1.10)_1$] $Q = \delta\left(\frac{\partial h^0}{\partial \mathbf{u}}\right)\delta u = \frac{\partial^2 h^0}{\partial \mathbf{u}\partial \mathbf{u}}\delta u \delta u < 0$; and

ii. by $\mathbf{u} = \frac{\partial h'^0}{\partial \mathbf{\Lambda}}$ [see $(1.15)_1$] $Q = \delta\Lambda\delta\left(\frac{\partial h'^0}{\partial \mathbf{\Lambda}}\right) = \frac{\partial^2 h'^0}{\partial \mathbf{\Lambda}\partial \mathbf{\Lambda}}\delta\mathbf{\Lambda}\delta\mathbf{\Lambda}$.

Thus we conclude

$$\frac{\partial^2 h'^0}{\partial \mathbf{\Lambda}\partial \mathbf{\Lambda}}\delta\mathbf{\Lambda}\delta\mathbf{\Lambda} = \frac{\partial^2 h^0}{\partial \mathbf{u}\partial \mathbf{u}}\delta\mathbf{u}\delta\mathbf{u} < 0.$$

This ensures that the system (1.19) is symmetric hyperbolic; it represents a particular case of symmetric hyperbolic systems defined by Friedrichs [59].

Even a merely hyperbolic system would guarantee finite speeds of propagation. The reason why we insist on having *symmetric* hyperbolic systems lies in their convenient and desirable mathematical properties, namely, the well-posedness of Cauchy problems, i.e., existence, uniqueness, and continuous dependence of the solutions on the data (see Fisher and Marsden [60]).

The fact that a system of balance-type field equations is symmetric hyperbolic if it is compatible with the entropy inequality and the concavity of the entropy density was discovered by Godunov [61] in a special case. In general this was proved Boillat [62] and Ruggeri and Strumia [63], the latter authors were motivated by a paper by Friedrichs and Lax [64].

The ingredients of the theory explained in this section are not only mathematically desirable, they also please the physicist. The attractive features include

— field equations of balance type,
— local constitutive relations,
— existence of an entropy inequality,
— concavity of the entropy density, and
— well-posedness of Cauchy problems.

We suggest, therefore, that thermodynamic theories ought to fit into the scheme explained in this section. Extended thermodynamics does that.

2 Galilean Invariance

2.1 Euclidean Transformations, Galilean Transformations, and Rotations

In classical thermodynamics we encounter three relevant types of transformations between frames, viz.

$$\begin{array}{ccc} \text{Euclidean} & \text{Galilean} & \text{pure rotations} \\ x_i^* = O_{ij}(t)x_j + b_i(t) & x_i^* = O_{ij}x_j + c_i t & x_i^* = O_{ij}x_j \\ t^* = t & t^* = t & t^* = t. \end{array} \tag{2.1}$$

A quantity is called a tensor of rank n, if its components in the two frames are related by the equation

$$T^*_{i_1 i_2 \cdots i_n} = O_{i_1 j_1} O_{i_2 j_2} \cdots O_{i_n j_n} T_{j_1 j_2 \cdots j_n}. \tag{2.2}$$

In the case of the Euclidean and Galilean transformations we call the tensors *objective* and *Galilean* respectively, and in the case of pure rotations we speak of tensors, without further qualification, or of tensors with respect to rotation.

In the present section we are mostly interested in Galilean transformations or, in fact, in the special case of a pure translation in which the velocity transforms according to

$$v_i^* = v_i + c_i. \tag{2.3}$$

In most specific cases in thermodynamics the generic field $\mathbf{u}$ contains, as some of its components, the velocity $\mathbf{v}$. We take this fact into account by decomposing the n vector $\mathbf{u}$ into the velocity $\mathbf{v}$ and an $(n-3)$ vector $\mathbf{w}$, which we assume to be composed of Galilei vectors. Thus the constitutive relations of extended thermodynamics in one Galilei frame read [cf. (1.2), (1.4)]

$$\mathbf{F}^0 = \mathbf{F}^0(\mathbf{v}, \mathbf{w}), \qquad \mathbf{F}^i = \mathbf{F}^i(\mathbf{v}, \mathbf{w}), \qquad \mathbf{f} = \mathbf{f}(\mathbf{v}, \mathbf{w}), \qquad h^0 = h^0(\mathbf{v}, \mathbf{w}), \qquad h^i = h^i(\mathbf{v}, \mathbf{w}). \tag{2.4}$$

In another Galilean frame these relations are assumed to have the forms

$$\overset{*}{\mathbf{F}^0} = \mathbf{F}^0(\overset{*}{\mathbf{v}}, \overset{*}{\mathbf{w}}), \qquad \overset{*}{\mathbf{F}^i} = \mathbf{F}^i(\overset{*}{\mathbf{v}}, \overset{*}{\mathbf{w}}), \overset{*}{\mathbf{f}^0} = \mathbf{f}^0(\overset{*}{\mathbf{v}}, \overset{*}{\mathbf{w}}), \qquad \overset{*}{h^0} = h^0(\overset{*}{\mathbf{v}}, \overset{*}{\mathbf{w}}), \qquad \overset{*}{h^i} = h^i(\overset{*}{\mathbf{v}}, \overset{*}{\mathbf{w}}). \tag{2.5}$$

Note that the constitutive *functions* are unchanged while both their values and the values of their variables may change. This amounts to the *assumption of material frame indifference* with respect to Galilei transformations. That assumption is part of the principle of relativity, because the field equations cannot have the same form in all Galilei frames unless that is first true for the constitutive relations.

We are noncommittal at this stage about the specific transformation properties of the densities $\mathbf{F}^0$. Later we shall assume that their components are scalars, vectors, and tensors, and we assume that the entropy density h^0 is a scalar.

2.2 Explicit Dependence of Constitutive Quantities on Velocity

We write the equations (1.1) and (1.3) in the slightly more explicit form

$$\frac{\partial \mathbf{F}^0}{\partial t} + \frac{\partial \mathbf{F}^i}{\partial x^i} = \mathbf{f} \quad \text{and} \quad \frac{\partial h^0}{\partial t} + \frac{\partial h^i}{\partial x^i} = \Sigma \geq 0 \tag{2.6}$$

and decompose the fluxes $\mathbf{F}^i$ and h^i into two parts each by splitting off the convective part $\mathbf{F}^0 v^i$ and $h^0 v_i$ respectively. Thus the relations read

$$\frac{\partial \mathbf{F}^0}{\partial t} + \frac{\partial (\mathbf{F}^0 v^i + \mathbf{G}^i)}{\partial x^i} = \mathbf{f} \quad \text{and} \quad \frac{\partial h^0}{\partial t} + \frac{\partial (h^0 v^i + \varphi^i)}{\partial x^i} = \Sigma \geq 0. \tag{2.7}$$

Here $\mathbf{G}^i$ and φ^i are called non-convective fluxes. Obviously they are constitutive quantities. The constitutive quantities $\mathbf{F}^0, \mathbf{G}^i, \mathbf{f}$ and h^0, φ^i, Σ may depend on the variables $\mathbf{u}$ or $(\mathbf{v}, \mathbf{w})$. We postulate Galilean invariance separately for the system of balance equations $(2.7)_1$ and for the entropy balance $(2.7)_2$. This implies that the constitutive quantities $\mathbf{F}^0, \mathbf{G}^i, \mathbf{f}$ and h^0, φ^i depend *explicitly* upon velocity. The purpose of this section is to determine that dependence.

First we consider the Galilean invariance of the balance equations $(2.7)_1$. The results about entropy will follow as a corollary. We make use of the requirement (2.4), (2.5) of material frame indifference and express the Galilei invariance of the field equations $(2.2)_1$ as

$$\frac{\partial \mathbf{F}^0(\mathbf{v},\mathbf{w})}{\partial t} + \frac{\partial \mathbf{F}^0(\mathbf{v},\mathbf{w})v^i + \mathbf{G}^i(\mathbf{v},\mathbf{w})}{\partial x^i} = \mathbf{f}^0(\mathbf{v},\mathbf{w}), \tag{2.8}$$

$$\frac{\partial \mathbf{F}^0(\overset{*}{\mathbf{v}},\overset{*}{\mathbf{w}})}{\partial t^*} + \frac{\partial \mathbf{F}^0(\overset{*}{\mathbf{v}},\overset{*}{\mathbf{w}})\,\overset{*}{v^i} + \mathbf{G}^i(\overset{*}{\mathbf{v}},\overset{*}{\mathbf{w}})}{\partial \overset{*}{x^i}} = \mathbf{f}^0(\overset{*}{\mathbf{v}},\overset{*}{\mathbf{w}}). \tag{2.9}$$

It is easy to see that

$$\frac{\partial}{\partial x_i^*} = \frac{\partial}{\partial x_i} \quad \text{and} \quad \frac{\partial}{\partial t^*} = \frac{\partial}{\partial t} - c_i\frac{\partial}{\partial x_i}.$$

Therefore (2.9) may be written in the form

$$\frac{\partial \mathbf{F}^0(\mathbf{v}+\mathbf{c},\mathbf{w})}{\partial t} + \frac{\partial \mathbf{F}^0(\mathbf{v}+\mathbf{c},\mathbf{w})v^i + \mathbf{G}^i(\mathbf{v}+\mathbf{c},\mathbf{w})}{\partial x_i} = \mathbf{f}(\mathbf{v}+\mathbf{c},\mathbf{w}). \tag{2.10}$$

Since this system of equations must be equivalent to the system (2.8), the terms in (2.10) must be linear combinations of the corresponding ones in (2.8). This means that there exists an $n \times n$ non-singular matrix $\mathbf{X}(\mathbf{c})$ such that

$$\mathbf{F}^0(\mathbf{v}+\mathbf{c},\mathbf{w}) = \mathbf{X}(\mathbf{c})\mathbf{F}^0(\mathbf{v},\mathbf{w}),\ \mathbf{G}^i(\mathbf{v}+\mathbf{c},\mathbf{w}) = \mathbf{X}(\mathbf{c})\mathbf{G}^i(\mathbf{v},\mathbf{w}),\ \mathbf{f}(\mathbf{v}+\mathbf{c},\mathbf{w}) = \mathbf{X}(\mathbf{c})\mathbf{f}(\mathbf{v},\mathbf{w}) \tag{2.11}$$

holds for all $\mathbf{v}$ and $\mathbf{c}$. In particular, we have for $\mathbf{v} = 0$ and for all $\mathbf{c}$

$$\mathbf{F}^0(\mathbf{c},\mathbf{w}) = \mathbf{X}(\mathbf{c})\mathbf{F}^0(0,\mathbf{w}), \qquad \mathbf{G}^i(\mathbf{c},\mathbf{w}) = \mathbf{X}(\mathbf{c})\mathbf{G}^i(0,\mathbf{w}), \qquad \mathbf{f}(\mathbf{c},\mathbf{w}) = \mathbf{X}(\mathbf{c})\mathbf{f}(0,\mathbf{w}), \tag{2.12}$$

where we call $\mathbf{F}^0(0,\mathbf{w}), \mathbf{G}^i(0,\mathbf{w}), \mathbf{f}(0,\mathbf{w})$ the intrinsic parts of the densities, fluxes, and productions. We denote these by $\hat{\mathbf{F}}^0, \hat{\mathbf{G}}^i, \hat{\mathbf{f}}$ and rewrite (2.12) as

$$\mathbf{F}^0(\mathbf{v},\mathbf{w}) = \mathbf{X}(\mathbf{v})\hat{\mathbf{F}}^0(\mathbf{w}), \qquad \mathbf{G}^i(\mathbf{v},\mathbf{w}) = \mathbf{X}(\mathbf{v})\hat{\mathbf{G}}^i(\mathbf{w}), \qquad \mathbf{f}(\mathbf{v},\mathbf{w}) = \mathbf{X}(\mathbf{v})\hat{\mathbf{f}}(\mathbf{w}). \tag{2.13}$$

From (2.11), (2.13) we conclude that

$$\mathbf{X}(\overset{(1)}{\mathbf{v}} + \overset{(2)}{\mathbf{v}}) = \mathbf{X}(\overset{(1)}{\mathbf{v}})\mathbf{X}(\overset{(2)}{\mathbf{v}}) = \mathbf{X}(\overset{(2)}{\mathbf{v}})\mathbf{X}(\overset{(1)}{\mathbf{v}}) \quad \text{and} \quad \mathbf{X}(0) = \mathbf{1} \tag{2.14}$$

hold, and this condition implies the differential equation

$$\frac{\partial \mathbf{X}}{\partial v_r}(\mathbf{v}) = \mathbf{A}^r\mathbf{X}(\mathbf{v}) = \mathbf{X}(\mathbf{v})\mathbf{A}^r \quad \text{with } \mathbf{X}(0) = 1 \text{ and } \mathbf{A}^r \equiv \frac{\partial \mathbf{X}}{\partial v_r}(0). \tag{2.15}$$

It follows that $\mathbf{X}(\mathbf{v})$ is an exponential operator

$$\mathbf{X}(\mathbf{v}) = e^{\mathbf{A}^r v_r} = \mathbf{1} + \mathbf{A}^r v_r + \frac{1}{2}\mathbf{A}^r\mathbf{A}^s v_r v_s + \cdots \tag{2.16}$$

where by differentiation of (2.15) with respect to $\mathbf{v}_s$ we obtain

$$\mathbf{A}^r\mathbf{A}^s = \mathbf{A}^s\mathbf{A}^r. \tag{2.17}$$

We note that (2.13) with (2.15) implies the differential equations

$$\frac{\partial \mathbf{F}^0}{\partial v_r} = \mathbf{A}^r \mathbf{F}^0, \qquad \frac{\partial \mathbf{G}^i}{\partial v_r} = \mathbf{A}^r \mathbf{G}^i, \qquad \frac{\partial \mathbf{f}}{\partial v_r} = \mathbf{A}^r \mathbf{f}. \tag{2.18}$$

Thus the symmetry conditions (2.17) are revealed as integrability conditions for $\mathbf{F}^0, \mathbf{G}^i$, or $\mathbf{f}$.

We return to the entropy balance $(2.7)_2$ and investigate the consequences—on h^0 and φ^i—of its Galilean invariance. Actually the arguments are identical to those for $\mathbf{F}^0$ and $\mathbf{G}^i$ and so are the results. In analogy to (2.12), (2.13), (2.16), and (2.18) we have

$$h^0(\mathbf{v}, \mathbf{w}) = x(\mathbf{v})\hat{h}^0(\mathbf{w}), \qquad \varphi^i(\mathbf{v}, \mathbf{w}) = x(\mathbf{v})\hat{\varphi}^i(\mathbf{w}), \qquad \Sigma(\mathbf{v}, \mathbf{w}) = x(\mathbf{v})\hat{\Sigma}(\mathbf{w}) \tag{2.19}$$

$$x(\mathbf{v}) = e^{a^r v_r} = 1 + a^r v_r + a^r a^s v_r v_s + \cdots, \tag{2.20}$$

$$\frac{\partial h^0}{\partial v_r} = a^r h^0, \qquad \frac{\partial \varphi^i}{\partial v_r} = a^r \varphi^i, \qquad \frac{\partial \Sigma}{\partial v_r} = a^r \Sigma. \tag{2.21}$$

As before the hat denotes values in the rest frame. $x(\mathbf{v})$ is now a scalar function of $\mathbf{v}$, i.e., it could possibly depend on v^2; but such a dependence is impossible by (2.20) and, therefore, x must be a constant, namely, 1, and a^r is equal to zero. Thus we conclude from (2.21) that the entropy density, the non-convective entropy flux, and the entropy production are all independent of $\mathbf{v}$. In other words, h^0, φ^i, and Σ consist only of intrinsic parts

$$h^0(\mathbf{v}, \mathbf{w}) = \hat{h}^0(\mathbf{w}), \qquad \varphi^i(\mathbf{v}, \mathbf{w}) = \hat{\varphi}^i(\mathbf{w}), \qquad \Sigma(\mathbf{v}, \mathbf{w}) = \hat{\Sigma}(\mathbf{w}). \tag{2.22}$$

It is clear from (2.22) and (2.7) that φ^i equals $\hat{h}^i$, and henceforth we shall write $\hat{h}^i$ instead of φ^i for the nonconvective *and* intrinsic part of the entropy flux.

We summarize these arguments in a theorem.

Theorem 2 *There exists an $(n \times n)$ matrix $\mathbf{X}(\mathbf{v})$ such that*

$$\mathbf{F}^0(\mathbf{v}, \mathbf{w}) = \mathbf{X}(\mathbf{v})\hat{\mathbf{F}}^0(\mathbf{w}), \qquad \mathbf{G}^i(\mathbf{v}, \mathbf{w}) = \mathbf{X}(\mathbf{v})\hat{\mathbf{G}}^i(\mathbf{w}), \qquad \mathbf{f}(\mathbf{v}, \mathbf{w}) = \mathbf{X}(\mathbf{v})\hat{\mathbf{f}}(\mathbf{w}), \tag{2.23}$$

where $\mathbf{X}(\mathbf{v})$ is an exponential matrix

$$\mathbf{X}(\mathbf{v}) = e^{\mathbf{A}^r v_r} = \mathbf{1} + \mathbf{A}^r v_r + \frac{1}{2}\mathbf{A}^r \mathbf{A}^s v_r v_s + \cdots \tag{2.24}$$

and $\mathbf{A}^r$ are constant $(n \times n)$-matrices with the property

$$\mathbf{A}^r \mathbf{A}^s = \mathbf{A}^s \mathbf{A}^r \qquad \forall r, s \in 1, 2, 3. \tag{2.25}$$

The density, nonconvective flux, and production of entropy are purely intrinsic, so

$$h^0(\mathbf{v}, \mathbf{w}) = \hat{h}^0(\mathbf{w}), \qquad \varphi^i(\mathbf{v}, \mathbf{w}) = \hat{h}^i(\mathbf{w}), \qquad \Sigma(\mathbf{v}, \mathbf{w}) = \hat{\Sigma}(\mathbf{w}). \tag{2.26}$$

We conclude from (2.23), (2.25) that the $\mathbf{v}$-dependence of the constitutive quantities $\mathbf{F}^0, \mathbf{G}^i$, and $\mathbf{f}$ is explicit, if only the three constant matrices $\mathbf{A}^r$ are known. The form of these matrices depends on the tensorial character of the densities $\mathbf{F}^0$ about which we have not made a commitment so far. The tensorial character of $\mathbf{F}^0$ is a constitutive property characteristic for a class of materials. In Section 3 we shall characterize such a class and calculate the matrices $\mathbf{A}^r$.

2.3 Galilean Invariance and Entropy Principle

Once the matrices $\mathbf{A}^r$ and, hence, $\mathbf{X}(\mathbf{v})$ are known for a particular class of materials we conclude from (2.23) that, within that class, only the intrinsic parts $\hat{\mathbf{F}}^0, \hat{\mathbf{G}}^i$, and $\hat{\mathbf{f}}$ of densities, fluxes, and productions remain as constitutive quantities and that $\mathbf{v}$ has dropped out as a variable. We shall now investigate how the remaining constitutive relations are restricted by the entropy principle. In other words, we wish to find out how the general restrictions [cf. (1.8), (1.9)]

$$dh^A = \mathbf{\Lambda} d\mathbf{F}^A \quad \text{and} \quad \mathbf{\Lambda f} \geq 0 \tag{2.27}$$

are reflected in restrictions on the intrinsic quantities.

By (2.27) and (2.23) we may write $(2.27)_1$ as

$$A = 0 : d\hat{h}^0 = \mathbf{\Lambda} d(\mathbf{X}\hat{\mathbf{F}}^0) \quad \Big| \quad A = i : d(\hat{h}^0 v^i + \hat{h}^i) = \mathbf{\Lambda} d[\mathbf{X}(\hat{\mathbf{F}}^0 v^i + \hat{\mathbf{G}}^i)]. \tag{2.28}$$

With $d\mathbf{X} = \mathbf{X}\mathbf{A}^r dv_r$ these two relations assume the forms

$$d\hat{h}^0 - \mathbf{\Lambda X} d\hat{\mathbf{F}}^0 = \mathbf{\Lambda X A}^r \hat{\mathbf{F}}^0 dv_r \quad \Big| \quad d\hat{h}^i - \mathbf{\Lambda X} d\hat{\mathbf{G}}^i = \left(\mathbf{\Lambda X A}^r \hat{\mathbf{G}}^i - [\hat{h}_0 - \mathbf{\Lambda X}\hat{\mathbf{F}}^0]\delta^{ri}\right) dv_r.$$

Since this must hold for all $d\mathbf{v}$, we have

$$d\hat{h}^0 = \mathbf{\Lambda X} d\hat{\mathbf{F}}^0, \ \mathbf{\Lambda X A}^r \hat{\mathbf{F}}^0 = 0 \quad \Big| \quad d\hat{h}^i = \mathbf{\Lambda X} d\hat{\mathbf{G}}^i, \ \mathbf{\Lambda X A}^r \hat{\mathbf{G}}^i = [\hat{h}_0 - \mathbf{\Lambda X}\hat{\mathbf{F}}^0]\delta^{ri}. \tag{2.29}$$

Moreover, $\hat{h}^0, \hat{\mathbf{F}}^0$ and $\hat{h}_i, \hat{\mathbf{G}}_i$ are independent of $\mathbf{v}$. Therefore $\mathbf{\Lambda X}$ is also independent of $\mathbf{v}$ and we write[2]

$$\hat{\mathbf{\Lambda}} = \hat{\mathbf{\Lambda}}(\mathbf{w}) = \mathbf{\Lambda X}, \tag{2.30}$$

which we may call the intrinsic Lagrange multipliers. Thus (2.30) reads

$$d\hat{h}^0 = \hat{\mathbf{\Lambda}} d\hat{\mathbf{F}}^0, \qquad \hat{\mathbf{\Lambda}}\mathbf{A}^r \hat{\mathbf{F}}^0 = 0, \qquad d\hat{h}^i - \hat{\mathbf{\Lambda}} d\hat{G}^i, \qquad \hat{\mathbf{\Lambda}}\mathbf{A}^r \hat{\mathbf{G}}^i = \left[\hat{h}_0 - \hat{\mathbf{\Lambda}}\hat{\mathbf{F}}^0\right]\delta^{ri}. \tag{2.31}$$

By $(2.23)_3$ and (2.31) the residual inequality $(2.28)_2$ may be written as

$$\hat{\mathbf{\Lambda}}\hat{\mathbf{f}} \geq 0. \tag{2.32}$$

The relations (2.32) and (2.33) represent the restrictions that were drawn from the entropy inequality and the requirement of Galilean invariance. The present form contains only intrinsic quantities; i.e., the velocity does not appear anymore. The differential form $(2.32)_1$, giving the differential of the entropy density, may be called the Gibbs equation of extended thermodynamics. Its equilibrium counterpart can identify the Lagrange mulitipliers $\mathbf{\Lambda}$ in equilibrium.

We may also reformulate the requirement of convexity in terms of the intrinsic quantities. If the fields $\mathbf{u}$ are chosen as densities $\mathbf{F}^0$, we recall that this requirement can be expressed as

$$Q = \delta\mathbf{\Lambda}\delta\mathbf{F}^0 = \delta\mathbf{\Lambda}\delta\mathbf{u} < 0. \tag{2.33}$$

Now $\mathbf{\Lambda} = \hat{\mathbf{\Lambda}}\mathbf{X}^{-1}$ and $\mathbf{F}^0 = \mathbf{X}\hat{\mathbf{F}}^0$, and therefore (2.34) reads

$$Q = \delta\hat{\mathbf{\Lambda}}\delta\hat{\mathbf{F}}^0 - 2\hat{\mathbf{\Lambda}}\mathbf{A}^r\delta\hat{\mathbf{F}}^0\delta v_r - \hat{\mathbf{\Lambda}}\mathbf{A}^r\mathbf{A}^s\hat{\mathbf{F}}^0\delta v_r\delta v_s < 0 \tag{2.34}$$

where (2.15) and $(2.32)_2$ were used. In particular, this implies

$$\delta\hat{\mathbf{\Lambda}}\delta\hat{\mathbf{F}}^0 < 0 \quad \text{and} \quad \hat{\mathbf{\Lambda}}\mathbf{A}^r\mathbf{A}^s\hat{\mathbf{F}}^0\delta v_r\delta v_s > 0. \tag{2.35}$$

---

[2]The conclusion (2.31) can obviously be drawn from $(2.30)_{1,3}$, if none of the components of $\hat{\mathbf{F}}^0$ and $\hat{\mathbf{G}}^i$ vanish. Later, however, we shall see that some of these components are zero, *but never the same ones*, and therefore the conclusion remains valid. In fact, if the same components of $\hat{\mathbf{F}}^0$ and $\hat{\mathbf{G}}^i$ were to vanish, the corresponding balance equations and the system of n equations would be underdetermined. We exlude that possibility.

We summarize the above results in a theorem.

Theorem 3 *The requirements of the entropy principle and of Galilean invariance provide restrictions on the intrinsic parts of the constitutive quantities* $\mathbf{F}^A, \mathbf{f}$ *and* $h^A, \mathbf{\Lambda}$. *These read*

$$\begin{aligned} d\hat{h}^0 &= \hat{\mathbf{\Lambda}} d\hat{\mathbf{F}}^0 \qquad d\hat{h}^i = \hat{\mathbf{\Lambda}} d\hat{\mathbf{G}}^i, \\ \hat{\mathbf{\Lambda}}\mathbf{A}^r\hat{\mathbf{F}}^0 &= 0, \\ \hat{\mathbf{\Lambda}}\mathbf{A}^r\hat{\mathbf{G}}^i &= [\hat{h}^0 - \hat{\mathbf{\Lambda}}\hat{\mathbf{F}}^0]\delta^{ri}. \end{aligned} \tag{2.36}$$

There remains the residual inequality

$$\hat{\mathbf{\Lambda}}\hat{\mathbf{f}} \geq 0, \tag{2.37}$$

and the requirement of convexity implies

$$\delta\hat{\mathbf{\Lambda}}\delta\hat{\mathbf{F}}^0 < 0, \qquad \hat{\mathbf{\Lambda}}\mathbf{A}^r\mathbf{A}^s\hat{\mathbf{F}}^0\delta v_r\delta v_s > 0. \tag{2.38}$$

Having identified the intrinsic parts $\hat{\mathbf{F}}^0, \hat{\mathbf{G}}^i$, and $\hat{\mathbf{f}}$ in the densities, non-convective fluxes, and productions we may introduce them into the balance equations $(2.7)_1$. Thus we obtain after a short calculation and by use of (2.15)

$$\mathbf{X}\left[\frac{d\hat{\mathbf{F}}^0}{dt} + \hat{\mathbf{F}}_0\frac{\partial v_i}{\partial x_i} + \frac{\partial \hat{\mathbf{G}}^i}{\partial x^i} + \mathbf{A}^r\left\{\hat{\mathbf{F}}^0\frac{dv_r}{dt} + \hat{\mathbf{G}}^i\frac{\partial v_r}{\partial x_i}\right\} - \hat{\mathbf{f}}\right] = 0. \tag{2.39}$$

Here d/dt is the material time derivative defined as $\partial/\partial t + v_i(\partial/\partial x_i)$. Obviously the square bracket in (2.40) has to vanish and that provides an alternative form of the equations of balance $(2.7)_1$, viz.

$$\frac{d\hat{\mathbf{F}}^0}{dt} + \hat{\mathbf{F}}^0\frac{\partial v_i}{\partial x_i} + \frac{\partial \hat{\mathbf{G}}^i}{\partial x_i} + \mathbf{A}^r\left\{\hat{\mathbf{F}}^0\frac{dv_r}{dt} + \hat{\mathbf{G}}^i\frac{\partial v_r}{\partial x_i}\right\} = \hat{\mathbf{f}}. \tag{2.40}$$

This form of the equations of balance is sometimes preferable because they contain the velocity explicitly and, apart from that, they contain only intrinsic quantities. The equations (2.3.1) are specific examples of these equations. Of course, the balance structure of the equations is disguised by the form (2.41) and from looking at them, without knowing the above derivation, it is not obvious that these equations can be combined to provide a set of equations of balance.

We recall the use of Lagrange multipliers $\mathbf{\Lambda}$ in (1.6). Since $\mathbf{F}^A{}_{,A} - \mathbf{f}$ in that inequality is equal to the left-hand side of the equation (2.40) we may write the inequality (1.6) as

$$\frac{\partial h^0}{\partial t} + \frac{\partial(h^0v^i + \varphi^i)}{\partial x^i} - \mathbf{\Lambda}\mathbf{X}\left[\frac{d\hat{\mathbf{F}}^0}{dt} + \hat{\mathbf{F}}_0\frac{\partial v_i}{\partial x_i} + \frac{\partial \hat{\mathbf{G}}^i}{\partial x^i} + \mathbf{A}^r\left\{\hat{\mathbf{F}}^0\frac{dv_r}{dt} + \hat{\mathbf{G}}^i\frac{\partial v_r}{\partial x^i}\right\} - \hat{\mathbf{f}}\right] \geq 0. \tag{2.41}$$

We therefore conclude that the intrinsic Lagrange multipliers $\hat{\mathbf{\Lambda}} = \mathbf{\Lambda}\mathbf{X}$ are those which take care of the constraints when the field equations are based on (2.41) rather than $(2.6)_1$ or $(2.7)_1$. We observe that the $\mathbf{\Lambda}$'s are independent variables, and therefore it was possible to choose them as field variables. In contrast to these the intrinsic Lagrange multipliers $\hat{\mathbf{\Lambda}}$ are constrained by (2.36), and therefore this set is only useful for the technique of exploiting the entropy inequality.

The constant matrices $\mathbf{A}^r$ in (2.41) have yet to be determined; only then will the dependence of $\mathbf{F}^0, \mathbf{F}^i$, and $\mathbf{f}$ on velocity be specific. The form of $\mathbf{A}^r$ depends on the tensor character of the variables, and we proceed to calculate $\mathbf{A}^r$ for the case in which the variables $\mathbf{u}$ are tensors of increasing rank. That case is realized by extended thermodynamics of 13 variables, as presented in Chapter 2, and by the kinetic theory of gases for tensors of arbitrary rank.

3 Variables of Increasing Tensorial Rank

3.1 Restrictive Conditions on the Matrices $\mathbf{X}(\mathbf{v})$ and $\mathbf{A}^r$

We consider the case where the densities, (non-convective) fluxes, and productions have the forms

$$\mathbf{F}^0 = \begin{bmatrix} F \\ F_{k_1} \\ F_{k_1k_2} \\ \vdots \\ F_{k_1k_2\cdots k_N} \end{bmatrix}, \qquad \mathbf{G}^i = \begin{bmatrix} G^i \\ G^i_{k_1} \\ G^i_{k_1k_2} \\ \vdots \\ G^i_{k_1k_2\cdots k_N} \end{bmatrix}, \qquad \mathbf{f} = \begin{bmatrix} f \\ f_{k_1} \\ f_{k_1k_2} \\ \vdots \\ f_{k_1k_2\cdots k_N} \end{bmatrix} \tag{3.1}$$

with $F_{k_1k_2\cdots k_j}, G^i_{k_1k_2\cdots k_j}, f_{k_1k_2\cdots k_j}$ as symmetric tensors. Thus the system of balance equations (1.1) or $(2.7)_1$ has a natural order induced by the increasing tensorial character of the equations of the system. Each block of tensorial equations of rank j governs the evolution of the new quantity $F_{k_1k_2\cdots k_j}$.

Lest the system of equations should not be determined, we must assume that the densities $F_{k_1k_2\cdots k_j}$ cannot be influenced by intrinsic densities $\hat{F}_{k_1k_2\cdots k_l}$ with $l > j$; therefore, the matrix $\mathbf{X}(\mathbf{v})$ in (2.23) must be a sub-triangular block matrix of the form

$$\mathbf{X}(\mathbf{v}) = \begin{bmatrix} x & 0 & 0 & & 0 \\ x_{k_1} & x^{h_1}_{k_1} & 0 & & 0 \\ x_{k_1k_2} & x^{h_1}_{k_1k_2} & x^{h_1h_2}_{k_1k_2} & \cdots & 0 \\ \vdots & \vdots & \vdots & & \vdots \\ x_{k_1k_2\cdots k_N} & x^{h_1}_{k_1k_2\cdots k_N} & x^{h_1h_2}_{k_1k_2\cdots k_N} & \cdots & x^{h_1h_2\cdots h_N}_{k_1k_2\cdots k_N} \end{bmatrix}. \tag{3.2}$$

The blocks $x_{jl} = x^{h_1\cdots h_j}_{k_1\cdots k_l}$ are fully symmetric with respect to the indices $h_1\cdots h_j$ and $k_1\cdots k_l$.
It is then possible to prove the following statement:

The matrices $\mathbf{A}^r$ are sub-triangular block matrices with blocks of zero in the main diagonal and they are nil-potent of degree $N+1$, i.e.,

$$\mathbf{A}^{k_1}\mathbf{A}^{k_2}\ldots\mathbf{A}^{k_{n+1}} = 0 \quad \textit{for all } k_1, k_2\cdots k_{N+1} = 1,2,3. \tag{3.3}$$

The matrix $\mathbf{X}(\mathbf{v})$ is a polynomial matrix in $\mathbf{v}$ of order N.

The proof of this statement proceeds as follows:
We call τ_m a triangular $M\times M$ block matrix such that the blocks a_{ij} are zero for all $i > j+m$ with $m \geq 0$. It is then trivially true that

i. τ_1 is a triangular block matrix with blocks of zero in the main diagonal,
ii. $\tau_h\tau_k$ is a sub-triangular block matrix τ_{h+k}, and
iii. The block matrices τ_1 are nil-potent of degree M.

Since $\mathbf{X}(\mathbf{v})$ is a triangular block matrix τ_0, so is $\mathbf{A}^r$ by (2.15); i.e., we have

$$\mathbf{A}^r = \begin{bmatrix} a^r & 0 & 0 & & 0 \\ a^r_{k_1} & a^{rh_1}_{k_1} & 0 & & 0 \\ a^r_{k_1k_2} & a^{rh_1}_{k_1k_2} & a^{rh_1h_2}_{k_1k_2} & \cdots & 0 \\ \vdots & \vdots & \vdots & & \vdots \\ a^r_{k_1k_2\cdots k_N} & a^{rh_1}_{k_1k_2\cdots k_N} & a^{rh_1h_2}_{k_1k_2\cdots k_N} & \cdots & a^{rh_1h_2\cdots h_N}_{k_1k_2\cdots k_N} \end{bmatrix}. \tag{3.4}$$

Now the block elements $a^{rh_1\cdots h_j}_{k_1\cdots k_l}$ must be constant tensors, and this means that they must be combinations of Kronecker symbols. Therefore the blocks with $j+l$ even must be zero and, in particular, the diagonal blocks vanish, so that $\mathbf{A}^r$ is a block matrix τ_1, which by iii. is nil-potent of degree $N+1$. It follows from (2.16) that $\mathbf{X}(\mathbf{v})$ is a polynomial matrix in $\mathbf{v}$ of order N.

We conclude that $\mathbf{A}^r$ has the form

$$\mathbf{A}^r = \begin{bmatrix} 0 & & & & \\ a^r_{k_1} & 0 & & & \\ 0 & a^{rh_1}_{k_1k_2} & 0 & & \\ a^r_{k_1k_2k_3} & 0 & a^{rh_1h_2}_{k_1k_2k_3} & 0 & \\ \vdots & \vdots & \vdots & \vdots & \\ a^r_{k_1k_2\cdots k_N} & 0 & a^{rh_1h_2}_{k_1k_2\cdots k_N} & 0\cdots & 0 \end{bmatrix} \tag{3.5}$$

with every second subdiagonal composed of zeros.

Hence follows by (2.16) that the matrices $\mathbf{A}^r$ do not contribute to the diagonal of $\mathbf{X}(\mathbf{v})$ and, therefore, that the elements of the diagonal of $\mathbf{X}(\mathbf{v})$ are blocks of Kronecker symbols. We have

$$\mathbf{X}(\mathbf{v}) = \begin{bmatrix} 1 & 0 & 0 & & \\ x_{k_1} & \delta^{h_1}_{k_1} & 0 & & \\ x_{k_1k_2} & x^{h_1}_{k_1k_2} & \delta^{h_1}_{k_1}\delta^{h_2}_{k_2} & & \\ \vdots & \vdots & \vdots & & \\ x_{k_1k_2\cdots k_N} & x^{h_1}_{k_1k_2\cdots k_N} & x^{h_1h_2}_{k_1k_2\cdots k_N} & \cdots & \delta^{h_1}_{k_1}\delta^{h_2}_{k_2}\cdots\delta^{h_N}_{k_N} \end{bmatrix}. \tag{3.6}$$

3.2 The Case $N = 3$: An Illustration

For illustration we elaborate the case $N = 3$. In this case we have

$$\mathbf{F}^0 = \begin{bmatrix} F \\ F_{k_1} \\ F_{k_1k_2} \\ F_{k_1k_2k_3} \end{bmatrix}, \qquad \mathbf{G}^i = \begin{bmatrix} G^i \\ G^i_{k_1} \\ G^i_{k_2k_3} \\ G^i_{k_1k_2k_3} \end{bmatrix}, \qquad \mathbf{f} = \begin{bmatrix} f \\ f_{k_1} \\ f_{k_1k_2} \\ f_{k_1k_2k_3} \end{bmatrix}. \tag{3.7}$$

The matrix $\mathbf{A}^r$ has the form

$$\mathbf{A}^r = \begin{bmatrix} 0 & 0 & 0 & 0 \\ a^r_{k_1} & 0 & 0 & 0 \\ 0 & a^{rh_1}_{k_1k_2} & 0 & 0 \\ a^r_{k_1k_2k_3} & 0 & a^{rh_1h_2}_{k_1k_2k_3} & 0 \end{bmatrix} \tag{3.8}$$

and $\mathbf{X}(\mathbf{v})$ reads

$$\mathbf{X}(\mathbf{v}) = \begin{bmatrix} 1 & 0 & 0 & 0 \\ 0 & \delta^{h_1}_{k_1} & 0 & 0 \\ 0 & 0 & \delta^{h_1}_{k_1}\delta^{h_2}_{k_2} & 0 \\ 0 & 0 & 0 & \delta^{h_1}_{k_1}\delta^{h_2}_{k_2}\delta^{h_3}_{k_3} \end{bmatrix} + \begin{bmatrix} 0 & 0 & 0 & 0 \\ a^r_{k_1} & 0 & 0 & 0 \\ 0 & a^{rh_1}_{k_1k_2} & 0 & 0 \\ a^r_{k_1k_2k_3} & 0 & a^{rh_1h_2}_{k_1k_2k_3} & 0 \end{bmatrix} v^r$$
$$+\frac{1}{2}\begin{bmatrix} 0 & 0 & 0\,0 \\ 0 & 0 & 0\,0 \\ a^{rj_1}_{k_1k_2}a^s_{j_1} & 0 & 0\,0 \\ 0 & a^{rj_1j_2}_{k_1k_2k_3}a^{sh_1}_{j_1j_2} & 0\,0 \end{bmatrix} v_rv_s + \frac{1}{6}\begin{bmatrix} 0 & 0\,0\,0 \\ 0 & 0\,0\,0 \\ 0 & 0\,0\,0 \\ a^{rj_1j_2}_{k_1k_2k_3}a^{sj_3}_{j_1j_2}a^t_{j_3} & 0\,0\,0 \end{bmatrix} v_rv_sv_t. \tag{3.9}$$

It follows that the relations between $\mathbf{F}^0$ and the intrinsic values $\hat{\mathbf{F}}^0$ have the form [see (2.23)]

$$\begin{aligned}
F &= \hat{F}, \\
F_{k_1} &= \hat{F}_{k_1} + a^r_{k_1} \hat{F} v_r, \\
F_{k_1k_2} &= \hat{F}_{k_1k_2} + a^{rh_1}_{k_1k_2} \hat{F}_{h_1} v_r + \tfrac{1}{2} a^{rj_1}_{k_1k_2} a^s_{j_1} \hat{F} v_r v_s, \\
F_{k_1k_2k_3} &= \hat{F}_{k_1k_2k_3} + (a^r_{k_1k_2k_3}\hat{F} + a^{rh_1h_2}_{k_1k_2k_3}\hat{F}_{h_1h_2})\, v_r + \tfrac{1}{2} a^{rj_1j_2}_{k_1k_2k_3} a^{sh_1}_{j_1j_2} \hat{F}_{h_1} v_r v_s \\
&\quad + \tfrac{1}{6} a^{rj_1j_2}_{k_1k_2k_3} a^{sj_3}_{j_1j_2} a^t_{j_3} \hat{F} v_r v_s v_t.
\end{aligned} \tag{3.10}$$

We see that the tensors $F_{k_1\cdots k_j}$ are polynomials in v^i of degree j. Analogous relations hold between $\mathbf{G}^i$ and $\hat{\mathbf{G}}^i$ and between $\mathbf{f}$ and $\hat{\mathbf{f}}$.

The a's are combinations of Kronecker symbols. Thus

$$\begin{aligned}
a^r_{k_1} &= \alpha\delta^r_{k_1}, \\
a^{rh_1}_{k_1k_2} &= \beta\delta^r_{(k_1}\delta^{h_1}_{k_2)} + \gamma\delta^{rh_1}\delta_{k_1k_2}, \\
a^r_{k_1k_2k_3} &= \delta\delta^r_{(k_1}\delta_{k_2k_3)}, \\
a^{rh_1h_2}_{k_1k_2k_3} &= \varepsilon\delta^r_{(k_1}\delta^{h_1}_{k_2}\delta^{h_2}_{k_3)} + \eta\delta^{r(h_1}\delta^{h_2)}_{(k_1}\delta_{k_2k_3)} + \zeta\delta^{h_1h_2}\delta^r_{(k_1}\delta_{k_2k_3)},
\end{aligned} \tag{3.11}$$

where the necessary symmetry with respect to the indices h and k has already been taken into account. The commutability condition (2.26) requires

$$\beta\left(\frac{1}{2}\eta - \zeta\right) = (\varepsilon + \eta + 3\zeta)\,\gamma. \tag{3.12}$$

Upon reflection it is clear that there is a redundance among the coefficients α through ζ. Indeed, the second equation among $(2.6)_1$ reads with $(3.11)_1$

$$\frac{\partial\left(\hat{F}_{k_1} + \alpha\hat{F}v_{k_1}\right)}{\partial t} + \frac{\partial\left((\hat{F}_{k_1} + \alpha\hat{F}v_{k_1})v^i + \hat{G}^i_{k_1} + \alpha\hat{G}^i v_{k_1}\right)}{\partial x_i} = \hat{f}_{k_1} + \alpha\hat{f}v_{k_1}.$$

Therefore α may be chosen as 1 without loss of generality. All this amounts to is a redefinition of $\hat{F}_{k_1}, \hat{G}^i_{k_1}$, and $\hat{f}_{k_1}$ by multiplication with the constant α. This does not change the only property, which so far characterize $\hat{F}_{k_1}, \hat{G}^i_{k_1}, \hat{f}_{k_1}$, namely, their Galilean tensor property. This argument applied to all irreducible parts of the tensors $\hat{F}_{k_1\cdots k_n}, \hat{G}_{k_1\cdots k_n}, \hat{f}_{k_1\cdots k_n}$ allows us to set

$$\alpha = 1, \qquad \beta = 2, \qquad \gamma = 0, \qquad \varepsilon = 3, \qquad \eta = 0, \qquad \zeta = 0 \tag{3.13}$$

without loss of generality. Thus (3.12) is identically satisfied. A little more subtle argument gets rid of the coefficient δ, which can be chosen as zero be redefinition of $\hat{F}_{k_1k_2k_3}$ as $\hat{F}_{k_1k_2k_3} + a^r_{k_1k_2k_3}\hat{F}_r$ and similarly for the fluxes and productions. Thus we obtain

$$\begin{aligned}
F &= \hat{F} \\
F_{k_1} &= \hat{F}_{k_1} + \hat{F} v_{k_1}, \\
F_{k_1k_2} &= \hat{F}_{k_1k_2} + 2\hat{F}_{(k_1} v_{k_2)} + \hat{F} v_{k_1}v_{k_2}, \\
F_{k_1k_2k_3} &= \hat{F}_{k_1k_2k_3} + 3\hat{F}_{(k_1k_2} v_{k_3)} + 3\hat{F}_{(k_1} v_{k_2}v_{k_3)} + \hat{F} v_{k_1}v_{k_2}v_{k_3}
\end{aligned} \tag{3.14}$$

and analogous relations for $\mathbf{G}^i$ and $\mathbf{f}$. There is some simplification, if F and F_{k_1} are mass and momentum density, because in this case $\hat{F}_{k_1}$ is zero and so is $\hat{G}^i$, because there is no non-convective mass flux nor is there an intrinsic momentum density. In this case the decomposition (3.14) is identical to the decompositions (1.3.8) of the kinetic theory or to the decomposition (2.1.6) of extended thermodynamics. The matrix $\mathbf{X}(\mathbf{v})$ reads for this case of $N = 3$

$$\mathbf{X}(\mathbf{v}) = \begin{bmatrix} 1 & 0 & 0 & 0 \\ v_{k_1} & \delta^{h_1}_{k_1} & 0 & 0 \\ v_{k_1}v_{k_2} & 2\delta^{h_1}_{(k_1}v_{k_2)} & \delta^{h_1}_{k_1}\delta^{h_2}_{k_2} & 0 \\ v_{k_1}v_{k_2}v_{k_3} & 3\delta^{h_1}_{(k_1}v_{k_2}v_{k_3)} & 3\delta^{h_1}_{(k_1}\delta^{h_2}_{k_2}v_{k_3)} & \delta^{h_1}_{k_1}\delta^{h_2}_{k_2} \end{bmatrix}. \tag{3.15}$$

3.3 The Matrix X(v) for Arbitrary N

For an arbitrary value of N it can be proved that the matrix $\mathbf{X}(\mathbf{v})$ has the form

$$\begin{bmatrix} 1 & & & & & & \\ v_{k_1} & \delta^{h_1}_{k_1} & & & & & \\ v_{k_1}v_{k_2} & 2\delta^{h_1}_{(k_1}v_{k_2)} & \cdots & & & & \\ v_{k_1}v_{k_2}v_{k_3} & 3\delta^{h_1}_{(k_1}v_{k_2}v_{k_3)} & \cdots & \delta^{h_1}_{k_1}\delta^{h_2}_{k_2}\delta^{h_3}_{k_3} & & & \\ \vdots & \vdots & \vdots & \vdots & & & \\ v_{k_1}v_{k_2}\cdots v_{k_n} & \binom{n}{1}\delta^{h_1}_{(k_1}v_{k_2}\cdots v_{k_n)} & \cdots & \vdots & & & \\ \vdots & \vdots & \vdots & \vdots & & \vdots & \vdots \\ v_{k_1}v_{k_2}\cdots v_{k_N} & \binom{N}{1}\delta^{h_1}_{(k_1}v_{k_2}\cdots v_{k_N)} & \cdots & \binom{N}{3}\delta^{h_1}_{(k_1}\delta^{h_2}_{k_2}\delta^{h_3}_{k_3}v_{k_4}\cdots v_{k_N)} & & \cdots & \delta^{h_1}_{k_1}\delta^{h_2}_{k_2}\cdots\delta^{h_N}_{k_N} \end{bmatrix}. \tag{3.16}$$

The rule by which this matrix is formed is evident by inspection, and we are thus able to decompose tensors of arbitrary rank into velocity dependent and intrinsic parts.

Chapter 7

Waves in Extended Thermodynamics

Waves represent an important probe into the constitutive properties of bodies, because their speeds and attenuations may be measured as well as, in the case of shock waves, the jumps in density, temperature, etc. Such observations permit the calculation of constitutive coefficients or alternatively, when the coefficients are known, they permit the evaluation of the range of validity of the theory.

Extended thermodynamics provides a rich field for the study of waves and, indeed, we have considered waves in previous chapters, viz.

— linear waves in mon-atomic gases and mixtures (Chapters 1 and 3),

— acceleration waves in relativistic gases (Chapter 4), and

— shock waves in the phonon and photon gas (Chapter 5).

This was done in an ad hoc fashion on the previous occasions, but now, having clarified the formal structure of extended thermodynamics in Chapter 6 we are able to approach the subject of wave propagation systematically.

The equations of extended thermodynamics are of balance type, see Paragraph 6.1.1

$$\mathbf{F}^A{}_{,A} = \mathbf{f} \quad \text{with } \mathbf{F}^A = \mathbf{F}^A(\mathbf{u}) \text{ and } \mathbf{f} = \mathbf{f}(\mathbf{u}).$$

It follows that the field equations form a first-order quasi-linear system of the generic form

$$\mathbf{A}^A(\mathbf{u})\mathbf{u}_{,A} = \mathbf{f}(\mathbf{u}) \quad \text{with } \mathbf{A}^A = \frac{\partial \mathbf{F}^A}{\partial \mathbf{u}}.$$

The purpose of this chapter is the study of the waves that this sytem allows to propagate. We shall consider linear waves, acceleration waves, and shock waves. In each case we illustrate the general results for the special system (2.4.3) of extended thermodynamics. We also determine the limits of hyperbolicity for that system.

1 Hyperbolicity and Symmetric Hyperbolic Systems

1.1 Hyperbolicity in the t-Direction

The system

$$\mathbf{A}^A(\mathbf{u})\mathbf{u}_{,A} = \mathbf{f}(\mathbf{u}) \tag{1.1}$$

generally permits the propagation of waves, in particular, of weak discontinuities waves, or acceleration waves, on which the fields $\mathbf{u}$ are continuous but their derivatives are not. The possible

$$\text{speeds } V \text{ and normal jumps } \delta\mathbf{u} = \left[\!\left[\frac{\partial \mathbf{u}}{\partial x^a}\right]\!\right] n_a$$

may be calculated from the eigenvalues λ and right eigenvectors $\mathbf{d}$ of the linear homogeneous system of algebraic equations (see Paragraph 4.1)

$$\left(\mathbf{A}^i n_i - \lambda \mathbf{A}^0\right)\mathbf{d} = 0. \tag{1.2}$$

n_i is the unit vector orthogonal to the wave front. The λ's are called the *characteristic speeds* and the polynomial

$$\det\left(\mathbf{A}^i n_i - \lambda \mathbf{A}^0\right) = 0 \tag{1.3}$$

is called the *characteristic polynomial.*

The system (1.1) is called *hyperbolic in the t-direction,* if it has the following two properties:

- $\det \mathbf{A}^0 \neq 0$, and
- for all unit vectors $\mathbf{n}$ the eigenvalue problem (1.3) admits only real eigenvalues λ and a set of linearly independent right eigenvectors $\mathbf{d}$.

It is well known from linear algebra that the left eigenvector $\mathbf{l}$, to be calculated from

$$\mathbf{l}\left(\mathbf{A}^i n_i - \lambda \mathbf{A}^0\right) = 0, \tag{1.4}$$

may be chosen such that

$$\mathbf{l}^{(i)} \cdot \mathbf{d}^{(j)} = \delta^{ij} \quad \text{for all } i, j = 1, 2, \ldots, n. \tag{1.5}$$

1.2 Symmetric Hyperbolic Systems

The system (1.1) is said to be *symmetric hyperbolic* (in the t-direction), or briefly symmetric, by a definition of Friedrichs, if

- $\mathbf{A}^A$ are symmetric, and
- $\mathbf{A}^0$ is positive definite.

By linear algebra every symmetric system is hyperbolic, but the reverse statement is not true.

Symmetric systems play an important role in extended thermodynamics, because that theory uses hyperbolic equations, and the entropy principle together with the concavity of the entropy density ensures that they form a *symmetric* hyperbolic suystem.

Indeed, as we have shown in Chapter 6, the compatibility of the balance laws

$$\mathbf{F}^A{}_{,A} = \mathbf{f} \quad \text{with } \mathbf{F}^A = \mathbf{F}^A(\mathbf{u}) \text{ and } \mathbf{f} = \mathbf{f}(\mathbf{u}) \tag{1.6}$$

with the entropy principle

$$h^A{}_{,A} \geq 0 \quad \text{and} \quad h^0(\mathbf{u}) \text{ - concave} \tag{1.7}$$

guarantees that the system (1.6) may be written in a symmetric form. We recall this from Paragraph 6.1.4.

The advantage of having symmetric hyperbolic systems lies in the fact that there exist theorems guaranteeing the well–posedness for the Cauchy problem (e.g., see [60]):

Theorem 1 *Each symmetric system, whose initial data are of class $H^s(R^N)$ with $s \geq 4$, has a unique solution $\in H^s(R^N)$ in the neighborhood of the initial manifold, even if the system is not strictly hyperbolic.*[1]

Since time is distinguished from the spatial variables in the definition of hyperbolicity and symmetric hyperbolicity, one might expect difficulties in the relativistic case. Indeed, such difficulties occur and Ruggeri [40] has discussed them. The results of that discussion are thought-provoking but not quite conclusive; therefore we do not introduce these arguments there.

2 Linear Waves

2.1 Plane Harmonic Waves: The Dispersion Relation

In extended thermodynamics the system (1.1) is symmetric hyperbolic so that without loss of generality it may be written in the form

$$\frac{\partial \mathbf{u}}{\partial t} + \mathbf{A}^a(\mathbf{u})\frac{\partial \mathbf{u}}{\partial x^a} = \mathbf{f}(\mathbf{u}). \tag{2.1}$$

If we consider the fields to depend on one space dimension only, we may write

$$\frac{\partial \mathbf{u}}{\partial t} + \mathbf{A}(\mathbf{u})\frac{\partial \mathbf{u}}{\partial x} = \mathbf{f}(\mathbf{u}). \tag{2.2}$$

We linearize by setting

$$\mathbf{u} = \tilde{\mathbf{u}} + \bar{\mathbf{u}}, \tag{2.3}$$

where $\bar{\mathbf{u}}$ is a small perturbation of an equilibrium state $\tilde{\mathbf{u}}$ for which we assume $\mathbf{f}(\tilde{\mathbf{u}}) = 0$. Thus we obtain the linearized equations

$$\frac{\partial \bar{\mathbf{u}}}{\partial t} + \tilde{\mathbf{A}}\frac{\partial \bar{\mathbf{u}}}{\partial x} = \tilde{\mathbf{B}}\bar{\mathbf{u}} \quad \text{where } \tilde{\mathbf{B}} = \left(\frac{\partial \mathbf{f}}{\partial \mathbf{u}}\right)_{\tilde{\mathbf{u}}}, \tag{2.4}$$

and we look for solutions of the form

$$\bar{\mathbf{u}} = \mathbf{w}e^{i(\omega t - kx)} \tag{2.5}$$

representing a plane harmonic wave of real frequency ω and complex wave number $k = k_r + ik_i$ with the complex amplitude $\mathbf{w}$. Equation (2.5) represents a wave travelling in the x-direction.

Substitution of (2.5) into (2.4) provides a homogeneous algebraic linear system of the form

$$\left(\mathbf{1} - z\tilde{\mathbf{A}} + \frac{i}{\omega}\tilde{\mathbf{B}}\right)\mathbf{w} = 0 \tag{2.6}$$

where z stands for k/ω.

[1]Strict hyperbolicity means that all characteristic speeds are different.

For non-trivial solutions the *dispersion relation* must be satisfied

$$\det\left(\mathbf{1} - z\tilde{\mathbf{A}} + \frac{i}{\omega}\tilde{\mathbf{B}}\right) = 0. \tag{2.7}$$

The dispersion relation permits the calculation of the phase velocity v_{ph} and of the attenuation α in terms of the frequency ω

$$v_{\mathrm{ph}} = \frac{\omega}{\mathrm{Re}(k)} = \frac{1}{\mathrm{Re}(z)}, \qquad \alpha = -\mathrm{Im}(k) = -\omega\mathrm{Im}(z). \tag{2.8}$$

For linear stability, $\alpha(\omega)$ must be positive for waves travelling to the right; it is negative for waves travelling to the left and, if we assume the waves to start at time $t = 0$ at $x = 0$, we have $\alpha(\omega)x > 0$. Since the path of the wave is given by $x = \tilde{\lambda}t$ under these circumstances, we may write the condition of linear stability as

$$\alpha(\omega)\tilde{\lambda} > 0. \tag{2.9}$$

2.2 The High Frequency Limit

Our objective in this paragraph is to estimate v_{ph} and α in the limit of high frequency, that is, for $\omega \to \infty$. For this purpose we consider the formal expansion of z and $\mathbf{w}$ in terms of powers of $1/\omega$, viz.

$$z = \sum_{\alpha\geq 0} \frac{z_\alpha}{\omega^\alpha}, \qquad \mathbf{w} = \sum_{\beta\geq 0} \frac{\mathbf{w}_\beta}{\omega^\beta}. \tag{2.10}$$

Insertion of (2.10) into (2.4) provides a recurrence formula (see [65])[2]

$$(\mathbf{1} - z_0\mathbf{A})\mathbf{w}_\alpha - (z_1\mathbf{A} - i\mathbf{B})\mathbf{w}_{\alpha-1} - \sum_{\beta=2}^{\alpha} z_\beta \mathbf{A}\mathbf{w}_{\alpha-\beta} = 0 \qquad (\alpha = 0, 1, 2, ...) \tag{2.11}$$

where $\mathbf{w}_\gamma = 0$ for $\gamma < 0$.

For $\alpha = 0$ we obtain from (2.11)

$$(\mathbf{1} - z_0\mathbf{A})\mathbf{w}_0 = 0, \qquad \text{hence} \qquad z_0 = \frac{1}{\lambda} \quad \text{and} \quad \mathbf{w}_0 = \mathbf{d}, \tag{2.12}$$

where λ and $\mathbf{d}$ are an eigenvalue and the corresponding right eigenvector of $\mathbf{A}$ evaluated at the constant state. We observe that z_0 and $\mathbf{w}_0$ are real.

For $\alpha = 1$ we have, recalling (2.12),

$$\left(\mathbf{1} - \frac{\mathbf{A}}{\lambda}\right)\mathbf{w}_1 - (z_1\mathbf{A} - i\mathbf{B})\mathbf{d} = 0. \tag{2.13}$$

Multiplication by the left eigenvector $\mathbf{l}$ gives with $\mathbf{l}\mathbf{A} = \lambda\mathbf{l}$

$$z_1 = \frac{i}{\lambda}(\mathbf{l}\cdot\mathbf{B}\cdot\mathbf{d}) \tag{2.14}$$

so that z_1 is purely imaginary.

Summarizing we have so far

$$z = \frac{k}{\omega} \approx z_0 + \frac{z_1}{\omega} = \frac{1}{\lambda}\left(1 + \frac{i}{\omega}(\mathbf{l}\cdot\mathbf{B}\cdot\mathbf{d})\right) \quad \text{for } \omega\text{-large} \tag{2.15}$$

[2] For simplicity we omit the tilde that characterizes the constant state.

and hence by (2.8) and (2.12)

$$\lim_{\omega\to\infty} v_{\text{ph}}(\omega) = \lambda \quad \text{and} \quad \lim_{\omega\to\infty} \alpha(\omega)\lambda = -\mathbf{l}\cdot\mathbf{B}\cdot\mathbf{d}. \tag{2.16}$$

The first equation in (2.16) provides the result that the phase velocity coincides, in the limit of high frequency, with the characteristic speeds. We recall that in Paragraph 1.2.3 we have anticipated that result when speaking of the maximum speed of propagation in extended TIP. The second result in (2.16) furnished a condition of linear stability, because by (2.9) for stability we must have

$$\mathbf{l}\cdot\mathbf{B}\cdot\mathbf{d} < 0. \tag{2.17}$$

Later we shall see that (2.17) guarantees non-linear stability as well (see Paragraph 3.2).

2.3 Higher-Order Terms

The exploitation of the recurrence formula (2.11) may be carried further and this has been done in [65], but we shall not present these calculations, which become increasingly difficult. Suffice it to list two of the general results in the form of a theorem.

Theorem 2 *The phase velocity and the attenuation depend on the frequency ω through ω^2, and therefore they are even functions of ω.*

The phase velocity and the attenuation corresponding to an eigenvalue λ are independent of frequency only if the left eigenvector of $\mathbf{A}$ is also left eigenvector of $\mathbf{B}$; in that case we have for all ω

$$v_{\text{ph}} = \lambda, \qquad \alpha = -\frac{\mathbf{l}\cdot\mathbf{B}\cdot\mathbf{d}}{\lambda}. \tag{2.18}$$

2.4 Linear Waves in Extended Thermodynamics

For an example of the foregoing calculations we refer to Chapter 2, specifically to the field equation (2.4.3) for the non-degenerate gas, in which we set $c = 0$ for consistency with the kinetic theory of gases. In a non-inertial frame we assume that $\mathbf{v}$ and $\mathbf{q}$ have only x_1-components and that $t_{<ij>}$ has only diagonal components with $t_{<11>} = \sigma$. All fields are assumed to depend only on x_1. Thus we obtain with $c = \sqrt{\frac{5}{3}\frac{k}{m}T}$ and

$$x = x_1, \qquad v = v_1, \qquad q = q_1, \qquad \sigma = t_{<11>} \tag{2.19}$$

$$\begin{bmatrix} \dot\rho \\ \dot v \\ \dot T \\ \dot\sigma \\ \dot q \end{bmatrix} + \begin{bmatrix} 0 & \rho & 0 & 0 & 0 \\ \frac{3c^2}{5\rho} & 0 & \frac{3c^2}{5T} & -\frac{1}{\rho} & 0 \\ 0 & \frac{2}{9}T(3-5\hat\sigma) & 0 & 0 & \frac{10}{9}\frac{T}{c^2\rho} \\ 0 & c^2\rho\left(\frac{7}{3}\hat\sigma - \frac{4}{5}\right) & 0 & 0 & -\frac{8}{15} \\ \frac{3}{5}c^4\hat\sigma & \frac{16}{5}c^3\rho\hat q & -\frac{3}{10}\frac{c^4\rho}{T}(5\hat\sigma-3) & -\frac{c^2}{5}(5\hat\sigma+3) & 0 \end{bmatrix} \cdot \begin{bmatrix} \frac{\partial\rho}{\partial x} \\ \frac{\partial v}{\partial x} \\ \frac{\partial T}{\partial x} \\ \frac{\partial\sigma}{\partial x} \\ \frac{\partial q}{\partial x} \end{bmatrix} = \begin{bmatrix} 0 \\ 0 \\ 0 \\ -\frac{3}{2\tau_q}\sigma \\ -\frac{1}{\tau_q}q \end{bmatrix} \tag{2.20}$$

We have assumed here that $\tau_q/\tau_t = 3/2$ holds, which is exactly true only for Maxwellian molecules, but it is very nearly true for all mon-atomic gases.

Note that the system (2.20) is not yet linearized. This fact is put in evidence by the occurrence of $\hat{\sigma}$ and $\hat{q}$ in the matrix. These quantities represent stress and heat flux in dimensionless form. We have

$$\hat{\sigma} = \frac{\sigma}{c^2 \rho} \quad \text{and} \quad \hat{q} = \frac{q}{c^3 \rho}. \tag{2.21}$$

The field vector $\mathbf{u}$ of the general theory must be identified as

$$\mathbf{u} = (\rho, v, T, \sigma, q), \tag{2.22}$$

and we linearize in $\bar{\mathbf{u}} = \mathbf{u} - \tilde{\mathbf{u}}$ about

$$\tilde{\mathbf{u}} = (\tilde{\rho}, 0, \tilde{T}, 0, 0). \tag{2.23}$$

The linearized system thus reads

$$\begin{bmatrix} \frac{\partial \bar{\rho}}{\partial t} \\ \frac{\partial \bar{v}}{\partial t} \\ \frac{\partial \bar{T}}{\partial t} \\ \frac{\partial \sigma}{\partial t} \\ \frac{\partial q}{\partial t} \end{bmatrix} + \begin{bmatrix} 0 & \tilde{\rho} & 0 & 0 & 0 \\ \frac{3\tilde{c}^2}{5\tilde{\rho}} & 0 & \frac{3\tilde{c}^2}{5\tilde{T}} & -\frac{1}{\tilde{\rho}} & 0 \\ 0 & \frac{2}{3}\tilde{T} & 0 & 0 & \frac{10}{9}\frac{\tilde{T}}{\tilde{c}^2\tilde{\rho}} \\ 0 & -\frac{4}{5}\tilde{c}^2\tilde{\rho} & 0 & 0 & -\frac{8}{15} \\ 0 & 0 & \frac{9}{10}\frac{\tilde{c}^4\tilde{\rho}}{\tilde{T}} & -\frac{3\tilde{c}^2}{5} & 0 \end{bmatrix} \cdot \begin{bmatrix} \frac{\partial \bar{\rho}}{\partial x} \\ \frac{\partial \bar{v}}{\partial x} \\ \frac{\partial \bar{T}}{\partial x} \\ \frac{\partial \sigma}{\partial x} \\ \frac{\partial q}{\partial x} \end{bmatrix} = \begin{bmatrix} 0 \\ 0 \\ 0 \\ -\frac{3}{2\tau_q}\sigma \\ -\frac{1}{\tau_q}q \end{bmatrix}. \tag{2.24}$$

Comparison with (2.4) identifies the matrices $\tilde{\mathbf{A}}$ and $\tilde{\mathbf{B}}$ and we may thus calculate the eigenvalues of $\tilde{\mathbf{A}}$, that is, the characteristic speeds in equilibrium. They result as

$$\lambda_1 \simeq -1.6503\,\tilde{c}, \qquad \lambda_2 \simeq -0.6297\,\tilde{c}, \qquad \lambda_3 = 0, \qquad \lambda_4 \simeq 0.6297\,\tilde{c}, \qquad \lambda_5 \simeq 1.6503\,\tilde{c}. \tag{2.25}$$

The dispersion relation (2.7) has an explicit form in this case. We obtain with $\Omega = \tilde{\tau}_q \omega$ as a dimensionless frequency

$$9\Omega(z\tilde{c})^4(5i - 6\Omega) + 3(z\tilde{c})^2(52\Omega^2 - 80i\Omega - 25) - 50\Omega^2 + 125i\Omega + 75 = 0. \tag{2.26}$$

From this equation we may calculate $z = k/\omega$ explicitly as a function of ω and hence v_{ph} and α. The Figures 1 and 2 show the results for the "mode" with $\lim_{\omega \to \infty} z = 1/\lambda_5$, i.e., the mode of the fastest wave (see the continuous lines in the figures).

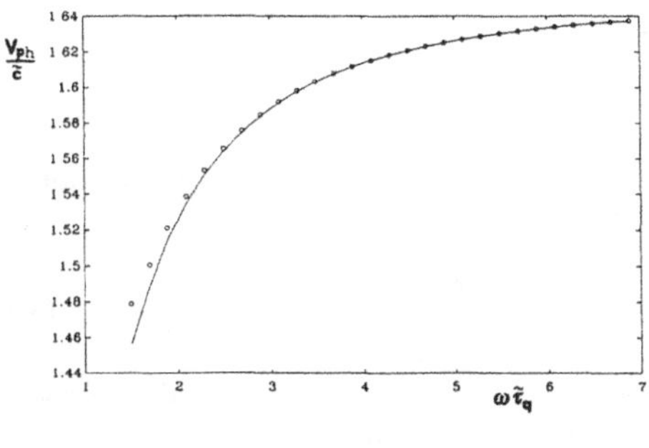

Figure 1. Phase speed as a function of ω.

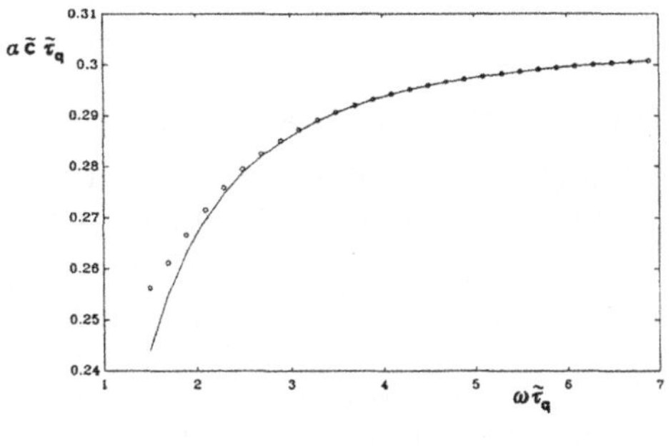

Figure 2. Attenuation as a function of ω.

Rather than solve (2.26) explicitly we may use the expansion (2.10). We have calculated the first four terms for the fastest mode and obtained

$$z\tilde{c} \sim 0.6060 - \frac{0.3046i}{\Omega} + \frac{0.2295}{\Omega^2} + \frac{0.1822i}{\Omega^3}, \tag{2.27}$$

whence follows by (2.8)

$$v_{ph} \simeq \left(1.6503 - \frac{0.6251}{\Omega^2}\right)\tilde{c} \quad \text{and} \quad \alpha \simeq \left(0.3046 - \frac{0.1822}{\Omega^2}\right)\frac{1}{\tilde{\tau}_q \tilde{c}}. \tag{2.28}$$

From (2.16) and (2.28) we obtain

$$\lim_{\omega \to \infty} \alpha(\omega)\lambda \simeq \frac{0.5027}{\tilde{\tau}_q}, \tag{2.29}$$

so extended thermodynamics satisfies the condition of linear stability.

Figures 1 and 2 show as dotted lines the phase speed and attenuation calculated from an expansion like (2.28), but with one more term. Thus we are able to get an idea about the rapid convergence of the series expansion.

3 Hyperbolicity in Extended Thermodynamics

3.1 The Characteristic Polynomial

We refer to the system (2.20), which is the non-linear system of field equations of extended thermodynamics. This system is not quite in the form (2.2) because of the occurrence of the dotted quantities, like $\dot{\rho} = \partial\rho/\partial t + v(\partial\rho/\partial x)$. The matrix $\mathbf{A}$ in (2.2) and the matrix $\overline{\mathbf{A}}$ in (2.20), therefore, are related by the equation

$$\overline{\mathbf{A}} = \mathbf{A} - v\mathbf{1}. \tag{3.1}$$

This matrix has the same eigenvectors as $\mathbf{A}$ and the eigenvalues are related by

$$\overline{\lambda} = \lambda - v \tag{3.2}$$

where λ are the characteristic speeds.

The characteristic polynomial from which the speeds may be calculated has the form

$$\hat{\lambda}\left\{\hat{\lambda}^4+\hat{\lambda}^2\left(\frac{62}{15}\hat{\sigma}-\frac{78}{25}\right)-\frac{96}{25}\hat{q}\hat{\lambda}+\frac{21}{5}\hat{\sigma}^2-\frac{18}{5}\hat{\sigma}+\frac{27}{25}\right\}=0, \tag{3.3}$$

where $\hat{\lambda}$ is defined as $\overline{\lambda}/\tilde{c}$ with $\tilde{c}=\sqrt{\frac{5}{3}\frac{k}{m}\tilde{T}}$.

From (3.3) we conclude that, apart from the trivial solution $\hat{\lambda}=0$, we have four waves with speeds $\hat{\lambda}$ that are roots of the fourth-order algebraic equation

$$\hat{\lambda}^4+\hat{\lambda}^2\left(\frac{62}{15}\hat{\sigma}-\frac{78}{25}\right)-\frac{96}{25}\hat{q}\hat{\lambda}+\frac{21}{5}\hat{\sigma}^2-\frac{18}{5}\hat{\sigma}+\frac{27}{25}=0. \tag{3.4}$$

We see that the speeds depend on the non-equilibrium state characterized by the values $\hat{\sigma}$ and $\hat{q}$. In equilibrium the equation (3.4) degenerates into a biquadratic equation, viz.

$$\hat{\lambda}^4-\frac{78}{25}\hat{\lambda}^2+\frac{27}{25}=0, \tag{3.5}$$

and according to the results of Paragraph 2.2 the corresponding speeds $\hat{\lambda}$ coincide with the phase speeds in the high-frequency limit, viz. by (2.25)

$$\hat{\lambda}\simeq\pm1.6503 \quad \text{and} \quad \hat{\lambda}\simeq\pm0.6297. \tag{3.6}$$

3.2 Region of Hyperbolicity

We see from (3.6) that in equilibrium the characteristic speeds of extended thermodynamics are real, and therefore the system of equations is hyperbolic. Actually we should like extended thermodynamics to be hyperbolic irrespective of how far from equilibrium the state is, and in order to achieve this we have required "convexity and causality", for example, see Paragraph 6.1.2. However, we were forced to make approximations in order to obtain specific results; thus the equations (2.4.3), hence (2.20) are derived from *linear* constitutive relations. It may therefore be suspected that the system (2.20) is hyperbolic only in the neighbourhood of equilibrium, i.e., in the range where the constitutive equations are valid. We proceed to investigate that suspicion. In fact, we shall find it confirmed and shall calculate the *region of hyperbolicity* in the $(\hat{\sigma},\hat{q})$-plane.

For $\hat{\sigma}=0$, $\hat{q}=0$ the system is hyperbolic as we have seen, since the four roots of the characteristic equation (3.4) are real [see (3.6)]. This means that the function on the left-hand side of (3.4) has four points of intersection with the $\hat{\lambda}$-axis for $\hat{\sigma}=0$, $\hat{q}=0$. That function depends on the values of $\hat{\sigma}$ and $\hat{q}$ and hyperbolicity is lost, if two of the four points of intersection merge into one. The mathematical condition for this to happen is that the characteristic equation has a double root μ_1 and two separate roots μ_2,μ_3, i.e., that the equation has the form

$$(\hat{\lambda}-\mu_1)^2(\hat{\lambda}-\mu_2)(\hat{\lambda}-\mu_3)=0. \tag{3.7}$$

Comparing this expression with (3.4) gives four conditions by equating the coefficients of $\hat{\lambda}$, viz.

$$\mu_1-\frac{1}{2}(\mu_2+\mu_3)=0, \tag*{(3.8)$_1$}$$

$$\frac{62}{15}\hat{\sigma}-\frac{78}{25}+\frac{1}{2}(\mu_2+\mu_3)^2+\frac{1}{4}(\mu_2-\mu_3)^2=0, \tag*{(3.8)$_2$}$$

$$\frac{21}{5}\hat{\sigma}^2-\frac{18}{5}\hat{\sigma}+\frac{27}{25}-\frac{1}{4}(\mu_2+\mu_3)^2\mu_2\mu_3=0, \tag*{(3.8)$_3$}$$

$$\frac{96}{25}\hat{q} = \frac{1}{4}(\mu_2 + \mu_3)(\mu_2 - \mu_3)^2. \tag{3.8$_4$}$$

From $(3.8)_{1,2,3}$ we obtain with some algebra

$$\mu_1 = -\frac{1}{15}\sqrt{117 - 155\hat{\sigma} + \sqrt{2}\Delta},$$
$$\mu_{2,3} = \frac{1}{15}\sqrt{117 - 155\hat{\sigma} + \sqrt{2}\Delta} \pm \frac{2}{15}\sqrt{117 - 155\hat{\sigma} - \frac{1}{\sqrt{2}}\Delta} \tag{3.9}$$

where Δ stands for

$$\Delta = \sqrt{47450\hat{\sigma}^2 - 48510\hat{\sigma} + 15957}. \tag{3.10}$$

If (3.9) and (3.10) are inserted into $(3.8)_4$ we obtain $\hat{q}$ as a function of $\hat{\sigma}$. That function—or rather $\hat{\sigma}$ as a function of $\hat{q}$—is shown in Figure 3. The area around $\hat{\sigma} = 0$, $\hat{q} = 0$, which is characterized as the hyperbolicity region, contains the pairs $(\hat{\sigma}, \hat{q})$ for which the characteristic speeds are real and, in fact, distinct.

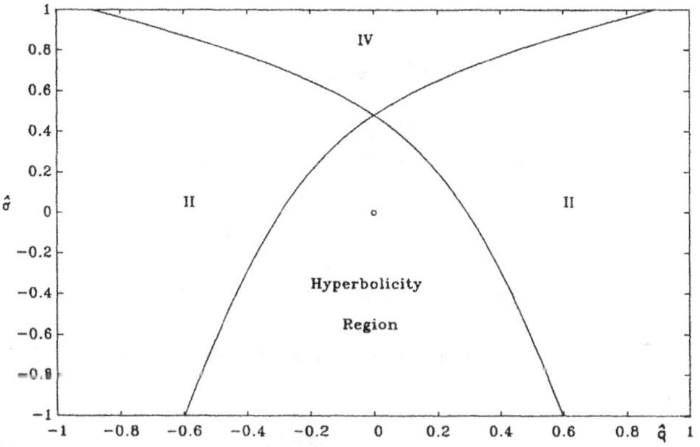

Figure 3. The region of hyperbolicity.

There are two more regions that can be identified in Figure 3. The region denoted by II corresponds to the case in which μ_2 and μ_3 are real, while in region IV all eigenvalues are complex. Needless to say, the system is non-hyperbolic in regions II and IV.

On the boundary of the hyperbolicity region two roots are real and distinct while two others are real and coincident. Thus on that boundary the system, while hyperbolic, is not strictly hyperbolic.

We have constructed the equation of the maximum circle centered in the origin of the $(\hat{\sigma}, \hat{q})$-plane and contained in the region of hyperbolicity (see Figure 4). The radius of that circle is given by

$$r = \sqrt{\hat{q}^2 + \hat{\sigma}^2} = \frac{1}{c^2\rho}\sqrt{\frac{q^2}{c^2} + \sigma^2} \simeq 0.2686, \tag{3.11}$$

it gives an estimate of the range of validity of extended thermodynamics concerning hyperbolicity.

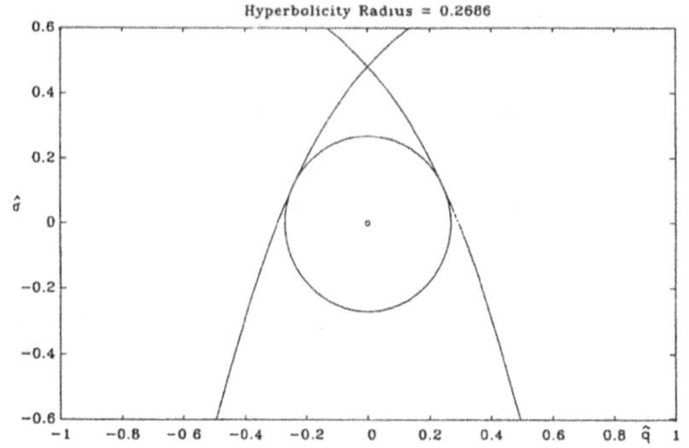

Figure 4. On the validity of extended thermodynamics for non-equilibrium processes.

4 Non-Linear Waves, Discontinuity Waves, or Acceleration Waves

4.1 Amplitude of Discontinuity Waves

We continue with the system (1.1) or (2.1) but we do not linearize now; nor do we consider harmonic waves. Rather we consider a moving surface $\varphi(\mathbf{x}, t) = 0$—the wave front—with unit normal $\mathbf{n}$, which separates the unperturbed field $\tilde{\mathbf{u}}$ in front from the perturbed field behind. The field itself is continuous, but its first derivatives have jumps. We may thus write

$$[\![\mathbf{u}]\!] = 0, \qquad \left[\!\left[\frac{\partial \mathbf{u}}{\partial x^a}\right]\!\right] n_a = \mathbf{\Pi} \neq 0, \tag{4.1}$$

where $\mathbf{\Pi}$ is called the *jump vector*. For the normal speed of the wave front we have

$$V = -\frac{\partial\varphi/\partial t}{|\,\mathrm{grad}\varphi\,|}. \tag{4.2}$$

For the investigation of the motion and the evolution of the wave front we assume that the unperturbed field is an equilibrium state with $\mathbf{f}(\mathbf{u}) = 0$. The more general case can be treated and has been treated in [66], [67], and [68].

The calculation proceeds in two steps: identification of speed and amplitude and determination of the temporal behaviour of the amplitude.

First of all we subtract the equations (2.1) on the two sides of the front and obtain with

$$\left[\!\left[\frac{\partial \mathbf{u}}{\partial t}\right]\!\right] = -V\,\mathbf{\Pi}, \qquad \left[\!\left[\frac{\partial \mathbf{u}}{\partial x^a}\right]\!\right] n_a = \mathbf{\Pi}: \quad (\mathbf{A}^a n_a - V\mathbf{1})\,\mathbf{\Pi} = 0. \tag{4.3}$$

Thus the possible speeds of the wave are given by the eigenvalues of $\mathbf{A}^a n_a$ and the jump vector is proportional to the right eigenvector of that matrix. We have

$$V = \tilde{\lambda} \equiv \lambda(\tilde{\mathbf{u}}) \quad \text{and} \quad \mathbf{\Pi} = \Pi\tilde{\mathbf{d}} \equiv \Pi\mathbf{d}(\tilde{\mathbf{u}}). \tag{4.4}$$

Π is called the *amplitude of the wave.*

Second, we calculate the evolution of the amplitude in time and, in doing this, we restrict the attention to the one-dimensional case where the wave moves in the x-direction. We obtain equation (4.6) below by a quick calculation:

We apply a left eigenvector $\mathbf{l}$ to the system (2.1) and obtain with $\mathbf{u}_t = \partial\mathbf{u}/\partial t$, $\mathbf{u}_x = \partial\mathbf{u}/\partial x$

$$\mathbf{l}\cdot(\mathbf{u}_t + \lambda\mathbf{u}_x) = \mathbf{l}\cdot\mathbf{f}.$$

This equation is differentiated with respect to x to give

$$\frac{\partial\mathbf{l}}{\partial\mathbf{u}}\cdot\mathbf{u}_x\,(\mathbf{u}_t + \lambda\mathbf{u}_x) + \mathbf{l}\cdot\left\{\mathbf{u}_{tx} + \lambda\mathbf{u}_{xx} + \mathbf{u}_x\frac{\partial\lambda}{\partial\mathbf{u}}\cdot\mathbf{u}_x\right\} - \frac{\partial(\mathbf{l}\cdot\mathbf{f})}{\partial\mathbf{u}}\cdot\mathbf{u}_x = 0.$$

We form the difference of this expression on the two sides of the front. The first term drops out, because by $(4.3)_{1,2}$ we have $[\![\mathbf{u}_t + \lambda\mathbf{u}_x]\!] = 0$ and $\mathbf{u}_t = 0$, $\mathbf{u}_x = 0$ before the front. What remains may be written as (with $d/dt = \partial/\partial t + \lambda(\partial/\partial x)$)

$$\mathbf{l}\cdot\frac{d\,[\![\mathbf{u}_x]\!]}{dt} + \mathbf{l}\cdot[\![\mathbf{u}_x]\!]\frac{\partial\lambda}{\partial\mathbf{u}}\cdot[\![\mathbf{u}_x]\!] - \frac{\partial(\mathbf{l}\cdot\mathbf{f})}{\partial\mathbf{u}}\cdot[\![\mathbf{u}_x]\!] = 0$$

or by $(4.3)_1$ and $(4.4)_2$ and with $\mathbf{f}(\mathbf{u}) = 0$ before the front

$$\frac{d\Pi}{dt} + \frac{\partial\lambda}{\partial\mathbf{u}}\cdot\mathbf{d}\,\Pi^2 - \mathbf{l}\cdot\frac{\partial\mathbf{f}}{\partial\mathbf{u}}\cdot\mathbf{d}\,\Pi = 0. \tag{4.5}$$

Equation (4.5) is a Bernoulli equation for the amplitude Π and we write it as

$$\frac{d\Pi}{dt} + \alpha\Pi^2 + \beta\Pi = 0 \quad \text{with } \alpha = \frac{\partial\lambda}{\partial\mathbf{u}}\cdot\mathbf{d},\ \ \beta = -\mathbf{l}\cdot\frac{\partial\mathbf{f}}{\partial\mathbf{u}}\cdot\mathbf{d}. \tag{4.6}$$

The coefficients α and β are constants in the present simple case where α and β must be evaluated for the constant state $\tilde{\mathbf{u}}$ in front of the wave. The Bernoulli equation can therefore be solved to give

$$\Pi(t) = \frac{\Pi(0)e^{-\beta t}}{1 - \Pi(0)\frac{\alpha}{\beta}\left(e^{-\beta t} - 1\right)}. \tag{4.7}$$

In the general case that the wave propagates into a region where the unperturbed field is not constant, α and β depend on time [68]. In that case the Bernoulli equation cannot be solved so easily. An instructive example is given by Ruggeri [68] and Muracchini and Seccia [69] for wave propagation in the gravitational atmosphere. Applications to continuum mechanics are given by Chen [70].

4.2 Growth and Decay

It is obvious from (4.6) and (4.7) that for a semi-linear system, where $\mathbf{A}^A$ and hence λ, $\mathbf{l}$, and $\mathbf{d}$ are constants, we have $\alpha = 0$, and therefore the amplitude will either grow or decay exponentially, depending on the sign of β.

Even for a non-linear system α may vanish, namely, if $\partial\lambda/\partial\mathbf{u}\cdot\mathbf{d} = 0$ holds. If this is not the case; i.e., if

$$\frac{\partial\lambda}{\partial\mathbf{u}}\cdot\mathbf{d} \neq 0 \tag{4.8}$$

for all eigenvalues, we say that the system is *genuinely non-linear*. We shall proceed on the assumption that (4.8) holds. Thus α is not zero and, without loss of generality, we assume $\alpha > 0$.

The equation (4.7) lends itself for a characterization of non-linear stability. Stability demands that $\Pi(t)$ is bounded for all times and inspection of (4.7) shows that this requires

$$1 + \Pi(0)\frac{\alpha}{\beta} > 0 \quad \text{and} \quad \beta > 0, \qquad \text{hence} \qquad \Pi(0) > -\frac{\beta}{\alpha}. \tag{4.9}$$

By $(4.6)_2$ and since $\mathbf{f}(\mathbf{u}) = 0$ in front of the wave, the condition $(4.9)_2$ provides a restriction on the material. If may be written in the form

$$\beta = -\mathbf{l} \cdot \frac{\partial \mathbf{f}}{\partial \mathbf{u}} \cdot \mathbf{d} > 0 \quad \text{or by (2.4)} \quad \beta = -\mathbf{l} \cdot \mathbf{B} \cdot \mathbf{d} > 0, \tag{4.10}$$

which confirms the earlier remark—made in Paragraph 2.2—that the linear stability condition guarantees non-linear stability as well. At least this is true, if the discontinuity wave travels into a region in equilibrium.

The condition $(4.9)_3$ provides a condition on the initial amplitude of the discontinuity. We may say that, if the amplitude $\mid \Pi(0) \mid$ is too big, the attenuation does not succeed to dissipate the discontinuity and, therefore, stability is lost. We proceed to investigate that case.

The denominator of (4.7) may become zero at some critical time t_{cr} between 0 and ∞, viz.

$$t_{\text{cr}} = \frac{1}{\beta} \ln \frac{1}{1 + (\beta/\alpha) \cdot (1/\Pi(0))}. \tag{4.11}$$

Since we must have $0 < t_{\text{cr}} < \infty$, this can only happen for

$$\Pi(0) < -\frac{\beta}{\alpha}. \tag{4.12}$$

The physical significance of this instability is clear: if the initial discontinuity in the derivatives is too strong, it cannot be damped; instead it grows to infinity and thus the acceleration wave develops into a shock wave.

4.3 Evolution of Amplitude in Extended Thermodynamics

We wish to calculate the amplitude $\Pi(t)$ from (4.7) for extended thermodynamics of 13 fields, the theory whose system of equations is given by (2.20), if we restrict attention to one space dimension. We do not linearize now, but, as before, we consider the gas before the wave to be in equilibrium and we shall assume that $\tilde{\mathbf{v}} = 0$ holds. We must calculate the coefficients α and β defined in (4.6); that is, we need to calculate the eigenvalues λ and the right and left eigenvectors $\mathbf{d}$ and $\mathbf{l}$ of the matrix $\mathbf{A}$ in (2.24). Fortunately the eigenvectors $\mathbf{d}$ and $\mathbf{l}$ may be evaluated in the equilibrium state; that is, they are eigenvalues of the matrix of the linearized system (2.24). An easy though cumbersome calculation provides

$$\mathbf{d} = \begin{bmatrix} 1 \\ c\frac{\hat{\lambda}}{\rho} \\ \frac{T}{27\rho}\left(25\hat{\lambda}^2 - 27\right) \\ -\frac{4c^2\hat{\lambda}^2}{9} \\ \frac{c^3\hat{\lambda}}{6}\left(5\hat{\lambda}^2 - 9\right) \end{bmatrix}, \quad \mathbf{l} = a \begin{bmatrix} \frac{1}{25}c^3\left(25\hat{\lambda}^2 - 33\right) \\ \frac{1}{15}c^2\hat{\lambda}\rho\left(25\hat{\lambda}^2 - 33\right) \\ \frac{1}{25T}c^3\rho\left(25\hat{\lambda}^2 + 12\right) \\ -\frac{1}{3}c\left(5\hat{\lambda}^2 - 3\right) \\ 2\hat{\lambda} \end{bmatrix}^T \quad \text{where } a = \frac{5}{4c^3\left(13\hat{\lambda}^2 - 9\right)}. \tag{4.13}$$

a was calculated from the requirement that $\mathbf{l} \cdot \mathbf{d} = 1$ must hold. We have given the value 1 to the first component of $\mathbf{d}$; therefore, the amplitude of the discontinuity wave is

$$\Pi = \delta\rho \quad \text{and hence} \quad \delta\mathbf{u} = \delta\rho\,\mathbf{d} \tag{4.14}$$

where δ stands for $[\![\partial/\partial x]\!]$, i.e., the jump of the normal derivative.

We calculate the coefficient α from $(4.6)_2$

$$\alpha = \frac{\partial \lambda}{\partial \mathbf{u}} \cdot \mathbf{d} = \frac{1}{\delta \rho} \frac{\partial \lambda}{\partial \mathbf{u}} \cdot \delta \mathbf{u} = \frac{\delta \lambda}{\delta \rho}, \tag{4.15}$$

and this must be evaluated in equilibrium. $\delta\lambda$ follows from the characteristic equation (3.4) by differentiation. We must take into account that

$$\lambda = c\hat{\lambda} + v \tag{4.16}$$

and obtain

$$\delta\lambda = \frac{c}{2T}\hat{\lambda}\delta T + c\delta\hat{\lambda} + \delta v \quad \text{or with (3.4)} \tag{4.17}$$

$$\delta\lambda = \delta v + \frac{c}{2T}\hat{\lambda}\delta T + \frac{1}{c\rho}\frac{18 - \frac{62}{5}\hat{\lambda}^2}{20\hat{\lambda}^3 - \frac{156}{5}\hat{\lambda}}\delta\sigma + \frac{1}{c^2\rho}\frac{96}{100\hat{\lambda}^2 - 156}\delta q. \tag{4.18}$$

From (4.14) and $(4.13)_1$ we conclude

$$\delta v = c\frac{\hat{\lambda}}{\rho}\delta\rho, \qquad \delta T = \frac{T}{\rho}\left(\frac{25}{27}\hat{\lambda}^2 - 1\right)\delta\rho, \qquad \text{etc.,} \tag{4.19}$$

and thus after a little calculation

$$\alpha = \frac{c}{\rho}\hat{\lambda}\frac{625\hat{\lambda}^4 + 1400\hat{\lambda}^2 - 3537}{54\left(25\hat{\lambda}^2 - 39\right)} \simeq 5.16061\frac{c}{\rho} \tag{4.20}$$

where $(4.20)_2$ refers to the "fast mode" with $\hat{\lambda} = 1.6503$ [see (3.6)].

Next we calculate β from $(4.6)_3$. Inspection of (2.20) shows that we have in equilibrium

$$\frac{\partial \mathbf{f}}{\partial \mathbf{u}} = \begin{bmatrix} 0 & 0 & 0 & 0 & 0 \\ 0 & 0 & 0 & 0 & 0 \\ 0 & 0 & 0 & 0 & 0 \\ 0 & 0 & 0 & -\frac{3}{2}\frac{1}{\tau_q} & 0 \\ 0 & 0 & 0 & 0 & -\frac{1}{\tau_q} \end{bmatrix}. \tag{4.21}$$

Multiplying this from left and right by $\mathbf{l}$ and $\mathbf{d}$, respectively, we obtain from (4.13)

$$\beta = \frac{1}{\tau_q}\frac{5}{4}\frac{5\hat{\lambda}^3 - 3}{13\hat{\lambda}^2 - 9} \simeq 0.5027\frac{1}{\tau_q}. \tag{4.22}$$

Note that $\beta\tau_q = 0.5027$ agrees with the high frequency limit of the attenuation of the "fast" harmonic wave that was calculated in (2.29). This fact reflects the general result about the equivalence of the criteria for linear and non-linear stability that we commented on in Paragraph 4.2 [see (4.10) and (2.17)].

If α and β from (4.20) and (4.22) are introduced into the evolution equation (4.7) of the jump—here, by (4.14), the jump $\delta\rho$ of the density gradient—we obtain

$$\delta\rho(t) = \frac{\delta\rho(0)e^{-0.5027\tau}}{1 - (\delta\rho(0)/\rho)c\tau_q 10.2658\left(e^{-0.5027\tau} - 1\right)} \quad \text{where } \tau = \frac{t}{\tau_q}. \tag{4.23}$$

We conclude that the discontinuity wave grows into a shock wave under the condition

$$\delta\rho(0) < -0.0974\frac{\rho}{c\tau_q}. \tag{4.24}$$

Introducing the jump in acceleration

$$\left[\left[\frac{\partial v}{\partial t}\right]\right] = -\lambda\delta v = -\lambda c\frac{\hat{\lambda}}{\rho}\delta\rho,$$

we obtain from (4.24)

$$\left[\left[\frac{\partial v}{\partial t}\right]\right]_{t=0} > 0.265\frac{c}{\tau_q}.$$

Under normal conditions this is an enormously big acceleration, which we may be sure will never be reached. If it were reached and surpassed by a factor 2 (say)—so that $\delta\rho(0)$ were equal to $-0.1948\frac{\rho}{c\tau_q}$—the acceleration wave would develop into a shock wave in the time

$$t_{cr} = 1.315\tau_q,$$

which is a very short time indeed under normal conditions.

5 Non-Linear Waves and Shock Waves

5.1 Weak Solutions

If the field $\mathbf{u}$ experiences a jump across the wave front we have a *shock wave.* Shock waves are possible only for systems of balance laws and they are a particular class of the so-called *weak solutions.* We recall the definition of weak solutions for systems of balance laws of the type

$$\mathbf{F}^A{}_{,A} = \mathbf{f} \quad \text{with } \mathbf{F}^A = \mathbf{F}^A(\mathbf{u}),\ \mathbf{f} = \mathbf{f}(\mathbf{u}). \tag{5.1}$$

This involves the test function $\Phi(x^A)$ having support in a four-dimensional domain $\mathcal{C}$ of space-time. We multiply (5.1) by Φ and integrate over $\mathcal{C}$

$$\int_{\mathcal{C}} \Phi\left(\mathbf{F}^A{}_{,A} - \mathbf{f}\right) d\mathcal{C} = 0 \quad \text{or} \quad \int_{\mathcal{C}} \left(\Phi\mathbf{F}^A\right)_{,A} d\mathcal{C} - \int_{\mathcal{C}} \left(\mathbf{F}^A\Phi_{,A} + \mathbf{f}\Phi\right) d\mathcal{C} = 0.$$

If Σ is the three-dimensional surface of $\mathcal{C}$ with the normal n_A, we obtain

$$\int_{\Sigma} n_A\Phi\mathbf{F}^A\, d\Sigma - \int_{\mathcal{C}} \left(\mathbf{F}^A\Phi_{,A} + \mathbf{f}\Phi\right) d\mathcal{C} = 0. \tag{5.2}$$

The first integral vanishes since $\Phi = 0$ on Σ and we have

$$\int_{\mathcal{C}} \left(\mathbf{F}^A\Phi_{,A} + \mathbf{f}\Phi\right) d\mathcal{C} = 0. \tag{5.3}$$

A field $\mathbf{u}(x^A)$ is said to be a *weak solution* of (5.1), if it satisfies (5.3) for every test function Φ.

The advantage of weak solutions lies in the fact that, in contrast to classical solutions, no differentiability conditions are required. Obviously all classical solutions are also weak solutions, while the reverse is not true.

5.2 Rankine–Hugoniot Equations

We consider a shock front with the normal ν_A as separating two regular solutions $\mathbf{u}_\pm(x^B)$ on the right and left, respectively, and we proceed to prove that the pair $(\mathbf{u}_+(x^B), \mathbf{u}_-(x^B))$ is a weak solution of (5.1), if and only if across the shock front it satisfies the *Rankine–Hugoniot equations*, viz.

$$\left[\!\left[\mathbf{F}^A(\mathbf{u})\right]\!\right] \nu_A = 0. \tag{5.4}$$

We refer to Figure 5 and write (5.2) for both subdomains with $\mathbf{F}^A_\pm = \mathbf{F}^A(\mathbf{u}_\pm)$

$$\int_{\Sigma^\pm \bigcup \sigma} \nu_A^\pm \Phi \mathbf{F}^A_\pm \, d\Sigma - \int_{\mathcal{C}^\pm} \left(\mathbf{F}^A \Phi,_A + \mathbf{f}\Phi\right) d\mathcal{C} = 0.$$

As Φ vanishes on Σ, we may write

$$\int_\sigma \nu_A^\pm \Phi \mathbf{F}^A_\pm \, d\Sigma - \int_{\mathcal{C}^\pm} \left(\mathbf{F}^A \Phi,_A + \mathbf{f}\Phi\right) d\mathcal{C} = 0. \tag{5.5}$$

We add those two equations to obtain with $\nu_A = \nu_A^+ = -\nu_A^-$

$$\int_\sigma \nu_A \left(\mathbf{F}^A_+ - \mathbf{F}^A_-\right) \Phi \, d\sigma - \int_{\mathcal{C}} \left(\mathbf{F}^A \Phi,_A + \mathbf{f}\Phi\right) d\mathcal{C} = 0. \tag{5.6}$$

The second integral vanishes, since $\mathbf{u}$ is a weak solution in $\mathcal{C}$, and we conclude the proof of (5.4), by writing

$$\nu_A \left(\mathbf{F}^A_+ - \mathbf{F}^A_-\right) = 0,$$

which follows from (5.6) since that relation holds for all domains $\mathcal{C}$.

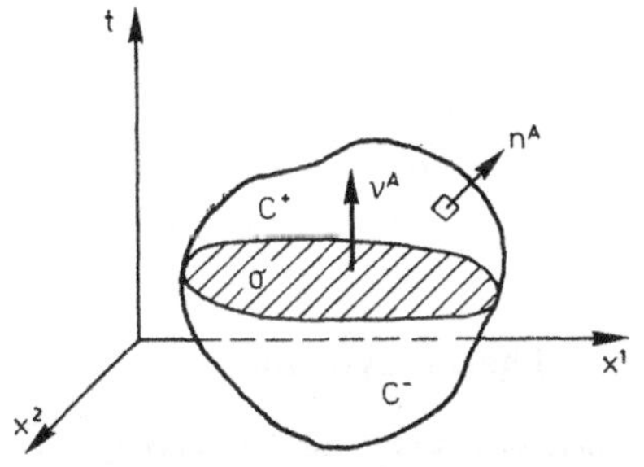

Figure 5. On the derivation of the Rankine–Hugoniot equations.

Therefore the Rankine–Hugoniot equations constitute the compatibility conditions for the existence of shocks in systems of balance laws of the type (5.1).

We now introduce more familiar notation. Let $\varphi(x^A)$ be the shock front, s its normal velocity, and $\mathbf{n}$ its unity normal vector in space so that we have

$$\nu^0 = -s \mid \mathrm{grad}\varphi \mid, \qquad \nu^a = n^a \mid \mathrm{grad}\varphi \mid . \tag{5.7}$$

Thus the Rankine–Hugoniot equations read

$$-\, s \left[\!\left[\mathbf{F}^o\right]\!\right] + \left[\!\left[\mathbf{F}^a\right]\!\right] n_a = 0. \tag{5.8}$$

If the unperturbed field $\mathbf{u}_+$ in front of the shock is known and we consider plane shocks with $\mathbf{n} = \text{const.}$, the Rankine–Hugoniot equations furnish a system of n equations for the $n+1$ unknowns $\mathbf{u}_-$ and s. Thus *one* among the $(n+1)$-tuple $(\mathbf{u}_-, s)$ may be chosen as the *shock parameter*, that is, the quantity that characterizes the *strength* of the shock, and that one quantity may be the speed s, or any components of $\mathbf{u}_-$, or a combination of these; generically we denote the shock parameter by μ.

In a schematic way we may represent the values $\mathbf{u}_-$ for given values $\mathbf{u}_+$ as a function of s as in Figure 6. The figure also indicates that *shock solutions may be seen as bifurcating from the trivial solution* $\mathbf{u}_+ = \mathbf{u}_-$ of (5.8) at the points where s is equal to one of the characteristic speeds λ. We proceed to prove this observation.

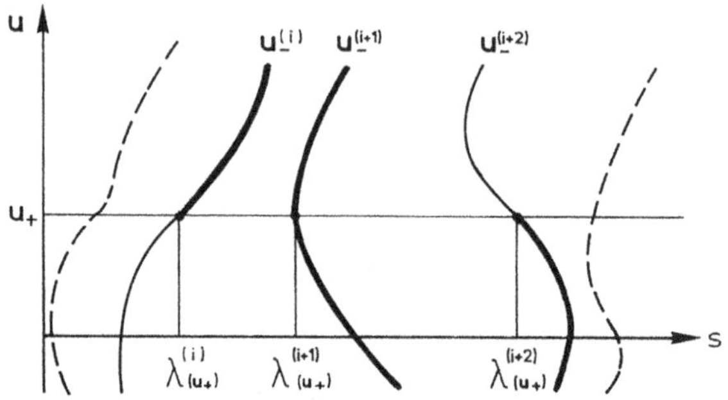

Figure 6. Shocks as bifurcating branches of the trivial solution with s as shock parameter.

We introduce the mapping

$$\boldsymbol{\Psi}_s(\mathbf{u}) = -s\mathbf{F}^o(\mathbf{u}) + \mathbf{F}^i(\mathbf{u})n_i$$

and observe that, by the Rankine–Hugoniot equations, $\Psi = \Psi(\mathbf{u})$ must be locally non-invertible. In fact by (5.8) we have $\boldsymbol{\Psi}_s(\mathbf{u}_-) = \boldsymbol{\Psi}_s(\mathbf{u}_+)$, and therefore $\det\partial\Psi_s/\partial\mathbf{u} = 0$ must hold at $\mathbf{u}_- = \mathbf{u}_+$. On the other hand, the Jacobian of the transformation $\Psi_s = \Psi_s(\mathbf{u})$, viz. (with $\mathbf{A}^A = \partial\mathbf{F}^A/\partial\mathbf{u}$)

$$\det\left(\frac{\partial\boldsymbol{\Psi}_s}{\partial\mathbf{u}}\right) = \det\left(\mathbf{A}^i n_i - s\mathbf{A}^o\right)$$

becomes singular for $s = \lambda(\mathbf{u})$ [see (1.3)]. Therefore the local non-invertibility occurs when s is equal to the characteristic speeds.

5.3 Shocks in Extended Thermodynamics

The equations of extended thermodynamics are of balance type as put in evidence by the system (2.1.2). We may therefore write Rankine–Hugoniot equations. We do this after using (2.1.6), (2.2.45), and $F(z) = \frac{k}{m}z$ in the system (2.1.2). Also we simplify that system by investigating the one-dimensional case with

$$x = x_1, \qquad v = v_1, \qquad q = q_1, \qquad \sigma = t_{<11>}$$

and by introducing the relative shock speed

$$u = v - s. \tag{5.9}$$

A little calculation provides the following explicit form of the Rankine–Hugoniot equations

$$\begin{aligned}
[\![\rho u]\!] &= 0,\\
\left[\!\!\left[\rho u^2 + p - \sigma\right]\!\!\right] &= 0,\\
\left[\!\!\left[2q + 5pu - 2\sigma u + \rho u^3\right]\!\!\right] &= 0,\\
\left[\!\!\left[\frac{8}{15}q + \frac{4}{3}pu - \frac{7}{3}\sigma u + \frac{2}{3}\rho u^3\right]\!\!\right] &= 0,\\
\left[\!\!\left[\rho u^4 + 5\frac{p^2}{\rho} - 7\frac{p\sigma}{\rho} + \frac{32}{5}uq + u^2(8p - 5\sigma)\right]\!\!\right] &= 0.
\end{aligned} \tag{5.10}$$

Before the shock we assume an equilibrium state; this is disturbed by the passage of the shock. Thus we have

$$\mathbf{u}_+ = (\rho_+, 0, T_+, 0, 0), \qquad \mathbf{u}_- = (\rho_-, v_-, T_-, \sigma_-, q_-) \quad \text{or}$$
$$\mathbf{u}_+ = (\rho_+, 0, p_+, 0, 0), \qquad \mathbf{u}_- = (\rho_-, v_-, p_-, \sigma_-, q_-). \tag{5.11}$$

For a simplified notation we introduce dimensionless fields behind the shock, viz.

$$r = \frac{\rho_-}{\rho_+}, \qquad w = \frac{u_-}{u_+}, \qquad \pi = \frac{p_-}{p_+}, \qquad \tau = \frac{\sigma_-}{p_+}, \qquad \chi = \frac{q_-}{p_+ c_+} \tag{5.12}$$

where $c_+ = \sqrt{\frac{5}{3}\frac{k}{m}T_+} = \sqrt{\frac{5}{3}\frac{p_+}{\rho_+}}$. Also we define a dimensionless speed of the shock, the Mach number

$$M_+ = \frac{s}{c_+}. \tag{5.13}$$

We choose M_+, that is, s as the shock parameter and obtain, after an easy though cumbersome calculation (see [71])

$$\begin{aligned}
r &= \frac{1}{w},\\
\pi &= 1 - \frac{5}{27}M_+^2\frac{(w-1)(7w-2)}{w},\\
\tau &= \frac{10}{27}M_+^2\frac{w^2-1}{w},\\
\chi &= -\frac{5}{18}M_+(w-1)\left(10M_+^2 w - 5M_+^2 - 9\right)
\end{aligned}$$

where $w = w(M_+)$ has to be inserted. This function results from solving the cubic equation

$$3470M_+^4 w^3 - \left(3550M_+^2 + 4212\right)M_+^2 w^2 + \left(715M_+^4 + 2106M_+^2 + 729\right)w + 40M_+^4 = 0. \tag{5.14}$$

By (5.9) and $p = \frac{k}{m}\rho T$ it is then also possible to calculate

$$V = \frac{v_-}{c_+} = M_+(1-w) \quad \text{and} \quad \Theta = \frac{T_-}{T_+} = \pi w. \tag{5.15}$$

We proceed to discuss some properties of the equations (5.14) through (5.16), which may be read off by inspection, such as that $w = 1$ corresponds to the continuous solution $\mathbf{u}_- = \mathbf{u}_+$, the null-shock.

In accordance with the general theory presented in Paragraph 5.2 the shocks are bifurcated branches of the continuous solution occurring when s approaches the characteristic speed of the system, and, indeed, for $w = 1$ (5.15) reduces to the equation

$$M_+^4 - \frac{78}{25}M_+^2 + \frac{27}{25} = 0,$$

which is identical to (3.5). From that equation we have calculated the characteristic speeds (3.6), and now we conclude that we have two families of shocks propagating to the right; they bifurcate from the null-shock when M_+ approaches [cf. (3.6)]

$$M_+^{(S)} = 0.6297 \quad \text{and} \quad M_+^{(F)} = 1.6503. \tag{5.16}$$

These are called slow shocks and fast shocks, respectively.

Another feature of the shocks, easily appreciated from $(5.14)_{3,4}$, is that behind the shock we always have a non-equilibrium state, because τ and χ are unequal to zero except in the null shock, where $w = 1$. This result must be contrasted with the classical one of the Navier-Stokes-Fourier theory. In that theory we have *shock structure* solutions in which two states—the ones before and behind the shock—are always in equilibrium, and they are connected by a continuous but very steep non-equilibrium process.

There exists an upper bound for the ratio r of densities ρ_-/ρ_+. This can be read off from (5.15), which may be written as a biquadratic equation for M_+, viz.

$$5M_+^4\left(694w^3 - 710w^2 + 143w + 8\right) - 2106M_+^2 w(2w-1) + 729w = 0. \tag{5.17}$$

According to this equation M_+ is a monotonically decreasing function of w, i.e., by $(5.14)_2$ an increasing function of r. Thus the biggest r corresponds to $M_+ \to \infty$ and then equation (5.19) can only be satisfied, if the coefficient of M_+^4, viz. $694w^3 - 710w^2 + 143w + 8$ is zero. This happens when $w = 0.7099$, that is, for

$$r_{\max} = \left.\frac{\rho_-}{\rho_+}\right|_{\max} = 1.4086. \tag{5.18}$$

We note that the corresponding result is $r_{\max} = 4$, if the mon-atomic gas is considered as an Eulerian fluid (e.g., see [72]), so here extended thermodynamics is in a marked contrast to the classical theory.

We calculate the two positive solutions $M_+ = M_+(w)$ of (5.18) corresponding to the slow and fast shocks [see (5.17)]. These solutions are introduced into the equations (5.14), (5.16) and thus provide

$$w = \frac{\rho_+}{\rho_-}, \quad \Theta = \frac{T_-}{T_+}, \quad \tau = \frac{\sigma_-}{p_+}, \quad \chi = \frac{q_-}{p_+c_+}, \quad V = \frac{v_-}{c_+} \tag{5.19}$$

as functions of s or M_+. These functions have been numerically calculated and they are represented in Figure 7 through Figure 11. The dashed lines correspond to the fast shock, while the solid lines correspond to the slow shock. All curves for the slow shock have been broken off at $M_+ = 0.908$; this is the value where w reaches the value 0.7099, i.e., where the fast shock creates the maximum jump in density.

Not all parts of these plots are physically relevant, because we have selection rules that imply that some parts are unphysical although they are perfectly valid solutions of the Rankine–Hugoniot equations. We proceed to discuss the selection criteria, of which there are two essentially equivalent ones. Afterward we return to the curves of Figure 7 through Figure 11 and discuss their admissible parts.

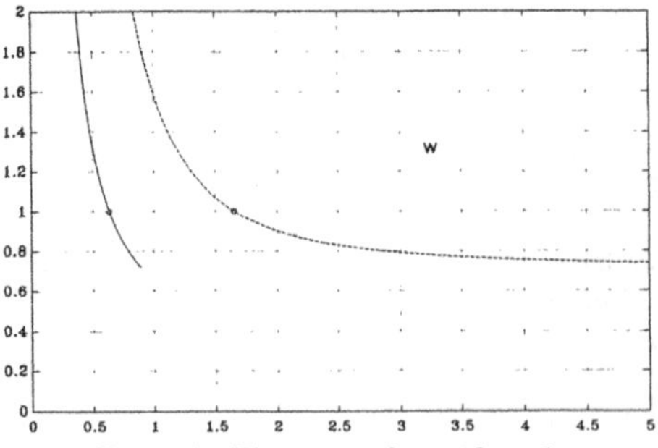

Figure 7. The ratio of specific volumes as a function of M_+.

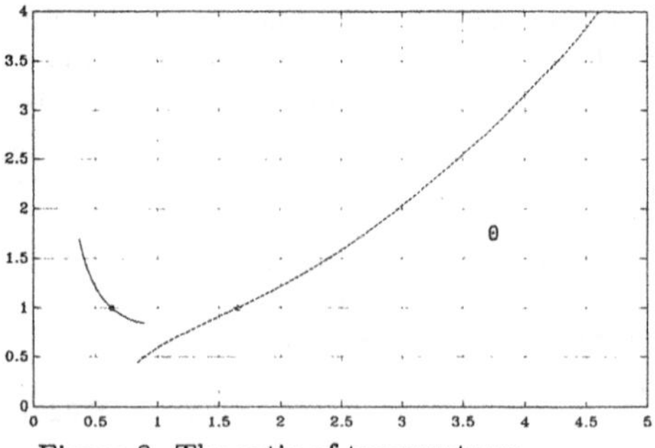

Figure 8. The ratio of temperatures as a function of M_+.

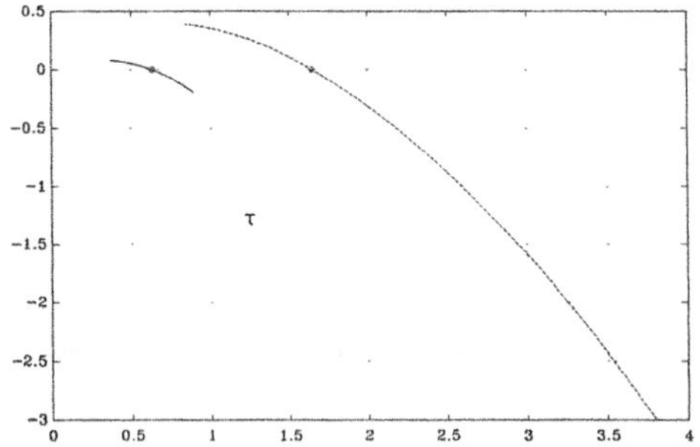

Figure 9. The stress behind the shock.

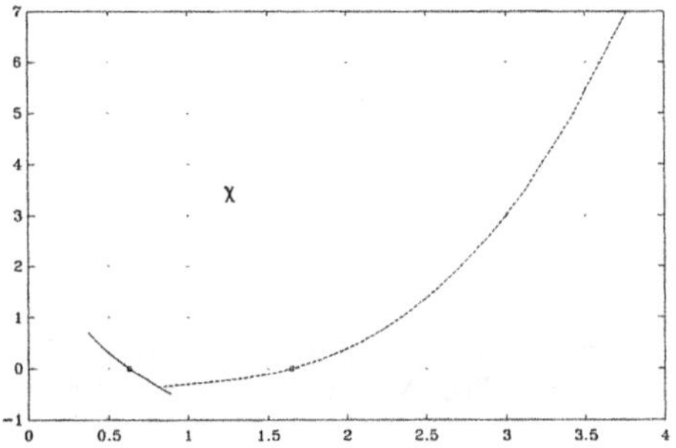

Figure 10. The heat flux behind the shock.

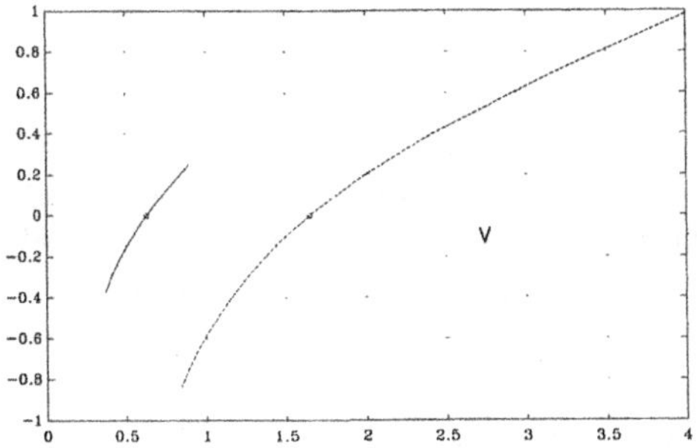

Figure 11. The velocity behind the shock as a function of M_+.

5.4 Selection Rules for Physical Shocks: The Entropy Growth Condition

In analogy to the derivation of the Rankine–Hugoniot conditions (5.8) from the equations of balance we may derive a jump condition for entropy. In this argument the entropy flux h^A replaces $\mathbf{F}^A$ and the entropy production Σ replaces $\mathbf{f}$, but there is one important difference: where $\mathbf{f}$ was supposed non-singular in the domain C, including the shock front σ, we cannot make that assumption for the entropy production. In fact, we must allow Σ to be singular on σ. Therefore the passage from (5.5) to (5.6) is impossible and, instead of (5.4), we obtain

$$\left[\!\left[h^A(\mathbf{u})\right]\!\right] \nu_A = \Sigma_S, \tag{5.20}$$

where Σ_S is the entropy production density on the shock front. Once the constitutive relation $h^A(\mathbf{u})$ is known, we may identify the shock parameter μ and introduce $\mathbf{u}_-(\mu)$ and $s(\mu)$ into (5.21). Thus we calculate $\Sigma_S(\mu)$. This quantity will generally be non-negative for some range of the shock parameter and negative elsewhere. The *entropy growth condition* requires (see [73] and [74])

$$\Sigma_S(\mu) \geq 0, \tag{5.21}$$

meaning that shocks are not admissible for those values of μ that violate (5.22).

In Chapter 5, Paragraph 3.2, we have made use of the entropy growth condition to prove that the passage of the shock will always increase the energy density. In classical gas dynamics it is well known that the entropy growth condition forbids expansive shocks.

5.5 Selection Rules for Physical Shocks: The Lax Conditions

Another selection rule for physical shocks is the Lax condition [75]. This conditon states that the speed of the shock lies between the characteristic speeds in front of the shock and behind it

$$\lambda(\mathbf{u}_+) < s < \lambda(\mathbf{u}_-). \tag{5.22}$$

Therefore in Figure 6 the admissible shocks are those lying on the branches that are emphasized by a thick line. Mostly this means that we can only have

$$\text{either} \quad \mu > \mu_0 \quad \text{or} \quad \mu < \mu_0 \tag{5.23}$$

where μ_0 is the value of the shock parameter for which $\mathbf{u}_+(\mu_0) = \mathbf{u}_-$. In exceptional cases—when the line $\mathbf{u}_-(s)$ intersects the values $\mathbf{u}_+$ vertically—no relation like (5.24) is dictated by the Lax condition.

It can be shown that for weak shocks the Lax condition and the entropy growth condition are equivalent (see [73] for a genuinely non-linear system where (4.8) holds for all eigenvalues).

We have written the Lax conditions (5.23) for a generic eigenvalue λ. Taking into account all n eigenvalues, which we suppose distinct, there are n Lax conditions, viz.

$$\begin{gathered}
\lambda_+^{(1)} < \lambda_+^{(2)} < \cdots < \lambda_+^{(k)} < s < \lambda_+{}^{(k+1)} < \cdots < \lambda_+^{(n)}, \\
\lambda_-^{(1)} < \lambda_-^{(2)} < \cdots < \lambda_-{}^{(k-1)} < s < \lambda_-{}^{(k)} < \cdots < \lambda_-^{(n)}.
\end{gathered} \tag{5.24}$$

For a fixed value k the shock that satisfies (5.25) is called a *k-shock* and, in particular, (5.23) applies to the k-shock in the form

$$\lambda_+^{(k)} < s < \lambda_-^{(k)}. \tag{5.25}$$

5.6 Lax Condition in Extended Thermodynamics

We return to Paragraph 5.3 and discuss the physical admissibility of the solutions of the Rankine–Hugoniot equations of extended thermodynamics.

We have four shocks in this case, which bifurcate from the four solutions of the characteristic equation (3.3). In front of the shock that solution is given by [see (3.6)]

$$\hat{\lambda}_+^{(1)} = -1.6503, \qquad \hat{\lambda}_+^{(2)} = -0.6297, \qquad \hat{\lambda}_+^{(3)} = 0.6297, \qquad \hat{\lambda}_+^{(4)} = 1.6503,$$

while behind the shock the four solutions are functions of σ and q, which in turn, by (5.12), (5.14), and (5.15), are functions of M_+, different ones for the fast and the slow shocks, because $w(M_+)$ is different. Thus $\hat{\lambda}_-^{(i)}$ $(i = 1, 2, 3, 4)$ are functions of M_+, which have been calculated and are represented in Figure 12.

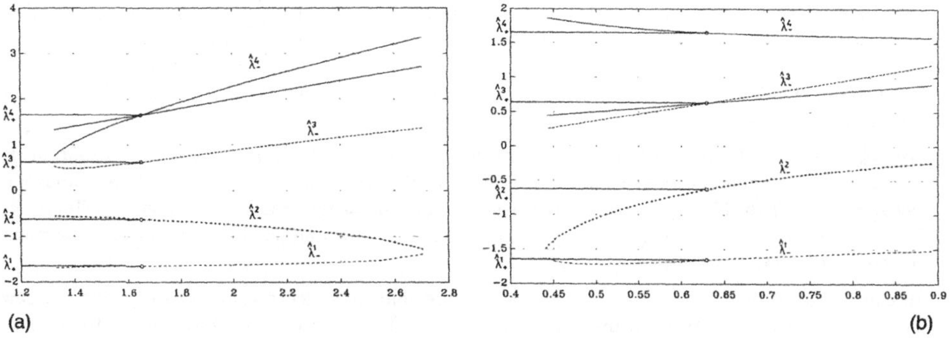

Figure 12. The characteristic speeds as functions of M_+. (a) Fast shock. (b) Slow shock.

The Lax condition (5.25) for the fastest wave $k = 4$ reads explicitly

$$\lambda_+^{(4)} < s \quad \text{and} \quad \lambda_-^{(3)} < s < \lambda_-^{(4)} \tag{5.26}$$

and of course, s/c_+ is the bisector of the $(\hat{\lambda}, M_+)$-plane shown in Figure 12(a). Inspection shows that the Lax conditions (5.27) are satisfied provided $M_+ > 1.6503$ holds. We therefore conclude that the dotted curves of Figures 7 through 11 are physically significant only in this range of the Mach number. In particular,

— the density jumps up,
— the temperature jumps up,
— the stress jumps down,
— the heat flux jumps up, and
— the velocity jumps up.

We shall now investigate the consequences of the Lax condition (5.25) for the slow wave $k = 3$. That condition (5.25) reads explicitly

$$\lambda_+^{(3)} < s < \lambda_+^{(4)} \quad \text{and} \quad \lambda_-^{(2)} < s < \lambda_-^{(3)} \tag{5.27}$$

and inspection of Figure 12(b) shows that the slow shock can only propagate in the range of Mach numbers $M_+ > 0.6297$. The drawn-out curves of the Figures 7 through 11 are physically relevant only in that range and we see that

— the density jumps up,
— the temperature jumps down,
— the stress jumps down,
— the heat flux jumps down, and
— the velocity jumps up.

From Figure 12 we also conclude that the system looses hyperbolicity after the passage of a fast shock with $M_+ \approx 2.7$, because two eigenvalues become equal in the disturbed region. We discussed this breakdown of hyperbolicity in Paragraph 3.2. We note that the loss of hyperbolicity at the Mach number 2.7 is not related to the critical Mach numbers found in [76] and [77] for shock structure solutions.

Anyway in the small range of Mach numbers where the slow shock can propagate, it is a *cold shock*, meaning that the temperature behind the shock is smaller than that in front of it. This is unusual, because we tend to think—motivated by the experience in gas dynamics—that a shock will always heat the gas up. This is not necessarily so, as we can see now and shall confirm in the treatment of the phonon gas in Chapter 8.

In conclusion of this section on shock waves we note that extended thermodynamics offers a new approach to shock waves when dissipation is present. The procedure is quite similar to the non dissipative case, since the equations of the system are balance laws. The analysis provides an alternative method to the usual shock structure approach. Moreover, the existence of a small set of Mach numbers for which there exists a cold shock may be a challenge for experimenters to check the results of extended thermodynamics.

Chapter 8

Second Sound in Solids at Low Temperature

The phenomenon of second sound in liquid He II was first predicted in the theoretical works of Tisza [78] and Landau [79]. Later Peshkov [80] detected this wave and confirmed that its temperature variations are uncoupled from the pressure variations: therefore the second sound is often referred to as a temperature wave. This phenomenon is now well understood. We described it in Chapter 3 following the work of Dreyer [81], who showed that the bifluid theory of Tisza and Landau is a special case of the thermodynamic theory of binary mixtures in general.

Prompted by a suggestion made by Peshkov [82] in 1947 scientists began to search for second sound phenomena in pure crystals at low temperatures. These were indeed observed, first in solid ^{4}He [83] and later in ^{3}He [84], NaF [85], and Bi [86]. The theoretical speculation about this new sound mode in solids immediately concentrated on phonons and the propagation of energy disturbances of the phonon gas; however, the speed of propagation was found to depend on temperature. Therefore the simple theory of phonons, as presented in Chapter 5, cannot explain the phenomenon, since it provides $c/\sqrt{3}$ as the only possible speed.

Guyer and Krumhansl [87] proposed a theory in which the conservation law of energy of the phonons is supplemented by a complicated generalization of the Cattaneo equation for the heat flux, viz.

$$\dot{\mathbf{q}} + \frac{1}{\tau_R}\mathbf{q} + \frac{1}{3}\varrho c_v c^2 \nabla T - \frac{3}{5}c^2 \tau_N \left[\Delta + 2\nabla(\nabla\cdot)\right]\mathbf{q}$$

where τ_N and τ_R are two relaxation times.

In extended thermodynamics we cannot have such a complicated equation, because we need the field equations to be of balance type. We therefore propose a different view, generalizing the equations of Chapter 5 but retaining their balance character. We are thus able to accommodate the temperature dependence of the (second) sound speed, and can analyse the propagation of shock waves. We predict the existence of a characteristic temperature $\tilde{T}_c$ below which the shock, if it exists, heats up the body, while for values of T above $\tilde{T}_c$ the shock cools the body. In this context we speak of a hot shock—the usual well-known phenomenon—and of a cold shock—a phenomenon discovered in [88] and described in detail in [89].

1 Extended Thermodynamics of Phonons Generalized

1.1 Equations of Balance and Constitutive Equations

We consider heat conduction in a rigid body at rest so that there are four basic fields only, viz.

temperature T and heat flux $\mathbf{Q}$.

The necessary equations are the conservation law of energy and a balance law for the heat flux. We write these equations in the forms

$$\frac{\partial e}{\partial t} + \frac{\partial Q_i}{\partial x_i} = 0, \qquad \frac{\partial w_i}{\partial t} + \frac{\partial T_{ij}}{\partial x_j} = b_i. \tag{1.1}$$

The energy density e, the vectors $\mathbf{w}$ and $\mathbf{b}$, and the tensor $\mathbf{T}$ are constitutive quantities. We allow them to depend on temperature T and heat flux Q in a manner that satisfies the principle of material frame indifference. Thus we may write

$$e = e(T, Q^2), \quad w_i = \alpha(T, Q^2)Q_i, \quad T_{ij} = \nu(T, Q^2)\delta_{ij} + \beta(T, Q^2)Q_iQ_j, \quad b_i = \mu(T, Q^2)Q_i. \tag{1.2}$$

We linearize the constitutive equations about an equilibrium state in which $\mathbf{Q}$ vanishes. Thus (1.2) assume the simplified forms

$$e = e(T) = \sigma T^4, \qquad w_i = \alpha(T)Q_i, \qquad T_{ij} = \nu(T)\delta_{ij}, \qquad b_i = \mu(T)Q_i, \tag{1.3}$$

where $(1.3)_1$ is due to the Stefan–Boltzmann equation [see (V.1.22)].

We introduce the heat conductivity

$$\kappa = -\frac{\nu'}{\mu} \quad \text{where }' \text{ stands for } \frac{d}{dT} \tag{1.4}$$

and obtain the field equations by insertion into (1.1)

$$\frac{\partial \sigma T^4}{\partial t} + \frac{\partial Q_i}{\partial x_i} = 0, \qquad \frac{\partial \alpha Q_i}{\partial t} + \nu' \frac{\partial T}{\partial x_i} = -\frac{\nu'}{\kappa} Q_i. \tag{1.5}$$

In order to make these equations specific field equations for T and Q, we must know the three functions $\alpha(T), \nu'(T)$, and $\kappa(T)$. The entropy principle will help to determine these.

Note that there are three not so subtle differences between this set of equations and the equations of Chapter 5 that were based on the kinetic theory of phonons, namely,

— energy is now assumed conserved,

— previously α was a constant which may be set equal to 1, and

— Previously we had $T_{ii} = e c^2$, that is, $\nu = \frac{1}{3} e c^2$.

In the present theory it is essential that α depends on T and that ν is unrelated to e; this theory was developed in [90] through [92] without regard to the kinetic theory. We present it here, because it has interesting implications for shock wave propagation.

1.2 Entropy Principle

We rely on the entropy principle to restrict the functions

$$\alpha = \alpha(T), \qquad \nu = \nu(T), \qquad \kappa = \kappa(T). \tag{1.6}$$

The entropy inequality

$$\frac{\partial h}{\partial t} + \frac{\partial h_i}{\partial x_i} \geq 0 \tag{1.7}$$

must hold for all solutions $T(\mathbf{x}, t), \mathbf{Q}(\mathbf{x}, t)$ of the equations (1.5). h and h_i are the entropy density and the entropy flux respectively; both are constitutive quantities and we have by material frame indifference

$$h = h(T, Q^2), \qquad h_i = \varphi(T, Q^2)Q_i. \tag{1.8}$$

We collect the results of the exploitation of the entropy principle in the following statement:

Statement 1. *The entropy principle restricts the constitutive functions* (1.6) *and* (1.8) *as follows:*

$$\alpha = \frac{\gamma}{\nu' T^2}, \qquad \kappa \geq 0, \qquad h = \frac{4}{3}\sigma T^3 - \frac{\gamma}{2(\nu' T^2)^2}Q^2, \qquad h_i = \frac{Q_i}{T} \tag{1.9}$$

where γ is a constant.

As always these results are most easily proved by the use of Lagrange multipliers. In the present case we choose λ and λ_i and write a new inequality

$$\frac{\partial h}{\partial t} + \frac{\partial h_i}{\partial x_i} - \lambda\left(\frac{\partial e}{\partial t} + \frac{\partial Q_i}{\partial x_i}\right) - \lambda_i\left(\frac{\partial \alpha Q_i}{\partial t} + \frac{\partial \nu}{\partial x_i} + \frac{\nu'}{\kappa}Q_i\right) \geq 0, \tag{1.10}$$

which must hold for all fields $T(\mathbf{x}, t), Q_i(\mathbf{x}, t)$. The Lagrange multipliers may depend on T and Q_i.

The left-hand side of the inequality is linear in

$$\frac{\partial T}{\partial t}, \ \frac{\partial Q_i}{\partial t}, \ \frac{\partial T}{\partial x_i}, \ \frac{\partial Q_i}{\partial x_j}.$$

Therefore we must have

$$\begin{aligned} dh &= \lambda\, de + \lambda_i d(\alpha Q_i), \\ dh_i &= \lambda\, dQ_i + \lambda_i d\nu, \end{aligned} \tag{1.11}$$

and there remains the residual inequality

$$\frac{\nu'}{\kappa}\lambda_i Q_i \leq 0. \tag{1.12}$$

From $(1.11)_2$ we obtain by $(1.8)_2$

$$\varphi = \varphi(T), \qquad \lambda = \varphi(T), \qquad \lambda_i = \frac{\varphi'}{\nu'}Q_i. \tag{1.13}$$

We use this in $(1.11)_1$ and obtain by use of $(1.3)_1$, (1.6), and $(1.8)_1$

$$\frac{\partial h}{\partial T} = \varphi(T)\frac{de}{dT} + \frac{\varphi'}{\nu'}\lambda' Q^2, \qquad \frac{\partial h}{\partial Q^2} = \frac{1}{2}\frac{\varphi'}{\nu'}\alpha. \tag{1.14}$$

From the equilibrium part of $(1.14)_1$ we conclude by comparison with the Gibbs equation that we have

$$\varphi(T) = \frac{1}{T} \qquad \begin{aligned} &\text{hence by } (1.3)_1 \quad h_E = \frac{4}{3}\sigma T^3 \\ &\text{and by } (1.8)_2 \quad h_i = \frac{Q_i}{T}. \end{aligned} \tag{1.15}$$

Also the two conditions (1.14) imply an integrability condition for h, viz.

$$\frac{\alpha'}{\alpha} = \frac{(1/T^2\nu')'}{1/T^2\nu'} \quad \text{hence} \quad \alpha = \gamma\frac{1}{T^2\nu'}, \tag{1.16}$$

where γ is a positive constant. Insertion of (1.15) and $(1.16)_2$ into (1.14) provides

$$h = \frac{4}{3}\sigma T^3 - \frac{\gamma}{2}\frac{1}{(T^2\nu')^2}Q^2. \tag{1.17}$$

By use of $(1.13)_3$ and with $\varphi = \frac{1}{T}$ it is easily seen that the residual inequality (1.12) requires $\kappa \geq 0$. This completes the proof of Statement 1.

1.3 Adjusting ν' to Observed Wave Speeds

By $(1.9)_1$ there is a relation between $\alpha(T)$ and $\nu'(T)$; therefore, the only unknown functions in the field equations (1.5) are $\nu'(T)$ and $\kappa(T)$ and, of course, there is the unknown constant γ. The heat conductivity κ may be determined experimentally and we shall assume that it is known as a function of T. It remains therefore to determine $\nu'(T)$ and this may be done by adjusting that function to the measured pulse speed or characteristic speed $U_E(\tilde{T})$.

For plane harmonic waves of small amplitude with $\mathbf{Q} = (Q, 0, 0)$, frequency ω, and wave vector $\mathbf{k} = (k, 0, 0)$ we may write (1.5) in the form

$$\begin{bmatrix} 4\sigma\tilde{T}^3 & -\frac{k}{\omega} \\ -\tilde{\nu}\frac{k}{\omega} & \tilde{\alpha} - \frac{i}{\omega}\frac{\tilde{\nu}'}{\tilde{\kappa}} \end{bmatrix} \begin{bmatrix} \bar{T} \\ \bar{Q} \end{bmatrix} = 0 \tag{1.18}$$

where the wiggles characterize the ground state of constant $\tilde{T}$ and $\tilde{Q} = 0$, while $\bar{T}, \bar{Q}$ are the complex amplitudes. The phase speed results from setting the determinant equal to zero and we obtain

$$\frac{1}{v_{\text{ph}}} = \text{Re}\,\frac{k}{\omega} = 2\sqrt{\sigma}\text{Re}\sqrt{\gamma\frac{\tilde{T}}{\tilde{\nu}'^2} - \frac{i}{\omega}\frac{\tilde{T}^3}{\tilde{\kappa}}}.$$

The pulse speed is the limit of v_{ph} for $\omega \to \infty$ (see Chapter 7). Therefore we have

$$U_E(T) = \frac{1}{2\sqrt{\sigma}}\frac{\nu'}{\sqrt{\gamma T}}, \quad \text{hence} \quad \nu' = 2\sqrt{\sigma}\sqrt{\gamma T}U_E(T) \quad \text{and by (1.16)} \quad \alpha = \frac{1}{2\sqrt{\sigma}}\sqrt{\gamma T}\frac{1}{T^3 U_E(T)}. \tag{1.19}$$

Provided the pulse speed is known, or measured, as a function of T, we may thus determine ν' and α as functions of T to within a common multiplicative constant $\sqrt{\gamma}$. What is more, inspection of $(1.5)_2$ shows that the constant drops out of the equation so that without loss of generality we may set γ equal to 1.

Thus we conclude that the field equations (1.5) are quite explicit, if only $U_E(T)$ and $\kappa(T)$ are measured. Also, for the study of shock waves the right-hand sides of the field equations do not matter. Therefore the behaviour of shocks is only influenced by $U_E(T)$.

1.4 An Analytic Form of $U_E(T)$

The pulse speed $U_E(T)$ has been measured by Jackson et al. [85] for NaF and by Narayanamurti and Dynes [86] for Bi. Coleman and Newman [93] fitted an analytic function to the measured data and obtained

$$\frac{1}{U_E^2} = A + BT^n. \tag{1.20}$$

Values of A, B, and n appropriate to NaF and Bi are as follows:

$$\begin{array}{llllll} \text{NaF}: & n=3,10 & A=9,09\cdot 10^{-12}, & B=2,22\cdot 10^{-15} & 10K\leq T\leq 18,5K, \\ \text{Bi}: & n=3,75 & A=9,07\cdot 10^{-11}, & B=7,58\cdot 10^{-13} & 1,4K\leq T\leq 4K. \end{array} \quad (1.21)$$

With these values U_E results in $\frac{\text{cm}}{\text{sec}}$, if T is inserted in K.

2 Shock Waves in a Rigid Conductor

2.1 Rankine–Hugoniot Equations

Efforts to create and monitor the second sound in a solid proceed as follows: the temperature of one side of a rod is suddenly raised or lowered. Thus the original profile of the "heat pulse" may be considered, in good approximation, as an upward or downward step constituting a shock front. We proceed to investigate whether such shocks are able to propagate according to the theory presented in Section 1.

We consider the system (1.5) of field equations. According to (7.5.4) the corresponding Rankine–Hugoniot equations read

$$\begin{aligned} -s\,[\![e]\!] + [\![Q_i]\!]\, n_i &= 0, \\ -s\,[\![\alpha Q_i]\!] + [\![\nu]\!]\, n_i &= 0. \end{aligned} \quad (2.1)$$

We restrict the attention to the one–dimensional case with $\mathbf{Q} = (Q, 0, 0)$ and $\mathbf{n} = (1, 0, 0)$ in which the shock propagates into an equilibrium state with a constant $\tilde{T}$ and $\tilde{Q} = 0$. Behind the shock the body is generally not in equilibrium and we denote its state by T_-, Q_-.

As the shock parameter we choose the perturbed temperature T_- so that the general solution of the Rankine–Hugoniot equation gives the speed of the shock wave and the heat flux behind it as function of T_- with $\tilde{T}$ as a parameter.

In the present case we use (1.19) and obtain from (2.1) after a little calculation

$$s = 2\sqrt{\frac{T_-^{5/2} U_E(T_-)}{T_-^4 - \tilde{T}^4} \int_{\tilde{T}}^{T_-} \sqrt{T}\, U_E(T)\, dT}, \qquad Q_- = \sigma\, s \left(T_-^4 - \tilde{T}^4\right). \quad (2.2)$$

We proceed to evaluate the Lax condition and the entropy growth condition for the shock (see Paragraphs 7.5.4 and 7.5.5).

2.2 Entropy Growth Condition

We use (1.9) and (2.2) to write the entropy production Σ_S in the form

$$\Sigma_S(T_-, \tilde{T}) = -s\frac{4}{3}\sigma\left(T_-^3 - \tilde{T}^3\right) + \frac{s^3}{8T_-^5 U_E^2(T_-)}\sigma\left(T_-^4 - \tilde{T}^4\right)^2 + \frac{1}{T_-} s\sigma\left(T_-^4 - \tilde{T}^4\right). \quad (2.3)$$

For three values $\tilde{T} = 11K$, $\tilde{T} = 15,36K$, and $\tilde{T} = 17K$ the entropy production is plotted in Figure 1 as a function of T_- by use of the analytic form (1.20) of $U_E(T)$ and for the values (1.21) corresponding to NaF.

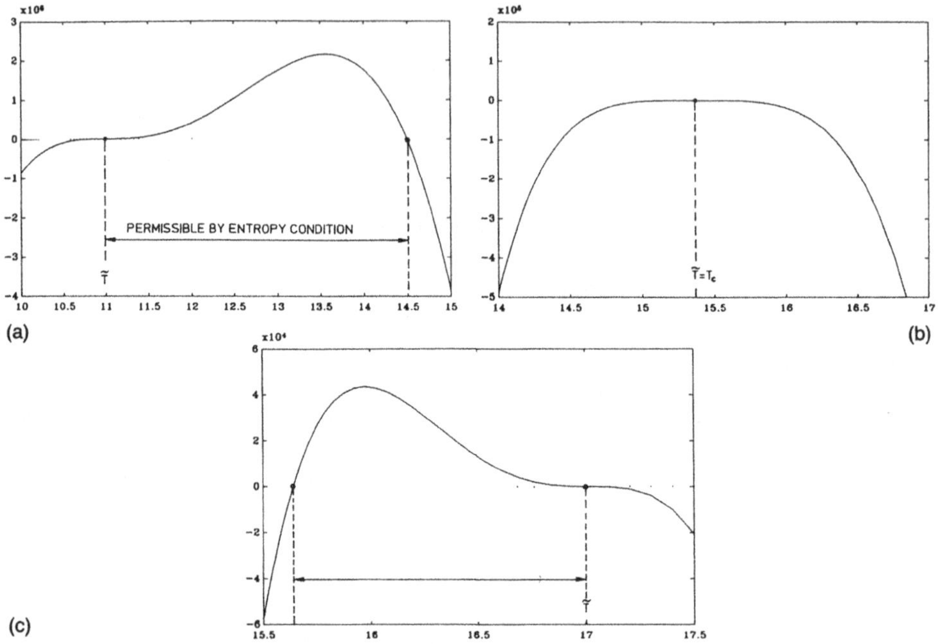

Figure 1. Entropy production for (a) $\tilde{T} = 11K$, (b) $\tilde{T} = 15,36K$, and (c) $\tilde{T} = 17K$.

The most remarkable feature in Figure 1 is the fact that the three cases permit different temperatures T_- behind the shock. In particular,

— If the temperature $\tilde{T}$ of the undisturbed region is equal to 11 K, $\Sigma_S > 0$ is satisfied only for values T_- that are bigger than 11 K. This is the usual situation—well-known from gas dynamics—in which the shock heats up the body. We speak of a *hot shock*, but this shock cannot be "hotter" than $3.5K$, because beyond $14.5K$ the entropy production is negative.

— The reverse happens for $\tilde{T} = 17K$. Here the admissible values of T_- are all smaller than $\tilde{T}$, that is, the shock cools the body and we speak of a *cold shock*. That situation is unusual, because it never occurs in gas dynamics; again, there is a limit on the size of the shock. The entropy condition forbids it to be as cold than $\simeq 2.3K$.

— The switch from a hot shock to a cold shock occurs at $\tilde{T} = 15,36K$ in NaF, and here no shock is possible, because Σ_S is negative for all $T_- \neq \tilde{T}$ as illustrated in Figure 1(b). We call that temperature the critical temperature $T_{\rm C}$.

2.3 Lax Conditions

We shall now make use of the considerations laid down in Paragraph VII.5.5. In order to formulate the Lax conditions, we need to determine the characteristic speeds λ_i. We compare the system (1.5) with the generic system (7.1.1) and obtain for the one-dimensional case

$$\mathbf{A}^0 = \begin{bmatrix} 4\sigma T^3 & 0 \\ \alpha' Q & \alpha \end{bmatrix}, \qquad \mathbf{A}^1 = \begin{bmatrix} 0 & 1 \\ \nu' & 0 \end{bmatrix}. \tag{2.4}$$

The characteristic polynomial therefore reads

$$\det\left(\mathbf{A}^1 - \lambda \mathbf{A}^0\right) = 4\sigma T^3 \alpha \lambda^2 + \alpha' Q \lambda - \nu' = 0, \tag{2.5}$$

whose positive solutions are

before the shock: $\tilde{\lambda} = U_E(\tilde{T})$

behind the shock: $$\lambda_- = \sqrt{U_E^2(T_-) + \left(\frac{Q_-}{8\sigma T_-^3}\right)^2 \left(\frac{5}{2T_-} + \frac{U_E'(T_-)}{U_E(T_-)}\right)^2} + \frac{Q_-}{8\sigma T_-^3}\left(\frac{5}{2T_-} + \frac{U_E'(T_-)}{U_E(T_-)}\right). \tag{2.6}$$

Since $U_E(T)$ is known by (1.20) and, since $Q_-(T_-, \tilde{T})$ and $s(T_-, \tilde{T})$ are known by (2.2), we may now investigate the restriction imposed on the shock by the Lax condition (7.5.23). In the present case this condition reads

$$\tilde{\lambda}(\tilde{T}) < s(\tilde{T}, T_-) < \lambda_-(\tilde{T}, T_-). \tag{2.7}$$

Figure 2 shows plots of s and λ_- as function of T_- for given values of $\tilde{T}$. We emphasize several salient features implied by these figures; which all refer to NaF.

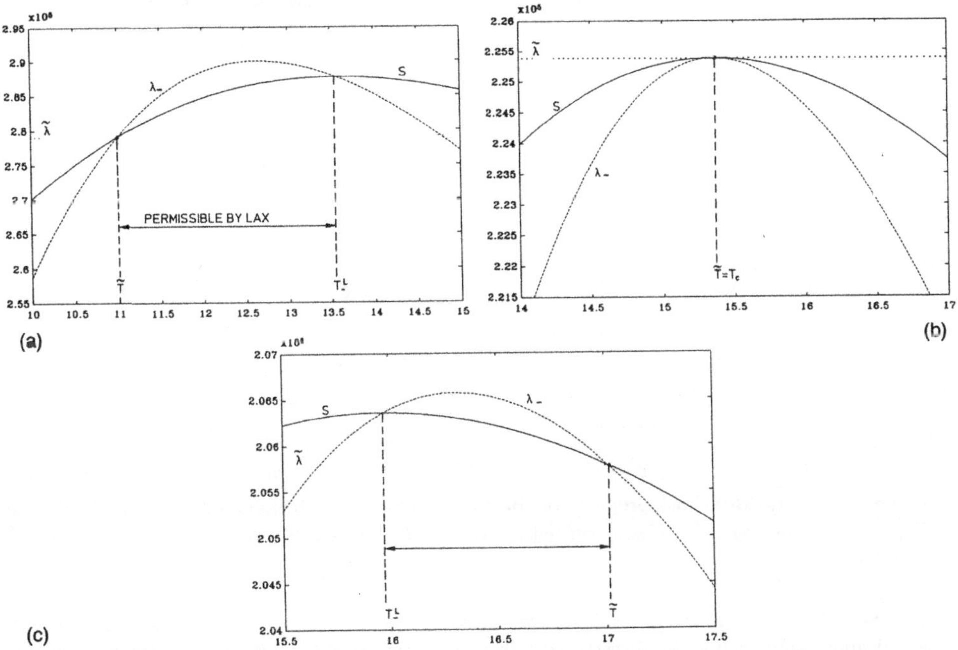

Figure 2. Lax conditions for (a) $\tilde{T} = 11K$, (b) $\tilde{T} = T_c = 15.36K$, and (c) $\tilde{T} = 17K$.

- For $\tilde{T} = 11K$ the Lax condition is satisfied for $T_- > \tilde{T}$, provided T_- is smaller than about 13,5 K. Thus only a hot shock is permitted.
- For $\tilde{T} = 17K$ the Lax condition permits only cold shocks.
- The switch from hot to cold shocks occurs at $\tilde{T} = T_c = 15.36K$ where no shocks are permissable (see Figure 2(b)).

All this is compatible with the evaluation of the entropy growth condition; however, the two conditions are not equivalent. A comparison of the plots of Figures 1 and 2 shows that the range of possible values of T_- allowed by the entropy condition is greater than for the Lax condition.

Further research is needed in order to determine which of the conditions is to be relied on. In that context it may be significant that the Lax condition is valid as long as the entropy growth grows. This can be seen by comparing the curves of Figures 1 and 2. The fact has led Ruggeri et al. [89] to suggest a *generalized entropy growth condition.*

2.4 Shape Changes of Second Sound Wave

We have seen in Paragraph 2.1 that second sound experiments may generate shocks, when the heater or cooler is turned on.

When a heater is turned on we may create the hot shock studied in the previous two paragraphs. That shock can propagate provided that

$$\tilde{T} < T_c \quad \text{and} \quad T_- < T_-^L \quad \text{[see Figure 2]}. \tag{2.8}$$

If (2.8) is violated we cannot expect a shock to propagate. The solution will be regularized as shown schematically in Figure 3. The shock also cannot propagate, if $\tilde{T} > T_c$ holds, we expect regularization in this case as well (see Figure 3).

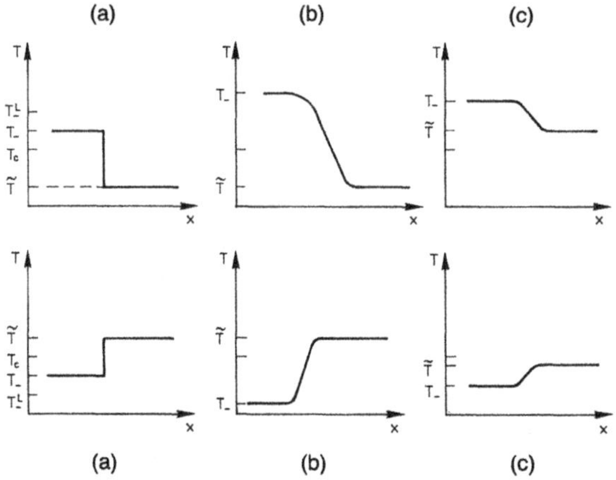

Figure 3. Top: (a) Hot shock propagation for $\tilde{T} < T_c$. (b) and (c) Regularization for $T_- > T_-^L$ and $\tilde{T} > T_c$. Bottom: (a) Cold shock propagation for $\tilde{T} > T_c$. (b) and (c) Regularization for $T_- < T_-^L$ and $\tilde{T} < T_c$.

This behavior seems to be supported by the oscilloscope trace in the case of ^4He [94], and for NaF [85] it may also be deduced indirectly from the observation that the difference between the arrival times for leading edges and peaks increases with temperature.

Chapter 9

Molecular Extended Thermodynamics

On previous occasions we have been surprised that extended thermodynamics furnished the same results as the kinetic theory of gases in the Grad approximation. The present chapter shows that this analogy goes far beyond 13 moments. Indeed, if we make sure that the phenomenological and the molecular theories choose the same variables, irrespective of their number, we can prove that extended thermodynamics and molecular extended thermodynamics lead to identical results. This was first shown by Dreyer [95]. To be sure, the methods are different in the two fields and, in particular, the role of entropy is different: In extended thermodynamics the entropy inequality is exploited, while in the molecular theory we maximize the kinetic expression for the entropy.

Weiss [96] has used the method of maximizing the entropy extensively. He was thus able to provide explicit field equations for moment theories of up to 35 moments. These are listed in Section 1 of this chapter.

There is a serious drawback in molecular thermodynamics of moments, because as we increase the number of moments, we need not only additional equations, but we must change the existing equations in subtle, non-transparent ways. On the other hand, if instead of moments we use mean eigenfunctions of the linear collision operator, the existing equations remain unchanged and the closure problem is most elegantly handled.

Section 2 gives a brief account of eigenfunctions, in particular, for Maxwellian molecules. It relies heavily on results obtained by Waldmann [97]. In Section 3 we identify some mean eigenfunctions and write the corresponding field equations in an explicit form. We discuss the closure problem.

1 Field Equations of Molecular Extended Thermodynamics of Many Moments

1.1 Densities, Fluxes, and Productions as Moments of the Phase Density

We make use of the generic notation introduced in Chapter 6 and consider the case in which the densities $\mathbf{F}^0$, the fluxes $\mathbf{F}^i$, and the production $\mathbf{f}$ are blocks of tensors

$$\mathbf{u} = \mathbf{F}^0 = \begin{bmatrix} F \\ F_{i_1} \\ F_{i_1 i_2} \\ \vdots \\ F_{i_1 i_2 \cdots i_N} \end{bmatrix}, \quad \mathbf{F}^i = \begin{bmatrix} F^i \\ F^i_{i_1} \\ F^i_{i_1 i_2} \\ \vdots \\ F^i_{i_1 i_2 \cdots i_N} \end{bmatrix}, \quad \mathbf{f} = \begin{bmatrix} f \\ f_{i_1} \\ f_{i_1 i_2} \\ \vdots \\ f_{i_1 i_2 \cdots i_N} \end{bmatrix}. \tag{1.1}$$

This choice characterizes the special case that is dictated by the kinetic theory of gases, see Section 1.3. Thus the flux in one equation equals the density in the following equation. In addition we assume that $\mathbf{u}, \mathbf{F}^i$, and $\mathbf{f}$ are related to the phase density f of the mon-atomic gas by requiring

$$F_{i_1 i_2 \cdots i_l} = m \int c_{i_1} \cdots c_{i_l} f \, d\mathbf{c}, \quad F^i_{i_1 \cdots i_l} = m \int c_i c_{i_1} \cdots c_{i_N} f \, d\mathbf{c}, \quad f_{i_1 \cdots i_l} = m \int c_{i_1} \cdots c_{i_l} S \, d\mathbf{c} \tag{1.2}$$

where S is the collision production in the Boltzmann equation [see (1.3.2.)].

We simplify the notation in a self-explanatory manner by introducing the multi-index $\alpha = i_1 i_2 \cdots i_l$ and write

$$u_\alpha = m \int c_\alpha f \, d\mathbf{c}, \qquad F^i_\alpha = m \int c^i c_\alpha f \, d\mathbf{c}, \qquad f_\alpha = m \int c_\alpha S \, d\mathbf{c}. \tag{1.3}$$

The balance equations assume the forms

$$\frac{\partial u_\alpha}{\partial t} + \frac{\partial F^i_\alpha}{\partial x^i} = f_\alpha. \tag{1.4}$$

Extended thermodynamics makes this a system of quasi-linear field equations by introducing constitutive relations of the forms

$$F^i_\alpha = \tilde{F}^i_\alpha(u_\beta) \quad \text{and} \quad f_\alpha = \tilde{f}_\alpha(u_\beta). \tag{1.5}$$

The constitutive functions $\tilde{F}^i_\alpha$ and $\tilde{f}_\alpha$ are restricted by the principle of relativity and the entropy principle as explained in Chapter 6.

Molecular extended thermodynamics assumes a different method for the determination of the constitutive functions; it calculates these functions by insertion, into $(1.3)_{2,3}$, of a phase density $f(u_\alpha(\mathbf{x}, t), \mathbf{c})$ that has been found by maximizing the entropy

$$\rho s = -k \int \ln f f \, d\mathbf{c} \tag{1.6}$$

under the constraints of fixed values of

$$u_\alpha = m \int c_\alpha f d\mathbf{c} \qquad (\alpha = 1, 2, \ldots, n). \tag{1.7}$$

1.2 Maximizing Entropy

In order to maximize the entropy density $\rho s = k \int f \ln f \, d\mathbf{c}$ under the constraints $u_\alpha = m \int c_\alpha f \, d\mathbf{c}$ we introduce Lagrange multipliers Λ_α and determine f such that the expression

$$\rho s - \sum_{\alpha=1}^{n} \Lambda_\alpha u_\alpha = -\int \left(k \ln f + \sum_{\alpha=1}^{n} \Lambda_\alpha m c_\alpha \right) f \, d\mathbf{c} \tag{1.8}$$

assumes a maximum without constraint. This phase density is given by

$$f = e^{-1-\frac{m}{k}\sum_{\alpha=1}^{n} \Lambda_\alpha c_\alpha}. \tag{1.9}$$

The Lagrange multipliers Λ_α must be calculated as functions of u_β by insertion of (1.9) into the constraints (1.7) so that we have

$$u_\beta = m \int c_\beta e^{-1-\frac{m}{k}\sum_{\alpha=1}^{n} \Lambda_\alpha c_\alpha} d\mathbf{c} \qquad \text{hence} \qquad u_\beta = u_\beta(\Lambda_\alpha) \quad \text{and} \quad \Lambda_\alpha = \Lambda_\alpha(u_\beta). \tag{1.10}$$

Thus we have obtained Λ_α as functions of u_β, at least in principle, and we may proceed to calculate the fluxes F^i_β by

$$F^i_\beta = m \int c^i c_\beta e^{-1-\frac{m}{k}\sum_{\alpha=1}^{n} \Lambda_\alpha c_\alpha} d\mathbf{c} \qquad \text{hence} \qquad F^i_\beta = \tilde{F}^i_\beta(\Lambda_\alpha) \quad \text{and} \quad F^i_\beta = \tilde{F}^i_\beta(u_\alpha). \tag{1.11}$$

1.3 Molecular Extended Thermodynamics Implics Extended Thermodynamics

We recall the condition (6.1.17) as the essential restriction that extended thermodynamics imposes upon the constitutive relations. With $\mathbf{F}^0 = \mathbf{u}$ this condition may be decomposed into two, namely,

$$\frac{\partial u_\alpha}{\partial \Lambda_\beta} = \frac{\partial u_\beta}{\partial \Lambda_\alpha} \quad \text{and} \quad \frac{\partial F^i_\alpha}{\partial \Lambda_\beta} = \frac{\partial F^i_\beta}{\partial \Lambda_\alpha}. \tag{1.12}$$

The two symmetry conditions (1.12) are also valid in molecular thermodynamics. Indeed by $(1.10)_1$ and $(1.11)_1$ we have

$$\frac{\partial u_\beta}{\partial \Lambda_\alpha} = m \int c_\beta c_\alpha e^{-1-\frac{m}{k}\sum_{\gamma=1}^{n} \Lambda_\gamma c_\gamma} d\mathbf{c} \quad \text{and} \quad \frac{\partial F^i_\beta}{\partial \Lambda_\alpha} = m \int c^i c_\beta c_\alpha e^{-1-\frac{m}{k}\sum_{\gamma=1}^{n} \Lambda_\gamma c_\gamma} d\mathbf{c}, \tag{1.13}$$

and these expressions are clearly symmetric with respect to α and β. We conclude that the Lagrange multipliers of extended thermodynamics may be identified with the Lagrange multipliers of molecular extended thermodynamics; also the results of molecular extended thermodynamics imply those of extended thermodynamics. If we know, or assume, that the variables and fluxes are moments, the two theories are equivalent. This is good to know, because molecular extended thermodynamics is able to handle the closure problem in an elegant manner as we shall see later.

Both theories have one difficulty in common: The inversion of the transformation $\Lambda_\beta = \Lambda_\beta(u_\alpha)$. Indeed, inspection of $(1.10)_1$ shows that neither $u_\alpha = u_\alpha(\Lambda_\beta)$ is easy to calculate, nor is that relation easy to invert. In this situation we are forced to linearize just as in extended thermodynamics. We shall explain the procedure in Paragraph 1.5.

1.4 Galilean Invariance and Central Moments

The matrix $\mathbf{X}(\mathbf{v})$ relates $\hat{\mathbf{F}}^0$ and $\mathbf{F}^0$ [see (6.2.23)], that is, $\hat{\mathbf{u}}$ and $\mathbf{u}$ in the present case. By (1.3) it must then be true that

$$c_\alpha = \sum_{\beta=1}^{n} X_{\alpha\beta}(\mathbf{v}) C_\beta$$

holds, a relation that can easily be confirmed directly from the fact that $C_i = c_i - v_i$ holds. Insertion into (1.9) gives

$$f = e^{-1-\frac{m}{k}\sum_{\alpha=1}^{n}\Lambda_\alpha c_\alpha} = e^{-1-\frac{m}{k}\sum_{\alpha=1}^{n}\hat{\Lambda}_\alpha C_\alpha}, \tag{1.14}$$

where (6.2.31) has been used to introduce $\hat{\Lambda}_\alpha$, the intrinsic Lagrange multipliers. It follows from $(1.10)_1$ that we have

$$\hat{u}_\beta = m \int C_\beta e^{-1-\frac{m}{k}\sum_{\alpha=1}^{n}\hat{\Lambda}_\alpha C_\alpha}\, d\mathbf{c}. \tag{1.15}$$

From this relation we may calculate the intrinsic variables $\hat{u}_\alpha$ as a function of $\hat{\Lambda}_\beta$ in an approximate manner, appropriate to processes that are close to equilibrium. We proceed to do this.

1.5 Linear Relation Between Lagrange Multipliers Λ_α and Fields u_α

We decompose the Lagrange multipliers $\hat{\Lambda}_\alpha$ into equilibrium parts $\hat{\Lambda}_\alpha^E$ and non-equilibrium parts λ_α and linearize the phase density (1.14) in the λ's.

In equilibrium the state of a gas at rest is characterized by five moments representing the densities of mass, momentum, and energy

$$\varrho = m \int f\, d\mathbf{c}, \qquad 0 = m \int C_i f_{\mathrm{E}}\, d\mathbf{c}, \qquad 3kT = m \int c^2 f_{\mathrm{E}}\, d\mathbf{c}. \tag{1.16}$$

These are the only constraints and accordingly there are only five non-vanishing Lagrange multipliers, viz. $\hat{\Lambda}_1^{\mathrm{E}}, \hat{\Lambda}_i^{\mathrm{E}}$, and $\hat{\Lambda}_5^{\mathrm{E}}$ so that by (1.14) f_{E} has the form

$$f_{\mathrm{E}} = e^{-1-\frac{m}{k}(\hat{\Lambda}_1^E + \hat{\Lambda}_i^E C_i + \hat{\Lambda}_5^E C^2)}. \tag{1.17}$$

In this case it is easy to calculate $\hat{\Lambda}_1^E, \hat{\Lambda}_i^E$, and $\hat{\Lambda}_5^E$ from (1.16), and we obtain

$$e^{-1-\frac{m}{k}\hat{\Lambda}_1^E} = \frac{\varrho}{m}\frac{1}{\sqrt{2\pi(k/m)T}^3}, \qquad \hat{\Lambda}_i^{\mathrm{E}} = 0, \qquad \hat{\Lambda}_5^{\mathrm{E}} = \frac{1}{2T}, \tag{1.18}$$

so that the equilibrium phase density is given by the Maxwell distribution

$$f_{\mathrm{M}} = \frac{\varrho}{m}\frac{1}{\sqrt{2\pi(k/m)T}^3} e^{-mC^2/2kT}. \tag{1.19}$$

In non-equilibrium we have n constraints rather than 5. Accordingly we have n multipliers of which, by (1.18), only two have non-vanishing equilibrium parts. We write

$$\begin{aligned}
\hat{\Lambda}_1 &= -\frac{k}{m}\ln\left[\frac{\varrho}{m}\frac{e}{\sqrt{2\pi\frac{k}{m}T}^3}\right] + \hat{\lambda}_1, \\
\hat{\Lambda}_i &= \hat{\lambda}_i, \\
\hat{\Lambda}_5 &= \frac{1}{2T} + \hat{\lambda}_5, \\
\hat{\Lambda}_\alpha &= \hat{\lambda}_\alpha \quad \text{for } 4 > 5.
\end{aligned} \tag{1.20}$$

Therefore f reads by (1.14)

$$f = f_{\mathrm{M}} e^{-\frac{m}{k}\sum_{\alpha=1}^{n}\hat{\lambda}_\alpha C_\alpha} \tag{1.21}$$

or by linearization in terms of $\hat{\lambda}_\alpha$

$$f = f_{\mathrm{M}}\left(1 - \frac{m}{k}\sum_{\alpha=1}^{n}\hat{\lambda}_\alpha C_\alpha\right). \tag{1.22}$$

It is now an easy matter in principle to calculate the $\hat{\lambda}$'s from the equations (1.15), which read

$$\hat{u}_\beta = m\int C_\beta\left(1 - \frac{m}{k}\sum_{\alpha=1}^{n}\hat{\lambda}_\alpha C_\alpha\right) f_{\mathrm{M}}\, d\mathbf{c} \tag{1.23}$$

or

$$\sum_{\alpha=1}^{n}\frac{m^2}{k}\left(\int C_\beta C_\alpha f_{\mathrm{M}} d\mathbf{c}\right)\hat{\lambda}_\alpha = \hat{u}_\beta^E - \hat{u}_\beta. \tag{1.24}$$

This is a set of *linear* algebraic equations for the determination of the functions $\hat{\lambda}_\alpha(\hat{u}_\beta)$. Once these functions have been found we obtain $f = f(\hat{u}_\beta(\mathbf{x},t),\mathbf{C})$ by insertion into (1.22), and thus we may calculate fluxes and productions.

1.6 The Case $n = 20$: An Illustration

As an illustration we consider the fairly simple case $n = 20$. This is the case that has already been considered in Paragraph 6.3.2. The densities (and variables) $\mathbf{u}$, fluxes $\mathbf{F}^i$, and productions $\mathbf{f}$ are given by

$$\mathbf{u} = \begin{bmatrix} F \\ F_i \\ F_{ij} \\ F_{ijk} \end{bmatrix} = \begin{bmatrix} m\int & f d\mathbf{c} \\ m\int c_i & f d\mathbf{c} \\ m\int c_i c_j & f d\mathbf{c} \\ m\int c_i c_j c_k & f d\mathbf{c} \end{bmatrix}, \quad F^e = \begin{bmatrix} F^e \\ F^e_i \\ F^e_{ij} \\ F^e_{ijk} \end{bmatrix} = \begin{bmatrix} m\int c_e & f d\mathbf{c} \\ m\int c_e c_i & f d\mathbf{c} \\ m\int c_e c_i c_j & f d\mathbf{c} \\ m\int c_e c_i c_j c_k & f d\mathbf{c} \end{bmatrix}, \tag{1.25$_1$}$$

$$\mathbf{f} = \begin{bmatrix} f \\ f_i \\ f_{ij} \\ f_{ijk} \end{bmatrix} = \begin{bmatrix} m\int & S d\mathbf{c} \\ m\int c_i & S d\mathbf{c} \\ m\int c_i c_j & S d\mathbf{c} \\ m\int c_i c_j c_k & S d\mathbf{c} \end{bmatrix}. \tag{1.25$_2$}$$

Accordingly the *intrinsic* variables, fluxes, and productions are

$$\hat{\mathbf{u}} = \begin{bmatrix} \hat{F} \\ 0 \\ \hat{F}_{ij} \\ \hat{F}_{ijk} \end{bmatrix} = \begin{bmatrix} m\int & f d\mathbf{c} \\ m\int C_i & f d\mathbf{c} \\ m\int C_iC_j & f d\mathbf{c} \\ m\int C_iC_jC_k & f d\mathbf{c} \end{bmatrix}, \qquad \hat{\mathbf{F}}^e = \begin{bmatrix} 0 \\ \hat{F}_i^e \\ \hat{F}_{ij}^e \\ \hat{F}_{ijk}^e \end{bmatrix} = \begin{bmatrix} m\int C^e & f d\mathbf{c} \\ m\int C^eC_i & f d\mathbf{c} \\ m\int C^eC_iC_j & f d\mathbf{c} \\ m\int C^eC_iC_jC_k & f d\mathbf{c} \end{bmatrix},$$

$$\hat{\mathbf{f}} = \begin{bmatrix} \hat{f} \\ \hat{f}_i \\ \hat{f}_{ij} \\ \hat{f}_{ijk} \end{bmatrix} = \begin{bmatrix} m\int & S d\mathbf{c} \\ m\int C_i & S d\mathbf{c} \\ m\int C_iC_j & S d\mathbf{c} \\ m\int C_iC_jC_k & S d\mathbf{c} \end{bmatrix}. \tag{1.26}$$

The field equations read, according to (6.2.6)

$$\frac{\partial \mathbf{u}}{\partial t} + \frac{\partial \mathbf{F}^e}{\partial x^e} = \mathbf{f} \tag{1.27}$$

or, in their intrinsic form (6.2.41)

$$\frac{d\hat{\mathbf{u}}}{dt} + \hat{\mathbf{u}}\frac{\partial v_n}{\partial x_n} + \frac{\partial \hat{\mathbf{F}}^n}{\partial x_n} + \mathbf{A}^r\left\{\hat{\mathbf{u}}\frac{dv_r}{dt} + \hat{\mathbf{F}}^n\frac{\partial v_r}{\partial x_n}\right\} = \hat{\mathbf{f}}, \tag{1.28}$$

where the matrix $\mathbf{A}^r$ is given by [see (6.3.8), (6.3.11), and (6.3.13)]

$$\mathbf{A}^r = \begin{bmatrix} 0 & 0 & 0 & 0 \\ \delta_k^r & 0 & 0 & 0 \\ 0 & 2\delta^r_{(k}\delta^h_{l)} & 0 & 0 \\ 0 & 0 & 3\delta^r_{(k}\delta^h_l\delta^i_{n)} & 0 \end{bmatrix}. \tag{1.29}$$

Thus we may write the equations (1.28) explicitly in the form

$$\begin{aligned}
&\frac{d\hat{F}}{dt} + \hat{F}\frac{\partial v_n}{\partial x_n} && = 0, \\
&\frac{\partial \hat{F}_{in}}{\partial x_n} + \delta_i^r\left\{F\frac{dv_r}{dt}\right\} && = 0, \\
&\frac{d\hat{F}_{ij}}{dt} + \hat{F}_{ij}\frac{\partial v_n}{\partial x_n} + \frac{\partial \hat{F}_{ijn}}{\partial x_n} + 2\delta^r_{(i}\delta^h_{j)}\left\{ + \hat{F}_{hn}\frac{\partial v_r}{\partial x_n}\right\} && = \hat{f}_{\langle ij\rangle}, \\
&\frac{d\hat{F}_{ijk}}{dt} + \hat{F}_{ijk}\frac{\partial v_n}{\partial x_n} + \frac{\partial \hat{F}_{ijkn}}{\partial x_n} + 3\delta^r_{(i}\delta^h_j\delta^m_{k)}\left\{\hat{F}_{hm}\frac{dv_r}{dt} + \hat{F}_{hmn}\frac{\partial v_r}{\partial x_n}\right\} && = \hat{f}_{ijk}.
\end{aligned} \tag{1.30}$$

The first two of these equations and the trace of the third represent the conservation laws of mass, momentum, and energy. Accordingly, the corresponding productions have been set equal to zero. Thus among the productions only $\hat{f}_{\langle ij\rangle}$, $\hat{f}_{ijk}$ remain to be determined and $\hat{F}_{ijkn}$ is the only flux that does not belong to the set of variables. These quantities must be calculated in order to close the system. We have

$$\begin{aligned}
\hat{f}_{\langle ij\rangle} &= m\int C_{\langle i}C_{j\rangle}(f'f^{1'} - ff^1)\sigma g \sin\theta \, d\theta \, d\varepsilon \, d\mathbf{c}^1 \, d\mathbf{c}, \\
\hat{f}_{ijk} &= m\int C_iC_jC_k(f'f^{1'} - ff^1)\sigma g \sin\theta \, d\theta \, d\varepsilon \mathbf{c}\mathbf{c}^1 \, d\mathbf{c} \quad \text{where } f = f_{\mathrm{M}}\left(1 - \frac{m}{k}\sum_{\alpha=1}^{20}\hat{\lambda}_\alpha C_\alpha\right), \\
\hat{F}_{ijkl} &= m\int C_iC_jC_kC_l f \, d\mathbf{c}.
\end{aligned} \tag{1.31}$$

More explicitly f may be written in the form

$$f = f_{\mathrm{M}}\left[1 - \frac{m}{k}\left(\hat{\lambda} + \hat{\lambda}_i C_i + \hat{\lambda}_{ij} C_i C_j + \hat{\lambda}_{ijk} C_i C_j C_k\right)\right], \tag{1.32}$$

and the $\hat{\lambda}$'s must be determined from the linear system (1.24), which in the present case reads explicitly

$$\begin{array}{llll} m\int & f_{\mathrm{M}}\,d\mathbf{c}\,\hat{\lambda} & +m\int C_pC_q & f_{\mathrm{M}}\,d\mathbf{c}\,\hat{\lambda}_{pq} = 0, \\ m\int C_iC_p & f_{\mathrm{M}}\,d\mathbf{c}\,\hat{\lambda}_p & +m\int C_iC_pC_qC_r & f_{\mathrm{M}}\,d\mathbf{c}\,\hat{\lambda}_{pqr} = 0, \\ m\int C_iC_j & f_{\mathrm{M}}\,d\mathbf{c}\,\hat{\lambda} & +m\int C_iC_jC_pC_q & f_{\mathrm{M}}\,d\mathbf{c}\,\hat{\lambda}_{pq} = -\frac{k}{m}\hat{F}_{\langle ij\rangle}, \\ m\int C_iC_jC_kC_p & f_{\mathrm{M}}\,d\mathbf{c}\,\hat{\lambda}_p & +m\int C_iC_jC_kC_pC_qC_r & f_{\mathrm{M}}\,d\mathbf{c}\,\hat{\lambda}_{pqr} = -\frac{k}{m}\hat{F}_{ijk}. \end{array} \tag{1.33}$$

Note that $\hat{F}_i = 0$ and that there is no distinction between mass and energy in equilibrium and non-equilibrium. Also $\hat{F}^{\mathrm{E}}_{ijk}$ is obviously zero.

We use standard methods to solve the system (1.33) for the $\hat{\lambda}$'s and obtain

$$\hat{\lambda} = 0, \qquad \hat{\lambda}_i = \frac{1}{2}\frac{1}{\varrho\frac{k}{m}T^2}\hat{F}_{ijj}, \qquad \hat{\lambda}_{ij} = -\frac{1}{2}\frac{1}{\varrho\frac{k}{m}T^2}\hat{F}_{\langle ij\rangle}, \qquad \hat{\lambda}_{ijk} = -\frac{1}{6}\frac{1}{\varrho\left(\frac{k}{m}\right)^2T^3}\hat{F}_{ijk}. \tag{1.34}$$

Hence follows f in terms of $\hat{F}_{\langle ij\rangle}$, $\hat{F}_{ijj}$, and $\hat{F}_{\langle ijk\rangle}$ by insertion into $(1.31)_4$

$$f = f_{\mathrm{M}}\left(1 - \frac{1}{2}\frac{1}{\varrho\left(\frac{k}{m}T\right)^2}\hat{F}_{\langle ij\rangle}C_iC_j - \frac{1}{2}\frac{1}{\varrho\left(\frac{k}{m}T\right)^2}C_i\left(1 - \frac{1}{5}\frac{C^2}{\frac{k}{m}T}\right)\hat{F}_{ijj} - \frac{1}{6}\frac{1}{\varrho\left(\frac{k}{m}T\right)^3}\hat{F}_{\langle ijk\rangle}C_iC_jC_k\right) \tag{1.35}$$

and $\hat{F}_{ijkl}$ by insertion into $(1.31)_3$

$$\hat{F}_{ijkl} = \varrho\left(\frac{k}{m}T\right)^2 3\delta_{(ij}\delta_{kl)} - \frac{k}{m}T6\delta_{(ij}\hat{F}_{kl)}. \tag{1.36}$$

$\hat{f}_{\langle ij\rangle}$, $\hat{f}_{ijk}$ may be calculated by insertion of f into $(1.31)_{1,2}$, but this is cumbersome and we prefer a shortcut that is valid for Maxwellian atoms. Indeed, for those atoms we have (see [14])

$$\begin{aligned} \hat{f}_{\langle ij\rangle} &= \sigma\,\hat{F}_{\langle ij\rangle}, \\ \hat{f}_{ijk} &= \sigma\left(\tfrac{3}{2}\hat{F}_{\langle ijk\rangle} + \tfrac{2}{5}\delta_{(ij}\hat{F}_{k)ll}\right) \end{aligned} \qquad \text{where } \sigma = -3\frac{\varrho}{m}\sqrt{\frac{\kappa}{2m}}1.370347\ldots. \tag{1.37}$$

Here κ is the coefficient that determines the strength of the interatomic potential $\Phi(r) = \frac{\kappa}{4}\frac{1}{r^4}$.

With (1.36) and (1.37) we may now write the system of field equations (1.30) for the 20-moment case in the *closed* form; in particular,

$$\begin{aligned}
&\frac{d\hat{F}}{dt} + \hat{F}\frac{\partial v_n}{\partial x_n} && = 0, \\
&\frac{\partial \hat{F}_{in}}{\partial x_n} + \left\{F\frac{dv_i}{dt}\right\} && = 0, \\
&\frac{d\hat{F}_{ij}}{dt} + \hat{F}_{ij}\frac{\partial v_n}{\partial x_n} + \frac{\partial \hat{F}_{ijn}}{\partial x_n} + \left\{\hat{F}_{n(i}\frac{\partial v_{j)}}{\partial x_n}\right\} && = \sigma\hat{F}_{\langle ij\rangle}, \\
&\frac{d\hat{F}_{ijk}}{dt} + \hat{F}_{ijk}\frac{\partial v_n}{\partial x_n} + \frac{\partial}{\partial x_n}\{\ \} + \left\{3\hat{F}_{(ik}\frac{dv_{j)}}{dt} + \hat{F}_{n(ik}\frac{\partial v_{j)}}{\partial x_n}\right\} && = \sigma\left(\tfrac{3}{2}\hat{F}_{\langle ijk\rangle} + \tfrac{2}{5}\delta_{(ij}\hat{F}_{k)ll}\right)
\end{aligned}$$

where

$$\{\quad\} \equiv \varrho\left(\frac{k}{m}T\right)^2 3\delta_{(ij}\delta_{kn)} - \frac{k}{m}T6\delta_{(ij}\hat{F}_{kn)}. \tag{1.38}$$

1.7 Field Equations for n = 13, 14, 20, 21, 26, 35

The derivation of the closed system (1.38) of 20 moments was typical for the procedure of molecular extended thermodynamics. We can apply this method to an arbitrary finite number of moments and in this paragraph we write the result for several values of n. In doing so we drop the synthetic notation of the F's and introduce the conventional notation by which we have

$$\hat{F} = \varrho, \qquad \hat{F}_{ii} = 2\varrho\varepsilon = 3kT, \qquad \hat{F}_{\langle ij\rangle} = -t_{\langle ij\rangle}, \qquad \hat{F}_{ijj} = 2q_i \tag{1.39}$$

with ϱ and T as mass–density and temperature and $t_{\langle ij\rangle}$ and q_i as stress deviator and heat flux. Of course, there is no conventional notation for the higher moments, but among the people who work in this field, a convenient nomenclature has evolved. By this we have

$$\hat{F}_{\langle ijk\rangle} = \varrho_{\langle ijk\rangle}, \qquad \hat{F}_{iijj} = 15\varrho\left(\frac{k}{m}T\right)^2 + \Delta, \qquad \hat{F}_{\langle ij\rangle kk} = \varrho_{\langle ij\rangle kk}, \qquad \hat{F}_{\langle ijkl\rangle} = \varrho_{\langle ijkl\rangle}. \tag{1.40}$$

Δ is the non-equilibrium part of the double trace of $\hat{F}_{ijkl}$.

The following equations are all linearized in the non-equilibrium parts of the moments, in velocity, and in the derivatives of ϱ and T. The uniform and time independent fields of ϱ and T underlying the process will be called $\tilde{\varrho}$ and $\tilde{T}$.

n = 13 Variables: F, F_i, F_{ij}, F_{ijj} or $\varrho, v_i, T, t_{\langle ij\rangle}, q_i$

$$\begin{aligned}
&\frac{\partial \varrho}{\partial t} + \tilde{\varrho}\frac{\partial v_i}{\partial x_i} = 0,\\
&\frac{\partial v_i}{\partial t} + \frac{\left(\frac{k}{m}\tilde{T}\right)}{\tilde{\varrho}}\frac{\partial \varrho}{\partial x_i} + \frac{\partial}{\partial x_i}\left(\frac{k}{m}T\right) - \frac{1}{\tilde{\varrho}}\frac{\partial t_{\langle ij\rangle}}{\partial x_j} = 0,\\
&\frac{\partial\left(\frac{k}{m}T\right)}{\partial t} + \frac{2}{3}\left(\frac{k}{m}\tilde{T}\right)\frac{\partial v_k}{\partial x_k} + \frac{2}{3}\frac{1}{\tilde{\varrho}}\frac{\partial q_k}{\partial x_k} = 0,\\
&\frac{\partial t_{\langle ij\rangle}}{\partial t} - \frac{4}{5}\frac{\partial q_{\langle i}}{\partial x_{j\rangle}} - 2\tilde{\varrho}\left(\frac{k}{m}\tilde{T}\right)\frac{\partial v_{\langle i}}{\partial x_{j\rangle}} = \sigma t_{\langle ij\rangle},\\
&\frac{\partial q_i}{\partial t} - \left(\frac{k}{m}\tilde{T}\right)\frac{\partial t_{\langle ik\rangle}}{\partial x_k} + \frac{5}{2}\tilde{\varrho}\left(\frac{k}{m}\tilde{T}\right)\frac{\partial\left(\frac{k}{m}T\right)}{\partial x_i} = \frac{2}{3}\sigma q_i.
\end{aligned} \tag{1.41}$$

n = 14 Variables: $F, F_i, F_{ij}, F_{ijj}F_{iijj}$ or $\varrho, v_i, T, t_{\langle ij\rangle}, q_i, \Delta$

$$\begin{aligned}
&\frac{\partial \varrho}{\partial t} + \tilde{\varrho}\frac{\partial v_i}{\partial x_i} = 0,\\
&\frac{\partial v_i}{\partial t} + \frac{\left(\frac{k}{m}\tilde{T}\right)}{\tilde{\varrho}}\frac{\partial \varrho}{\partial x_i} + \frac{\partial}{\partial x_i}\left(\frac{k}{m}T\right) - \frac{1}{\tilde{\varrho}}\frac{\partial t_{\langle ij\rangle}}{\partial x_j} = 0,\\
&\frac{\partial\left(\frac{k}{m}T\right)}{\partial t} + \frac{2}{3}\left(\frac{k}{m}\tilde{T}\right)\frac{\partial v_k}{\partial x_k} + \frac{2}{3}\frac{1}{\tilde{\varrho}}\frac{\partial q_k}{\partial x_k} = 0,\\
&\frac{\partial t_{\langle ij\rangle}}{\partial t} - \frac{4}{5}\frac{\partial q_{\langle i}}{\partial x_{j\rangle}} - 2\tilde{\varrho}\left(\frac{k}{m}\tilde{T}\right)\frac{\partial v_{\langle i}}{\partial x_{j\rangle}} = \sigma t_{\langle ij\rangle},
\end{aligned} \tag{1.42}$$

$$\frac{\partial q_i}{\partial t} - \left(\frac{k}{m}\tilde{T}\right)\frac{\partial t_{\langle ik\rangle}}{\partial x_k} + \frac{5}{2}\tilde{\varrho}\left(\frac{k}{m}\tilde{T}\right)\frac{\partial\left(\frac{k}{m}T\right)}{\partial x_i} + \frac{1}{6}\frac{\partial\Delta}{\partial x_i} = \frac{2}{3}\sigma q_i,$$
$$\frac{\partial\Delta}{\partial t} + 8\left(\frac{k}{m}\tilde{T}\right)\frac{\partial q_i}{\partial x_i} = \frac{2}{3}\sigma\Delta.$$

n = 20 Variables: F, F_i, F_{ij}, F_{ijk} or $\varrho, v_i, T, t_{\langle ij\rangle}, q_i, \varrho_{\langle ijk\rangle}$

$$\begin{aligned}
&\frac{\partial\varrho}{\partial t} + \tilde{\varrho}\frac{\partial v_i}{\partial x_i} = 0,\\
&\frac{\partial v_i}{\partial t} + \frac{\left(\frac{k}{m}\tilde{T}\right)}{\tilde{\varrho}}\frac{\partial\varrho}{\partial x_i} + \frac{\partial}{\partial x_i}\left(\frac{k}{m}T\right) - \frac{1}{\tilde{\varrho}}\frac{\partial t_{\langle ij\rangle}}{\partial x_j} = 0,\\
&\frac{\partial\left(\frac{k}{m}T\right)}{\partial t} + \frac{2}{3}\left(\frac{k}{m}\tilde{T}\right)\frac{\partial v_k}{\partial x_k} + \frac{2}{3}\frac{1}{\tilde{\varrho}}\frac{\partial q_k}{\partial x_k} = 0,\\
&\frac{\partial t_{\langle ij\rangle}}{\partial t} - \frac{4}{5}\frac{\partial q_{\langle i}}{\partial x_{j\rangle}} - 2\tilde{\varrho}\left(\frac{k}{m}\tilde{T}\right)\frac{\partial v_{\langle i}}{\partial x_{j\rangle}} - \frac{\partial\varrho_{\langle ijk\rangle}}{\partial x_k} = \sigma t_{\langle ij\rangle},\\
&\frac{\partial q_i}{\partial t} - \left(\frac{k}{m}\tilde{T}\right)\frac{\partial t_{\langle ik\rangle}}{\partial x_k} + \frac{5}{2}\tilde{\varrho}\left(\frac{k}{m}\tilde{T}\right)\frac{\partial\frac{k}{m}T}{\partial x_i} = \frac{2}{3}\sigma q_i,\\
&\frac{\partial\varrho_{\langle ijk\rangle}}{\partial t} - 3\frac{k}{m}\tilde{T}\left\{\frac{\partial t_{(\langle ij\rangle}}{\partial x_{k)}} - \frac{2}{5}\frac{\partial t_{\langle r(i\rangle}}{\partial x_r}\delta_{jk)}\right\} = \frac{3}{2}\sigma\varrho_{\langle ijk\rangle}.
\end{aligned} \tag{1.43}$$

n = 21 Variables: $F, F_i, F_{ij}, F_{ijk}F_{iijj}$ or $\varrho, v_i, T, t_{\langle ij\rangle}, q_i, \varrho_{\langle ijk\rangle}, \Delta$

$$\begin{aligned}
&\frac{\partial\varrho}{\partial t} + \tilde{\varrho}\frac{\partial v_i}{\partial x_i} = 0,\\
&\frac{\partial v_i}{\partial t} + \frac{\left(\frac{k}{m}\tilde{T}\right)}{\tilde{\varrho}}\frac{\partial\varrho}{\partial x_i} + \frac{\partial}{\partial x_i}\left(\frac{k}{m}T\right) - \frac{1}{\tilde{\varrho}}\frac{\partial t_{\langle ij\rangle}}{\partial x_j} = 0,\\
&\frac{\partial\left(\frac{k}{m}T\right)}{\partial t} + \frac{2}{3}\left(\frac{k}{m}\tilde{T}\right)\frac{\partial v_k}{\partial x_k} + \frac{2}{3}\frac{1}{\tilde{\varrho}}\frac{\partial q_k}{\partial x_k} = 0,\\
&\frac{\partial t_{\langle ij\rangle}}{\partial t} - \frac{4}{5}\frac{\partial q_{\langle i}}{\partial x_{j\rangle}} - 2\tilde{\varrho}\left(\frac{k}{m}\tilde{T}\right)\frac{\partial v_{\langle i}}{\partial x_{j\rangle}} - \frac{\partial\varrho_{\langle ijk\rangle}}{\partial x_k} = \sigma t_{\langle ij\rangle},\\
&\frac{\partial q_i}{\partial t} - \left(\frac{k}{m}\tilde{T}\right)\frac{\partial t_{\langle ik\rangle}}{\partial x_k} + \frac{5}{2}\tilde{\varrho}\left(\frac{k}{m}\tilde{T}\right)\frac{\partial\left(\frac{k}{m}T\right)}{\partial x_i} + \frac{1}{6}\frac{\partial\Delta}{\partial x_i} = \frac{2}{3}\sigma q_i,\\
&\frac{\partial\varrho_{\langle ijk\rangle}}{\partial t} - 3\frac{k}{m}\tilde{T}\left\{\frac{\partial t_{(\langle ij\rangle}}{\partial x_{k)}} - \frac{2}{5}\frac{\partial t_{\langle r(i\rangle}}{\partial x_r}\delta_{jk)}\right\} = \frac{3}{2}\sigma\varrho_{\langle ijk\rangle},\\
&\frac{\partial\Delta}{\partial t} + 8\frac{k}{m}\tilde{T}\frac{\partial q_i}{\partial x_i} = \frac{2}{3}\sigma\Delta.
\end{aligned} \tag{1.44}$$

n = 26 Variables: $F, F_i, F_{ij}, F_{ijk}F_{ijkk}$ or $\varrho, v_i, T, t_{\langle ij\rangle}, q_i, \varrho_{\langle ijk\rangle}, \Delta, \varrho_{kk\langle ij\rangle}$

$$\frac{\partial\varrho}{\partial t} + \tilde{\varrho}\frac{\partial v_i}{\partial x_i} = 0,$$
$$\frac{\partial v_i}{\partial t} + \frac{\left(\frac{k}{m}\tilde{T}\right)}{\tilde{\varrho}}\frac{\partial\varrho}{\partial x_i} + \frac{\partial}{\partial x_i}\left(\frac{k}{m}T\right) - \frac{1}{\tilde{\varrho}}\frac{\partial t_{\langle ij\rangle}}{\partial x_j} = 0,$$
$$\frac{\partial\left(\frac{k}{m}T\right)}{\partial t} + \frac{2}{3}\left(\frac{k}{m}\tilde{T}\right)\frac{\partial v_k}{\partial x_k} + \frac{2}{3}\frac{1}{\tilde{\varrho}}\frac{\partial q_k}{\partial x_k} = 0,$$

$$\frac{\partial t_{\langle ij\rangle}}{\partial t}-\frac{4}{5}\frac{\partial q_{\langle i}}{\partial x_{j\rangle}}-2\tilde{\varrho}\left(\frac{k}{m}\tilde{T}\right)\frac{\partial v_{\langle i}}{\partial x_{j\rangle}}-\frac{\partial \varrho_{\langle ijk\rangle}}{\partial x_k}=\sigma t_{\langle ij\rangle},\tag{1.45}$$
$$\frac{\partial q_i}{\partial t}-\frac{5}{2}\left(\frac{k}{m}\tilde{T}\right)\frac{\partial t_{\langle ik\rangle}}{\partial x_k}+\frac{5}{2}\tilde{\varrho}\left(\frac{k}{m}\tilde{T}\right)\frac{\partial \frac{k}{m}T}{\partial x_i}+\frac{1}{6}\frac{\partial \Delta}{\partial x_i}+\frac{1}{2}\frac{\partial \varrho_{kk\langle ij\rangle}}{\partial x_r}=\frac{2}{3}\sigma q_i,$$
$$\frac{\partial \varrho_{\langle ijk\rangle}}{\partial t}+\frac{3}{7}\left\{\frac{\partial \varrho_{rr(\langle ij\rangle}}{\partial x_{k)}}-\frac{2}{5}\frac{\partial \varrho_{nn\langle r(i\rangle}}{\partial x_r}\delta_{jk)}\right\}=\frac{3}{2}\sigma\varrho_{\langle ijk\rangle},$$
$$\frac{\partial \Delta}{\partial t}+8\left(\frac{k}{m}\tilde{T}\right)\frac{\partial q_i}{\partial x_i}=\frac{2}{3}\sigma\Delta,$$
$$\frac{\partial \varrho_{kk\langle ij\rangle}}{\partial t}+14\tilde{\varrho}\left(\frac{k}{m}\tilde{T}\right)^2\frac{\partial v_{\langle i}}{\partial x_{j\rangle}}+\frac{56}{6}\frac{k}{m}\tilde{T}\frac{\partial q_{\langle i}}{\partial x_{j\rangle}}+9\frac{k}{m}\tilde{T}\frac{\partial \varrho_{\langle ijk\rangle}}{\partial x_k}=\frac{7}{6}\sigma\left(\varrho_{kk\langle ij\rangle}+7t_{\langle ij\rangle}\right).$$

n = 35 Variables: $F, F_i, F_{ij}, F_{ijk}F_{ijkl}$ or $\varrho, v_i, T, t_{\langle ij\rangle}, q_i, \varrho_{\langle ijk\rangle}, \Delta, \varrho_{kk\langle ij\rangle}, \varrho_{\langle ijkl\rangle}$

$$\frac{\partial \varrho}{\partial t}+\tilde{\varrho}\frac{\partial v_i}{\partial x_i}=0,$$
$$\frac{\partial v_i}{\partial t}+\frac{\left(\frac{k}{m}\tilde{T}\right)}{\tilde{\varrho}}\frac{\partial \varrho}{\partial x_i}+\frac{\partial}{\partial x_i}\left(\frac{k}{m}T\right)-\frac{1}{\tilde{\varrho}}\frac{\partial t_{\langle ij\rangle}}{\partial x_j}=0,$$
$$\frac{\partial\left(\frac{k}{m}T\right)}{\partial t}+\frac{2}{3}\left(\frac{k}{m}\tilde{T}\right)\frac{\partial v_k}{\partial x_k}+\frac{2}{3}\frac{1}{\tilde{\varrho}}\frac{\partial q_k}{\partial x_k}=0,$$
$$\frac{\partial t_{\langle ij\rangle}}{\partial t}-\frac{4}{5}\frac{\partial q_{\langle i}}{\partial x_{j\rangle}}-2\tilde{\varrho}\left(\frac{k}{m}\tilde{T}\right)\frac{\partial v_{\langle i}}{\partial x_{j\rangle}}-\frac{\partial \varrho_{\langle ijk\rangle}}{\partial x_k}=\sigma t_{\langle ij\rangle},$$
$$\frac{\partial q_i}{\partial t}-\frac{5}{2}\frac{k}{m}\tilde{T}\frac{\partial t_{\langle ik\rangle}}{\partial x_k}+\frac{5}{2}\tilde{\varrho}\left(\frac{k}{m}\tilde{T}\right)\frac{\partial \frac{k}{m}T}{\partial x_i}+\frac{1}{6}\frac{\partial \Delta}{\partial x_i}+\frac{1}{2}\frac{\partial \varrho_{kk\langle ij\rangle}}{\partial x_r}=\frac{2}{3}\sigma q_i,\tag{1.46}$$
$$\frac{\partial \varrho_{\langle ijk\rangle}}{\partial t}+\frac{\partial \varrho_{\langle ijkr\rangle}}{\partial x_r}+\frac{3}{7}\left\{\frac{\partial \varrho_{rr(\langle ij\rangle}}{\partial x_{k)}}-\frac{2}{5}\frac{\partial \varrho_{nn\langle r(i\rangle}}{\partial x_r}\delta_{jk)}\right\}=\frac{3}{2}\sigma\varrho_{\langle ijk\rangle},$$
$$\frac{\partial \Delta}{\partial t}+8\left(\frac{k}{m}\tilde{T}\right)\frac{\partial q_i}{\partial x_i}=\frac{2}{3}\sigma\Delta,$$
$$\frac{\partial \varrho_{kk\langle ij\rangle}}{\partial t}+14\tilde{\varrho}\left(\frac{k}{m}\tilde{T}\right)^2\frac{\partial v_{\langle i}}{\partial x_{j\rangle}}+\frac{56}{5}\left(\frac{k}{m}\tilde{T}\right)\frac{\partial q_{\langle i}}{\partial x_{j\rangle}}+9\left(\frac{k}{m}\tilde{T}\right)\frac{\partial \varrho_{\langle ijk\rangle}}{\partial x_k}=\frac{7}{6}\sigma\left(\varrho_{kk\langle ij\rangle}+7t_{\langle ij\rangle}\right),$$
$$\frac{\partial \varrho_{\langle ijkl\rangle}}{\partial t}+4\frac{k}{m}\tilde{T}\left\{\frac{\partial \varrho_{(\langle ijk\rangle}}{\partial x_{l)}}-\frac{3}{7}\frac{\partial \varrho_{\langle r(ij\rangle}}{\partial x_r}\delta_{kl)}\right\}=1.8731...\sigma\varrho_{\langle ijkl\rangle}.$$

All these equations were first given, to our knowledge, by Ikenberry and Truesdell [23]. Their derivation involved the repeated solution of the linear algebraic system (1.24) for increasing values of n. On the right-hand side of $(1.46)_9$ the production $\hat{f}_{\langle ijkl\rangle}$ has again been taken from [14]; this is a convenient alternative to the lengthy calculation of the collision production with the phase density (1.22).

There is an awkward feature in the equations (1.41) through (1.46) that eventually frustrates even the most diligent student: As n is increased it is not safe to assume that the equations that have been calculated for lower n's remain unchanged. Thus for instance, as we proceed from $n = 20$ to $n = 21$, we do not merely add another equation to the system, but we change the equation for $\partial q_i/\partial t$ in a previously unforeseeable manner.

Because of this feature, the method of moments is not the most useful manner for the improved description of a gas. In Section 3 we shall learn about a better procedure, but first we have to provide more knowledge about the kinetic theory of gases.

2 More Kinetic Theory

2.1 Boltzmann Equation and Transfer Equations

We recall that the kinetic theory describes the state of a gas by the phase density $f(\mathbf{x}, \mathbf{c}, t)$, which obeys the Boltzmann equation

$$\frac{\partial f}{\partial t} + c_i \frac{\partial f}{\partial x_i} = S, \tag{2.1}$$

where S is the collision operator given by

$$S = \int \left(f' f^{1'} - f f' \right) \sigma g \sin \vartheta \, d\vartheta \, d\varepsilon \, d\mathbf{c}^1. \tag{2.2}$$

Multiplication by a generic function $\psi(\mathbf{x}, \mathbf{c}, t)$ and integration over $\mathbf{c}$ leads to an equation of transfer of the form

$$\frac{\partial \varrho \bar{\psi}}{\partial t} + \frac{\partial \varrho \overline{\psi c_i}}{\partial x_i} = \varrho \overline{\left(\frac{\partial \psi}{\partial t} + c_i \frac{\partial \psi}{\partial x_i} \right)} + \frac{m}{4} \int \left(\psi + \psi^1 - \psi' - \psi^{1'} \right) S \, d\mathbf{c}. \tag{2.3}$$

Here $\bar{\psi}$ is the mean value of ψ, viz.

$$\bar{\psi} = \frac{m}{\varrho} \int \psi f \, d\mathbf{c}. \tag{2.4}$$

In particular, for $\psi = -\frac{k}{m} \ln f$ we have

$$\bar{\psi} = s = -\frac{1}{\varrho} k \int \ln f \, f \, d\mathbf{c} \quad \text{and} \quad \varrho \overline{\psi c_i} = h_i = -\frac{1}{\varrho} k \int c_i \ln f f \, d\mathbf{c}. \tag{2.5}$$

Here s is the specific entropy and h_i the entropy flux. The corresponding equation of transfer is

$$\frac{\partial \varrho s}{\partial t} + \frac{\partial h_i}{\partial x_i} \geq 0, \tag{2.6}$$

and it represents the entropy inequality.

Most macroscopic thermodynamic quantities are moments, or central moments of the phase density. Generically these have the forms

$$F_\alpha = m \int c_\alpha f d\mathbf{c} \quad \text{and} \quad \varrho_\alpha = m \int C_\alpha f \, d\mathbf{c} \tag{2.7}$$

where α is a multi-index. F_α stands for $F_{i_1 i_2 \cdots i_n}$ and c_α for $c_{i_1} c_{i_2} \cdots c_{i_n}$.

2.2 Linearized Kinetic Theory

Often we are interested in processes close to equilibrium and for those it is appropriate to use an approximate linearized form of the Boltzmann equation. We introduce

$$f = \tilde{f}(1 + \varphi) \quad \text{with} \; \tilde{f} = \frac{\tilde{\varrho}}{m} \sqrt{\frac{m}{2\pi k \tilde{T}}}^{3} e^{-mc^2/2k\tilde{T}} \tag{2.8}$$

so that $\tilde{f}$ is the Maxwell distribution for a gas at rest with constant and uniform fields $\tilde{\varrho}$ and $\tilde{T}$ of density and temperature. We linearize the collision operator in φ and obtain from (2.2)

$$S = \tilde{f} \int \tilde{f}^1(\varphi' + \varphi^{1'} - \varphi - \varphi^1)\sigma g \sin\vartheta \, d\vartheta \, d\varepsilon \, d\mathbf{c}^1, \tag{2.9}$$

so the Boltzmann equation assumes the form

$$\frac{\partial \varphi}{\partial t} + c_i \frac{\partial \varphi}{\partial x_i} = I(\varphi) \quad \text{where } I(\varphi) = \int \tilde{f}^1(\varphi' + \varphi^{1'} - \varphi - \varphi^1)\sigma g \sin\vartheta \, d\vartheta \, d\varepsilon \, d\mathbf{c}^1. \tag{2.10}$$

The approximate solution of this linearized integral equation proceeds by converting it into a system of partial differential equations for the coefficients $b^{(r)}_{\langle i_1 \cdots i_l \rangle}(\mathbf{x}, t)$ in the expansion

$$\varphi(\mathbf{x}, \mathbf{c}, t) = \sum_{r,l} b^{(r)}_{\langle i_1 \cdots i_l \rangle}(\mathbf{x}, t) \psi^{(r)}_{\langle i_1 \cdots i_l \rangle},$$
$$\varphi(\mathbf{x}, \mathbf{c}, t) = \sum_{r,l} b^{(r)}_{\langle i_1 \cdots i_l \rangle}(\mathbf{x}, t) \psi_{lr}(c) \frac{c_{\langle i_1 \cdots i_l \rangle}}{c^l} \tag{2.11}$$

of φ in terms of the eigenfunctions $\psi^{(r)}_{\langle i_1 \cdots i_l \rangle} = \psi_{lr}(c) \frac{c_{\langle i_1 \cdots i_l \rangle}}{c}$ of the linearized collision operator $I(\varphi)$. We have

$$I\left(\psi_{lr}(c) \frac{c_{\langle i_1 \cdots i_l \rangle}}{c^l}\right) = -\zeta_{rl} \psi_{lr}(c) \frac{c_{\langle i_1 \cdots i_l \rangle}}{c^e}. \tag{2.12}$$

Note that the dependence of the eigenfunctions on the direction of $\mathbf{c}$ is given by the trace–less tensors

$$\frac{c_{\langle i_1 \cdots i_l \rangle}}{c^l} \equiv \frac{(-1)^l}{1 \cdot 3 \cdot 5 \cdots (2l-1)} c^{l+1} \frac{\partial^l}{\partial c_{i_1} \cdots \partial c_{i_l}} \left(\frac{1}{c}\right), \tag{2.13}$$

which satisfy the orthogonality conditions

$$\int \frac{c_{\langle i_1 \cdots i_l \rangle}}{c^l} \frac{c_{\langle j_1 \cdots j_n \rangle}}{c^n} \sin\vartheta \, d\vartheta \, d\varphi = 0 \quad \text{for } l \neq n. \tag{2.14}$$

The general properties of the operator I guarantee that its eigenvalues ζ_{rl} are real and non-negative and that the functions ψ_{lr} form an orthonormal set (see [97]), so we have

$$\int \psi_{lr} \psi_{ls} \frac{m}{\tilde{\varrho}} \tilde{f} \, dc = \delta_{rs}. \tag{2.15}$$

Insertion of (2.11) into (2.10) and use of the relations (2.13), (2.14) provides the equation[1]

$$\sum_r \left(\frac{\partial b^{(r)}_{\langle i_1 \cdots i_l \rangle}}{\partial t} \psi_{lr} + \frac{\partial b^{(r)}_{\langle i_1 \cdots i_{l-1} \rangle}}{\partial x_{i_l \rangle}} c\psi_{l-1r} + \frac{l+1}{2l+3} \frac{\partial b^{(r)}_{\langle i_1 \cdots i_{l+1} \rangle}}{\partial x_{i_{l+1}}} c\psi_{l+1r} + \zeta_{rl} \psi_{lr} b^{(r)}_{\langle i_1 \cdots i_l \rangle} \right) = 0. \tag{2.16}$$

Formally we can further reduce this equation by expanding $c\psi_{l-1r}$ and $c\psi_{l+1r}$ in terms of the functions ψ_{lr}, namely,

$$c\psi_{l-1r} = \sum_{r'} c(l-1, r|l, r')\psi_{lr}, \qquad c\psi_{l+1r} = \sum_{r'} c(l+1, r|l, r')\psi_{lr'}.$$

[1] In (2.16) $b^{(r)}_{\langle i_1 \cdots i_{l-1} \rangle}$ is zero for $l = 0$.

Thus by use of the orthogonality conditions (2.15) we obtain from (2.16)

$$\frac{\partial b^{(s)}_{\langle i_1\cdots i_l\rangle}}{\partial t}+\sum_r\left[c(l-1,r|l,s)\frac{\partial b^{(r)}_{\langle i_1\cdots i_{l-1}}}{\partial x_{i_l\rangle}}+\frac{l+1}{2l+3}c(l+1,r|l,s)\frac{\partial b^{(r)}_{\langle i_1\cdots i_{l+1}\rangle}}{\partial x_{i_{l+1}}}\right]+\zeta_{sl}b^{(s)}_{\langle i_1\cdots i_l\rangle}=0. \quad (2.17)$$

This is the desired system of partial differential equations for the coefficients $b^{(r)}_{\langle i_1\cdots i_l\rangle}(\mathbf{x},t)$ in the expansion (2.11) of $\varphi(\mathbf{x},\mathbf{c},t)$. The system is equivalent to the linearized Boltzmann equation (2.10), but of course it is not specific until we know the eigenvalues and the coefficients $c(l-1,r|l,s)$ and $c(l+1,r|l,s)$. That knowledge depends on the potential energy of the atomic interaction and Waldmann [97] has calculated these values for Maxwellian atoms.

2.3 Eigenfunctions and Eigenvalues for Maxwellian Atoms

According to [97] for Maxwellian atoms the functions $\psi_{lr}(c)$ are Sonine polynomials, which we may write in the form

$$\psi_{lr}(c)=\sqrt{\frac{\sqrt{\pi}r!}{2(l+\frac{1}{2}+r)!}}S^{(r)}_{l+1/2}\left(\frac{c^2}{2\frac{k}{m}\tilde{T}}\right)\left(\frac{c}{\sqrt{2\frac{k}{m}\tilde{T}}}\right)^l$$

where

$$S^{(r)}_{l+1/2}=\sum_{p=0}^{r}\frac{\left(l+\frac{1}{2}\right)\left(l+\frac{3}{2}\right)\cdots\left(l+\frac{1}{2}+r\right)}{p!(r-p)!\left(l+\frac{1}{2}\right)\left(1+\frac{3}{2}\right)\cdots\left(l+\frac{1}{2}+p\right)}\left(-\frac{c^2}{2\frac{k}{m}\tilde{T}}\right)^p, \quad (2.18)$$

and the eigenvalues are given by

$$\zeta_{rl}=\frac{\tilde{\varrho}}{m}2\pi\int_0^{\pi/2}\left[1-\sin^{l+2r}\Theta P_l(\sin\Theta)-\cos^{l+2r}\Theta P_l(\cos\Theta)\right]\Gamma(\Theta)\sin\Theta\,d\Theta \quad (2.19)$$

where $P_l(\alpha)$ are Legendre polynomials and $\Gamma(\Theta)=4\cos\Theta\sigma g$.

In Table 1 we list the ratio ζ_{rl}/ζ_{02} for some pairs (r,l).

$r\backslash l$	0	1	2	3
0	0	0	1	3/2
1	0	2/3	7/6	1.570
2	2/3	1	1.342	1.667
3	1	1.228	1.491	1.763

Table 1. Some eigenvalues ζ_{rl}/ζ_{02}.

We note that ζ_{00},ζ_{10}, and ζ_{01} are all zero. This reflects the conservation laws of mass, momentum, and energy, because the corresponding equations (2.17) are the equations of balance of mass, momentum, and energy.

With the explicit knowledge of $\psi_{lr}(c)$ it is now possible to calculate the coefficients $c(l-1,r|l,s)$ and $c(l+1,r|l,s)$ in (2.17). We obtain

$$\begin{aligned} c(l-1,r|l,s) &= \sqrt{2\frac{k}{m}\tilde{T}}\left(\sqrt{l+\frac{1}{2}+s}\,\delta_{r,s}-\sqrt{s+1}\,\delta_{s+1,r}\right), \\ c(l+1,r|l,s) &= \sqrt{2\frac{k}{m}\tilde{T}}\left(\sqrt{l+\frac{3}{2}+s}\,\delta_{r,s}-\sqrt{s}\,\delta_{s-1,r}\right), \end{aligned} \tag{2.20}$$

so (2.17) now reads

$$\begin{aligned} \frac{\partial b^{(s)}_{\langle i_1\cdots i_l\rangle}}{\partial t} + \sqrt{2\frac{k}{m}\tilde{T}}\Bigg[\sqrt{l+\frac{1}{2}+s}\frac{\partial b^{(s)}_{\langle i_1\cdots i_{l-1}}}{\partial x_{i_l\rangle}} - \sqrt{s+1}\frac{\partial b^{(s+1)}_{\langle i_1\cdots i_{l-1}}}{\partial x_{i_l\rangle}} \\ +\frac{l+1}{2l+3}\left(\sqrt{l+\frac{3}{2}+s}\frac{\partial b^{(s)}_{\langle i_1\cdots i_{l+1}}}{\partial x_{i_{l+1}}} - \sqrt{s}\frac{\partial b^{(s-1)}_{\langle i_1\cdots i_{l+1}\rangle}}{\partial x_{i_{l+1}}}\right)\Bigg] = -\zeta_{sl} b^{(s)}_{\langle i_1\cdots i_l\rangle}. \end{aligned} \tag{2.21}$$

It is obvious from (2.11) and the orthogonality conditions (2.14), (2.15) that the coefficients $b^{(r)}_{\langle i_1\cdots i_e\rangle}$ are proportional to the mean values of the eigenfunctions $\psi^r_{\langle i_1\cdots i_l\rangle} = \psi_{lr}(c)\frac{c_{\langle i_1\cdots i_l\rangle}}{c^e}$. In fact we have

$$b^{(r)}_{\langle i_1\cdots i_l\rangle} = \frac{2^{l+1}\left(l+\frac{3}{2}\right)!}{\sqrt{\pi}l!}\,\overline{\psi^{(r)}_{\langle i_1\cdots i_l\rangle}} = \frac{2^{l+1}\left(l+\frac{3}{2}\right)!}{\sqrt{\pi}l!}\frac{m}{\tilde{\varrho}}\int \psi^{(r)}_{\langle i_1\cdots i_e\rangle} f dc. \tag{2.22}$$

Insertion of (2.22) into (2.21) thus provides us with a system of coupled equations for $\overline{\psi^{(r)}_{\langle i_1\cdots i_e\rangle}}$, viz.

$$\begin{aligned} \frac{\partial \overline{\psi^{(s)}_{\langle i_1\cdots i_l\rangle}}}{\partial t} + \sqrt{2\frac{k}{m}\tilde{T}}\Bigg[\frac{l}{2l+1}\left(\sqrt{l+\frac{1}{2}+s}\frac{\partial \overline{\psi^{(s)}_{\langle i_1\cdots i_{l-1}}}}{\partial x_{i_l\rangle}} - \sqrt{s+1}\frac{\partial \overline{\psi^{(s+1)}_{\langle i_1\cdots i_{l-1}}}}{\partial x_{i_l\rangle}}\right) \\ +\sqrt{l+\frac{3}{2}+s}\frac{\partial \overline{\psi^{(s)}_{\langle i_1\cdots i_{l+1}\rangle}}}{\partial x_{i_{l+1}}} - \sqrt{s}\frac{\partial \overline{\psi^{(s-1)}_{\langle i_1\cdots i_{l+1}\rangle}}}{\partial x_{i_{l+1}}}\Bigg] = -\zeta_{sl}\overline{\psi^{(s)}_{\langle i_1\cdots i_l\rangle}}. \end{aligned} \tag{2.23}$$

This is a system of increasing tensorial character. We shall be interested in the subsystem of scalar equations that results from the application of the operator $\partial^l/\partial x_{i_1}\cdots\partial x_{i_l}$ and obtain with

$$\frac{\partial^l \overline{\psi^{(s)}_{\langle i_1\cdots i_l\rangle}}}{\partial x_{i_1}\cdots\partial x_{i_l}} \equiv B^{(s)}_l \quad\text{and}\quad \frac{\partial^l}{\partial x_{i_1}\cdots\partial x_{i_l}}\left(\frac{\partial \overline{\psi^{(s)}_{\langle i_1\cdots i_{l-1}}}}{\partial x_{i_l\rangle}}\right) = \frac{l}{2l-1}\Delta B^{(s)}_{l-1}, \tag{2.24}$$

$$\begin{aligned} \frac{\partial B^{(s)}_l}{\partial t} + \sqrt{2\frac{k}{m}\tilde{T}}\Bigg\{\frac{l^2}{(2l+1)(2l-1)}\left(\sqrt{l+\frac{1}{2}+s}\Delta B^{(s)}_{l-1} - \sqrt{s+1}\Delta B^{(s+1)}_{l-1}\right) \\ +\sqrt{l+\frac{3}{2}+s}B^{(s)}_{l+1} - \sqrt{s}B^{(s-1)}_{l+1}\Bigg\} = -\zeta_{sl}B^{(s)}_l. \end{aligned} \tag{2.25}$$

Here Δ is the Laplace operator.

The system (2.25) is obviously uncoupled from the tensorial systems that follow from (2.23) by a multiple application the operators $\partial/\partial x_j$ and $\varepsilon_{ijk}(\partial/\partial x_k)$. It governs sonic waves and density fluctuations, and that is why we consider it.

3 Equations for Mean Eigenfunctions and Closure

3.1 Mean Eigenfunctions

In Section 1 we deplored the fact that the addition of another moment to the set of variables not only added an equation to the previously derived set but also changed that set in subtle and non–transparent ways.

This is not so, if we choose the variables, fluxes, and productions to be related to the mean eigenfunctions $\overline{\psi^r_{\langle i_1\cdots i_e\rangle}}$ rather than to the moments $\overline{c_{i_1\cdots i_n}}$, that is

$$\begin{array}{llll} v_\alpha = m\int \psi^r_{\langle i_1\cdots i_e\rangle} f\,dc & \text{rather than} & u_\alpha = m\int c_{i_1\cdots i_n} f\,dc, \\ G^k_\alpha = m\int c_k\psi^r_{\langle i_1\cdots i_e\rangle} f\,dc & \text{rather than} & f_{\alpha k} = m\int c_k c_{i_1\cdots i_n} f\,dc, \\ g_\alpha = m\int \psi^r_{\langle i_1\cdots i_e\rangle} S\,dc & \text{rather than} & f_\alpha = m\int c_{i_1\cdots i_n} S\,dc. \end{array} \tag{3.1}$$

There is the question of ordering the v_α's, because α stands for the double index (r, l). We consider it natural to proceed in such a manner, that α increases with the polynomial degree $2r+l$ in c_i and, within those v_α's that have the *same* degree, we let α increase with l. Thus the first nine v_α's are given by (with $\beta = 1/\sqrt{2\frac{k}{m}\tilde{T}}$)

$$\begin{array}{llll}
(r,l)=(0,0): & \psi^{(0)} = 1 & v^{00} = \varrho, \\
(r,l)=(0,1): & \psi^{(1)}_i = \sqrt{\tfrac{2}{3}}\beta c_i & v^{01}_i = \sqrt{\tfrac{2}{3}}\beta\varrho v_i, \\
(r,l)=(1,0): & \psi^{(1)} = \sqrt{\tfrac{2}{3}}\left(\tfrac{3}{2}-(\beta c)^2\right) & v^{10} = \sqrt{\tfrac{2}{3}}\left(\tfrac{3}{2}\varrho-\beta^2 F_{ii}\right), \\
(r,l)=(0,2): & \psi^{0}_{\langle ij\rangle} = \tfrac{2}{\sqrt{15}}c_{\langle i}c_{j\rangle} & v^{02}_{\langle ij\rangle} = \tfrac{2}{\sqrt{15}}\beta^2 F_{\langle ij\rangle}, \\
(r,l)=(1,1): & \psi^{(1)}_i = \tfrac{2}{\sqrt{15}}\left(\tfrac{5}{2}-(\beta c)^2\right)\beta c_i & v^{11}_i = \tfrac{2}{\sqrt{15}}\left(\tfrac{5}{2}\beta\varrho v_i-\beta^3 F_{ijj}\right), \\
(r,l)=(0,3): & \psi^{(0)}_{\langle ijk\rangle} = \sqrt{\tfrac{8}{105}}\beta^3 c_{\langle i}c_j c_{k\rangle} & v^{03}_{\langle ijk\rangle} = \sqrt{\tfrac{8}{105}}\beta^3 F_{\langle ijk\rangle}, \\
(r,l)=(2,0): & \psi^{(2)} = \sqrt{\tfrac{8}{15}}\left(\tfrac{15}{8}-\tfrac{5}{2}(\beta c)^2+\tfrac{1}{2}(\beta c)^4\right) & v^{20} = \sqrt{\tfrac{8}{15}}\left(\tfrac{15}{8}\varrho-\tfrac{5}{2}\beta^2 F_{ii}+\tfrac{1}{2}\beta^4 F_{iijj}\right), \\
(r,l)=(1,2): & \psi^{(1)}_{\langle ij\rangle} = \sqrt{\tfrac{8}{105}}\left(\tfrac{7}{2}-(\beta c)^2\right)\beta^2 c_{\langle i}c_{j\rangle} & v^{12}_{\langle ij\rangle} = \sqrt{\tfrac{8}{105}}\left(\tfrac{7}{2}\beta^2 F_{\langle ij\rangle}-\beta^4 F_{\langle ij\rangle kk}\right), \\
(r,l)=(0,4): & \psi^{(0)}_{\langle ijkl\rangle} = \tfrac{4}{\sqrt{945}}\beta^4 c_{\langle i}c_j c_k c_{l\rangle} & v^{04}_{\langle ijkl\rangle} = \tfrac{4}{\sqrt{945}}\beta^4 F_{\langle ijkl\rangle}.
\end{array} \tag{3.2}$$

This list may easily be continued indefinitely. Therefore, by insertion into (2.23) a system of field equations for the mean eigenfunctions may be derived in which each consecutive step on the (r,l)-ladder provides an additional equation without changing those previous.

The mean eigenfunctions are irreducible combinations of the moments. In particular, we conclude from $(3.2)_4$ that the equation for $(r,l) = (0,2)$ is the equation of balance of the deviator of the momentum flux. Therefore $1/\zeta_{02}$ is the relaxation time of the momentum flux or stress–deviator. By the considerations of Paragraph 2.3.2 that relaxation time is related to the viscosity μ. We have

$$\zeta_{02} = \frac{\mu}{p}. \tag{3.3}$$

$$\partial_t B_0^{(0)} + a\{ \sqrt{\tfrac{3}{2}} \quad B_1^{(0)} \} = 0$$

$$\partial_t B_1^{(0)} + a\{ \tfrac{1}{\sqrt{6}}\Delta B_0^{(0)} - \tfrac{1}{3}\Delta B_0^{(1)} + \sqrt{\tfrac{5}{2}} \quad B_2^{(0)} \} = 0$$

$$\partial_t B_0^{(1)} + a\{ - \quad B_1^{(0)} + \sqrt{\tfrac{5}{2}} \quad B_1^{(1)} \} = 0$$

$$\partial_t B_2^{(0)} + a\{ + \tfrac{4}{3\sqrt{10}}\Delta \quad B_1^{(0)} - \tfrac{4}{15}\Delta \quad B_1^{(1)} + \sqrt{\tfrac{7}{2}} \quad B_3^{(0)} \} = -\zeta_{02} B_2^{(0)}$$

$$\partial_t B_1^{(1)} + a\{ \sqrt{\tfrac{5}{18}}\Delta B_0^{(1)} - \quad B_2^{(0)} - \tfrac{\sqrt{2}}{3}\Delta B_0^{(2)} + \tfrac{7}{2} \quad B_2^{(1)} \} = -\zeta_{11} B_1^{(1)}$$

13

$$\partial_t B_3^{(0)} + a\{ + \tfrac{9}{5\sqrt{14}}\Delta \quad B_2^{(0)} - \tfrac{9}{35}\Delta \quad B_2^{(1)} + \sqrt{\tfrac{9}{2}} B_4^{(0)} \} = -\zeta_{03} B_3^{(0)}$$

20

$$\partial_t B_0^{(2)} + a\{ -\sqrt{2} \quad B_1^{(1)} + \sqrt{\tfrac{7}{2}} \quad B_1^{(2)} \} = -\zeta_{20} B_0^{(2)}$$

21

$$\partial_t B_2^{(1)} + a\{ + \tfrac{4\sqrt{7}}{15\sqrt{2}}\Delta \quad B_1^{(1)} - \quad B_3^{(0)} - \tfrac{4\sqrt{2}}{15}\Delta \quad B_1^{(2)} + \sqrt{\tfrac{9}{2}} \quad B_3^{(1)} \} = -\zeta_{12} B_2^{(1)}$$

26

$$\partial_t B_4^{(0)} + a\{ + \tfrac{16}{7\sqrt{18}}\Delta \quad B_3^{(0)} - \tfrac{16}{63}\Delta \quad B_3^{(1)} + \sqrt{\tfrac{11}{2}} B_5^{(0)} \} = -\zeta_{04} B_4^{(0)}$$

35

$$a = \sqrt{2\frac{k}{m}T} \qquad (3.4)$$

Thus the system (2.25) is entirely specific as a system of field equations for the functions $B_l^{(s)}$. The eigenfunctions ζ_{sl} may be taken from (2.19) or Table 1 and equation (3.3). We proceed to write down the first few equations explicitly.

3.2 Field Equations for Mean Eigenfunctions and Closure

In an explicit form the first 9 equations of the system (2.25) are listed on the previous page in equation (3.4). This system may easily be extended to an arbitrary number of equations. Infact, Weiss [96] has used hundreds of equations in his calculations whose results will be presented in subsequent chapters.

While the system (3.3) is explicit in the $B_l^{(s)}$'s, it is not closed at any stage. Thus as usual in the kinetic theory, and in macroscopic thermodynamics, we face a closure problem. This problem is handled most elegantly in the present context.

Indeed, let us be interested in a 13–moment theory with the objective of determining the fields

$$\varrho, v_i, F_{ii}, F_{\langle ij\rangle}, F_{ijj}, \quad \text{i.e.,} \quad v^{00}, v_i^{01}, v^{10}, v_{\langle ij\rangle}^{02}, v_i^{11}. \tag{3.5}$$

In that case the first five of the equations (3.4) are relevant as indicated by the appropriate frame in the system (3.4). We close that system by setting $B_3^{(0)}, B_0^{(2)}$, and $B_2^{(1)}$ equal to zero. By $(3.2)_{6-8}$ this means, in the rest frame of the gas

$$\varrho_{\langle ijk\rangle} = 0, \qquad \varrho_{iijj} = 5\frac{\varrho_{ii}}{\beta^2} - \frac{15}{4}\frac{\varrho}{\beta^4}, \qquad \varrho_{\langle ij\rangle kk} = \frac{7}{2}\frac{\varrho_{\langle ij\rangle}}{\beta^2} \tag{3.6}$$

or with $\varrho_{ii} = 3\varrho\frac{k}{m}T$, $\frac{1}{\beta} = \sqrt{2\frac{k}{m}T}$

$$\varrho_{\langle ijk\rangle} = 0, \qquad \varrho_{iijj} = 15\varrho\left(\frac{k}{m}T\right)^2, \qquad \varrho_{\langle ij\rangle kk} = 7\frac{k}{m}T\varrho_{\langle ij\rangle}. \tag{3.7}$$

This is exactly the result of extended thermodynamics [see (2.2.45)] with $F(z) = (k/m)z$ and $z = \varrho/T^{3/2}$. Also this is the same result as in equation (1.3.33), which was obtained from a closure by use of Grad's 13-moment method.

In this case of 13 fields we thus have three different but equivalent closure procedures, viz.

i. Chapter 1: (Kinetic theory of 13 moments)
The system of equations of balance of moments is closed by the calculation of $\varrho_{\langle ijk\rangle}$ and ϱ_{ijkk} with the Grad distribution.

ii. Chapter 2: (Extended thermodynamics)
The system of balance laws is closed by formulating constitutive relations for $\varrho_{\langle ijk\rangle}$ and ϱ_{ijkk}. These relations are calculated by the exploitation of the principle of material frame indifference and the entropy principle.

iii. Chapter 10: (Molecular thermodynamics of eigenfunctions)
The system of balance laws is closed by setting the "supernumerary" mean values $B_3^{(0)}, B_0^{(2)}$, and $B_2^{(1)}$ equal to zero.

Obviously the latter method is the most elegant one. We have no general proof that the procedures i. through iii. are equivalent for an arbitrary number of fields, but we strongly suspect that this is so; Weiss, in exploiting molecular thermodynamics of many fields, has proceeded on this assumption.

In the system (3.4) we have framed the closed subsystems that correspond to 13, 20, 21, 26, and 35 fields.

4 Molecular Extended Thermodynamics and Method of Eigenfunctions

4.1 Phase Density in Molecular Extended Thermodynamics

The arguments of molecular extended thermodynamics in Section 1 are largely independent of the form of the variables. In particular, the variables u_α, fluxes F^i_α, and productions f_α in (1.1), (1.2) may be replaced by the variables v_α, fluxes G^k_α, and productions g_α in (3.1). If we write

$$
\begin{aligned}
v_\alpha &= m \int \psi_\alpha f \, dc \quad \text{with} \quad \psi_\alpha = \psi^r_{\langle i_1 \cdots i_e \rangle}(\mathbf{c}), \\
\hat{v}_\alpha &= m \int \Psi_\alpha f \, dc \quad \text{with} \quad \Psi_\alpha = \psi^r_{\langle i_1 \cdots i_e \rangle}(\mathbf{C}),
\end{aligned} \tag{4.1}
$$

we may run through the analysis of Paragraphs 1.1 through 1.5 without any change other than replacing

$$
u_\alpha, F^i_\alpha, f_\alpha, c_\alpha, C_\alpha \quad \text{by} \quad v_\alpha, G^i_\alpha, g_\alpha, \psi_\alpha, \Psi_\alpha. \tag{4.2}
$$

Thus in particular we obtain, instead of (1.22),

$$
\begin{aligned}
f &= f_M \left(1 - \frac{m}{k} \sum_{\alpha=1}^{n} \hat{\lambda}_\alpha \Psi_\alpha \right), \\
f &= f_M \left(1 - \frac{m}{k} \sum_{\alpha=1}^{n} \lambda_\alpha \psi_\alpha \right).
\end{aligned} \tag{4.3}
$$

It is clear that the Lagrange multipliers here are different from those in Section 1, but we use the same symbols.

4.2 Lagrange Multipliers as Expansion Coefficients

The equation (4.3) has been obtained by a maximization of entropy and subsequent linearization in the Lagrange multipliers of non–equilibrium. That equation, however, is just what the kinetic theory obtains, according to Paragraph 2.2, when it expands the phase density in terms of eigenfunctions, viz. by (2.8) and (2.11)

$$
f = \tilde{f} \left(1 + \sum_\alpha b_\alpha \psi_\alpha \right). \tag{4.4}
$$

We conclude that the expansion coefficients

$$
b_\alpha = b^{(r)}_{\langle i_1 \cdots i_e \rangle} \tag{4.5}
$$

are equal to the Lagrange multipliers λ_α to within a constant factor.

This observation provides an interpretation for the Lagrange multipliers that goes beyond their interpretation of mere auxiliary quantities. Indeed, by this observation the Lagrange multipliers are proportional to the mean values of the eigenfunctions.

Chapter 10

Testing Extended Thermodynamics by Light

Light scattering is a powerful tool for the experimental determination of transport coefficients like heat conductivity and viscosity. The great advantage of this method is that it works in equilibrium; in fact, it makes use of the density fluctuations in equilibrium. By Onsager's hypothesis the auto-correlation function of such a fluctuation is related to the solution of the macroscopic field equations.

We apply light scattering to monatomic gases whose transport coefficients are already known and we use it to determine the ranges of validity of the Navier–Stokes–Fourier theory and of extended thermodynamics with differently many moments. The rationale is this:

Depending on the position of the detector that monitors the scattered light the scale of the density fluctuation that scatters the light is different. Thus a large scale fluctuation is seen in forward scattering and a small scale fluctuation is seen in backward scattering. For a large scale fluctuation the Navier–Stokes–Fourier theory is adequate, while we need extended thermodynamics for small scale fluctuations. We shall answer the following question: How many moments do we need in extended thermodynamics, if we wish to describe processes whose density changes significantly over a mean free path?

The chapter begins with a short description of the physics of light scattering and its evaluation. This part is abstracted from the book [98] by Berne and Pecora. The results concerning extended thermodynamics are due to Weiss [96].

1 Basic Electrodynamics

1.1 Distant Field Approximation

Light is scattered by a gas in equilibrium, because unavoidable fluctuations $\delta\varrho$ of density make the dielectric constant ε fluctuate. We proceed to describe the basic electrodynamics that governs the effect. For visualization we refer to Figure 1.

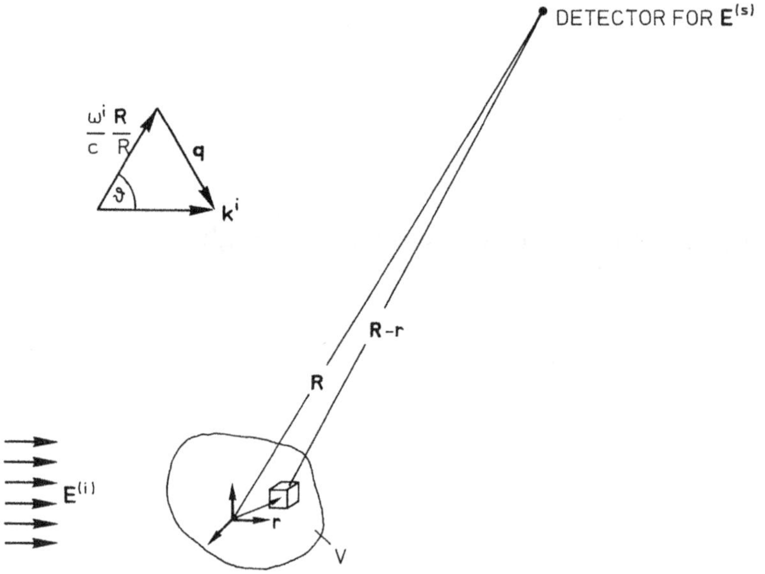

Figure 1. Incident light is scattered at point $\mathbf{r}$ in the scattering volume V.

The incident electromagnetic field is given by the fields

$$\begin{array}{llll} \mathbf{E}^i & \text{electric field,} & \mathbf{D}^i & \text{dielectric displacement,} \\ \mathbf{H}^i & \text{magnetic field,} & \mathbf{B}^i & \text{magnetic flux density.} \end{array} \tag{1.1}$$

Inside the scattering volume the electromagnetic field is a superposition of the incident field and the scattered field. We have

$$\begin{array}{ll} \mathbf{E} = \mathbf{E}^i + \mathbf{E}^s, & \mathbf{D} = \varepsilon\ (\mathbf{E}^i + \mathbf{E}^s) \approx \varepsilon_0(\mathbf{E}^i + \mathbf{E}^s) + \delta\varepsilon\mathbf{E}^i, \\ \mathbf{H} = \mathbf{H}^i + \mathbf{H}^s, & \mathbf{B} = \mu_0(\mathbf{H}^i + \mathbf{H}^s), \end{array} \tag{1.2}$$

provided we neglect $\delta\varepsilon \cdot \mathbf{E}^s$ and if no magnetization occurs in the gas. The fields $\mathbf{E}, \mathbf{D}, \mathbf{H}, \mathbf{B}$ and $\mathbf{E}^i, \mathbf{D}^i, \mathbf{H}^i, \mathbf{B}^i$ satisfy the Maxwell equations; therefore, $\mathbf{E}^s, \mathbf{D}^s, \mathbf{H}^s$, and $\mathbf{B}^s$ must also satisfy these equations so that we have

$$\frac{\partial \mathbf{B}^s}{\partial t} + \operatorname{curl} \mathbf{E}^s = 0, \quad \operatorname{div} \mathbf{B}^s = 0,$$

or by elimination of $\mathbf{B}^s, \mathbf{E}^s$

$$-\frac{\partial \mathbf{D}^s}{\partial t} + \operatorname{curl} \mathbf{H}^s = 0, \quad \operatorname{div} \mathbf{D}^s = 0$$

$$\begin{aligned} \mu_0 \frac{\partial \mathbf{H}^s}{\partial t} + \frac{1}{\varepsilon_0} \operatorname{curl} \mathbf{D}^s &= \frac{1}{\varepsilon_0} \operatorname{curl} (\delta\varepsilon \mathbf{E}^i), \qquad \operatorname{div} \mathbf{H}^s = 0, \\ -\frac{\partial D^s}{\partial t} + \quad \operatorname{curl} \mathbf{H}^s &= 0, \qquad \operatorname{div} \mathbf{D}^s = 0. \end{aligned} \tag{1.3}$$

By elimination of $\mathbf{H}^s$ we obtain with $c^2 = 1/\varepsilon_0\mu_0$

$$\frac{1}{c^2} \frac{\partial^2 \mathbf{D}^s}{\partial t^2} - \Delta \mathbf{D}^s = \operatorname{curl} \operatorname{curl} (\delta\varepsilon \mathbf{E}^i). \tag{1.4}$$

We introduce a potential $\boldsymbol{\pi}$ defined by $\mathbf{D}^s = \operatorname{curl} \operatorname{curl} \boldsymbol{\pi}$ and obtain the differential equation

$$\frac{1}{c^2} \frac{\partial^2 \boldsymbol{\pi}}{\partial t^2} - \Delta \boldsymbol{\pi} = \delta\varepsilon \cdot \mathbf{E}^{(i)} + \mathbf{F} \quad \text{with curl curl } \mathbf{F} = 0. \tag{1.5}$$

The solution that vanishes at infinity is the well-known *retarded potential* given by

$$\boldsymbol{\pi}(\mathbf{R}, t) = \frac{1}{4\pi} \int_V d\mathbf{r} \frac{\delta\varepsilon(\mathbf{r}, t')\mathbf{E}(\mathbf{r}, t')}{|\mathbf{R} - \mathbf{r}|} \quad \text{where } t' = t - \frac{|\mathbf{R} - \mathbf{r}|}{c} \tag{1.6}$$

is the time it takes the light to reach the detector from the scattering site in V. The function $\mathbf{F}$ in (1.5) can be set equal to zero without loss of generality.

We calculate $\mathbf{D}^s(\mathbf{R}, t)$ under two simplifying assumptions, namely,

i. *slow* fluctuation of $\delta\varepsilon(\mathbf{r}, t)$, and

ii. detector *far-away* from the scattering site.

Assumption i. allows us to set $\delta\varepsilon(\mathbf{r}, t') \approx \delta\varepsilon(\mathbf{r}, t)$; it implies that the size of the fluctuation $\delta\varepsilon = \varepsilon - \varepsilon_0$ does not appreciably change as the scattered light travels to the detector. By assumption ii. we retain only *the* term in $\mathbf{D}^s = \operatorname{curl} \operatorname{curl} \boldsymbol{\pi}$ that is of order $\frac{1}{R}$. Thus we obtain as the electric field $\mathbf{E}^s = \frac{1}{\varepsilon_0} \mathbf{D}^s$ of scattering on a "slow" fluctuation in the far-field approximation, i.e., at the site of the detector

$$E_p^s(\mathbf{R}, t) = \frac{1}{4\pi\varepsilon_0 c^2} \frac{1}{R} \left(\frac{R_p R_q}{R^2} - \delta_{pq} \right) \int_V d\mathbf{r} \, \delta\varepsilon(\mathbf{r}, t) \frac{\partial^2 E_q^i(\mathbf{r}, t')}{\partial t'^2}. \tag{1.7}$$

1.2 Incident Plane Harmonic Wave

If the incident electro-magnetic field is a plane harmonic light wave of frequency ω^i, wave length λ_i, and wave number k^i with the complex representation

$$\hat{E}_q^i(\mathbf{r}, t') = n_q^i \hat{E}_0^i e^{i(k_n^i r_n - \omega^i t')}, \tag{1.8}$$

we obtain

$$\frac{\partial^2 \hat{E}_q^i(\mathbf{r}, t')}{\partial t'^2} = -n_q^i \hat{E}_0^i \omega^{i2} e^{i\left(\left(k_e^i - \frac{\omega}{c} \frac{R_e}{R}\right) r_e + \frac{\omega^i}{c} R - \omega^i t\right)}. \tag{1.9}$$

In (1.9) $t' - t$ has been written as $\frac{1}{c}\frac{R_e}{R}r_e - \frac{1}{c}R$ as is appropriate for the far field. $\mathbf{n}^i$ is the unit vector in the direction of the incident electric field and $\hat{E}_0^i$ is the complex amplitude of the wave; the actual incident electric field is given by $\mathbf{E}^i = \mathrm{Re}\,\hat{\mathbf{E}}^i$.

We insert (1.9) into (1.7) and obtain

$$\hat{E}_p^s(\mathbf{R},t) = -\frac{\hat{E}_0^i}{4\pi\varepsilon_0 c^2}\frac{1}{R}\omega^{i2}\left(\frac{R_pR_q}{R^2} - \delta_{pq}\right)\left[\int_V d\mathbf{r}\,\underbrace{e^{i\left(k_e^i - \frac{\omega^i}{c}\frac{R_e}{R}\right)r_e}}_{q_e}\delta\varepsilon(\mathbf{r},t)\right] n_q^i e^{i\left(\frac{\omega^i}{c}R - \omega^i t\right)}, \tag{1.10}$$

$$\hat{E}_p^s(\mathbf{R},t) = -\frac{\hat{E}_0^i}{4\pi\varepsilon_0 c^2}\frac{1}{R}\omega^{i2}\left(\frac{R_pR_q}{R^2} - \delta_{pq}\right)\underbrace{\left[\int_V d\mathbf{r} e^{iq_e r_e}\delta\varepsilon(\mathbf{r},t)\right]}_{\delta\varepsilon(\mathbf{q},t)} n_q^i e^{i\left(\frac{\omega^i}{c}R - \omega^i t\right)}, \tag{1.11}$$

$$\boxed{\hat{E}_p^s(\mathbf{R},t) = -\frac{\hat{E}_0^i}{4\pi\varepsilon_0 c^2}\frac{1}{R}\omega^{i2}\left(\frac{R_pR_q}{R^2} - \delta_{pq}\right)\delta\varepsilon(\mathbf{q},t)n_q^i e^{i\left(\frac{\omega^i}{c}R - \omega^i t\right)}} \tag{1.12}$$

We note that $\delta\varepsilon(\mathbf{q},t)$, defined by (1.11), is the spatial Fourier component of the fluctuation $\delta\varepsilon(\mathbf{r},t)$ corresponding to the wave number $\mathbf{q}$ and we conclude

i. The scattered field at the position $\mathbf{R}$ is represented by an amplitude-modulated oscillation with the basic frequency ω^i.

ii. The amplitude modulation at $\mathbf{R}$ is determined by the fluctuation $\delta\varepsilon(\mathbf{r},t)$ in all points $\mathbf{r} \in V$ but only through the combination $\delta\varepsilon(\mathbf{q},t)$; i.e., the scattered field at the position $\mathbf{R}$ is determined by *the* spatial harmonic of the fluctuation field $\delta\varepsilon(\mathbf{r},t)$, which has the wave number $\mathbf{q} = \mathbf{k}^i - \frac{\omega^i}{c}\frac{\mathbf{R}}{R}$ (see Figure 1).

iii. The magnitude of $\mathbf{q}$ is given by

$$q = 2k^i \sin\frac{\vartheta}{2} = \frac{4\pi}{\lambda^i}\sin\frac{\vartheta}{2} = 2\frac{\omega^i}{c}\sin\frac{\vartheta}{2}. \tag{1.13}$$

That is, forward scattering with $\vartheta \ll \pi$, hence $q \ll 4\pi/\lambda^i$, represents the large scale fluctuations and backward scattering with $\vartheta \lesssim \pi$, hence $q \lesssim 4\pi/\lambda^i$, represents small scale scattering of the order of magnitude of the wave length of the incident light.

It follows that the position of the detector in the range $0 \le \vartheta \le \pi$ determines the scale of the fluctuation that has scattered the light that is being detected.

1.3 Intensity of Scattered Light

Generally the detector sees only the component $\hat{E}^s = \hat{E}_p^s n_p$ of the electric field in the direction of a unit vector $\mathbf{n}$ perpendicular to the beam of the scattered light. As that beam propagates approximately in the direction $\mathbf{R}$ we have by (1.12)

$$\hat{E}^s(\mathbf{R},t) = \underbrace{\frac{\hat{E}_0^i}{4\pi\varepsilon_0 c^2}\frac{1}{R}\omega^{i2}\mathbf{n}\cdot\mathbf{n}^i\delta\varepsilon(\mathbf{q},t)e^{i\frac{\omega^i}{c}R}}_{\hat{a}(t)} e^{i\omega^i t}. \tag{1.14}$$

At the point $\mathbf{R}$ this is an oscillation with the frequency ω^i whose amplitude is modulated by the fluctuating field $\varepsilon(\mathbf{r},t)$. Figure 2 gives a qualitative view of that oscillation. $\hat{a}(t)$ is the complex amplitude of this oscillation.

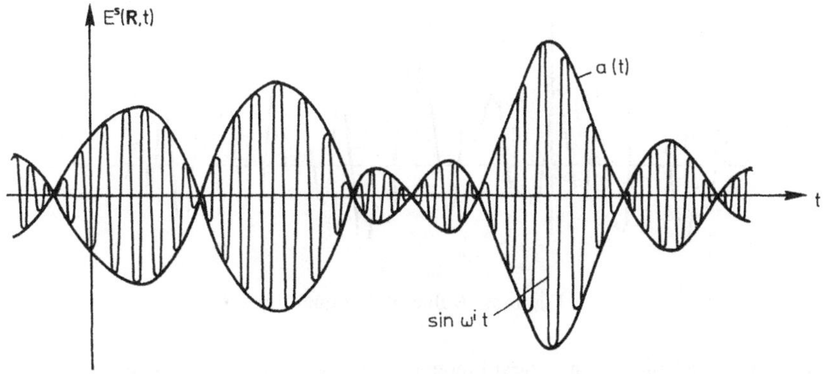

Figure 2. The amplitude modulated oscillation seen by the detector.

We must realize that the period of the scattered light is about 10^{-15} sec, whereas the mean duration of a fluctuation may be approximately 10^{-8} sec. Therefore, if we average the energy flux vector of the scattered light over an interval of time of 10^{-11} sec (say) around t, the amplitude will not significantly change during that interval in which 10^4 oscillations occur. The mean value of the energy flux "at time t" is proportional to the square of the absolute value of the complex amplitude and, suppressing a constant factor, we call

$$I^s(R,t) = \frac{|\hat{E}_0^i|^2}{R^2}\omega^{i4}(\mathbf{n}^i \cdot \mathbf{n})^2 \delta\varepsilon^*(\mathbf{q},t)\delta\varepsilon(\mathbf{q},t) \tag{1.15}$$

the *intensity* of the scattered light at time t. The star on $\delta\varepsilon$ denotes the complex conjugate of that function. $I^s(R,t)$ is a fluctuating quantity, its fluctuations are dictated by the fluctuations of $\varepsilon(\mathbf{r},t)$.

Equation (1.15) is a famous formula, because it allows us to conclude that waves of large frequencies are considerably more effectively scattered than those of small frequencies. The blue colour of the sky is attributed to that effect.

2 Auto-Correlation Function as the Characteristic Feature of a Fluctuating Quantity

2.1 The Simplest Case: One Real Fluctuating Quantity

We consider the fluctuating quantity $A(t)$ whose typical behavior in time is shown in Figure 3 such that the mean value over a long time is zero. This behavior is random and we ask for a characterization in which the random character is eliminated by averaging. *One* such characterization is the *mean regression* of a fluctuation of the value A (say). This function is formed in three steps by

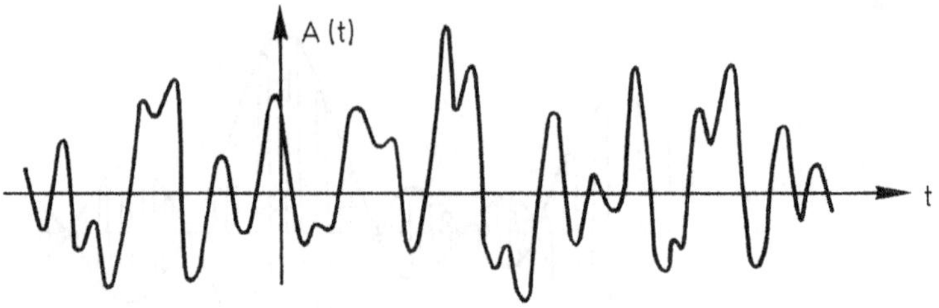

Figure 3. A fluctuating quantity.

i. determining all times t_i in a (large) interval T, for which $A(t_i) = A$ holds;

ii. reading off $A(t_i + \tau)$; and

iii. forming the average

$$R(\tau; A) = \frac{1}{N} \sum_{i=1}^{N} A(t_i + \tau). \tag{2.1}$$

The function R of τ is called the mean regression of the value A. One expects that R tends to zero as τ becomes bigger and that

$$R(\tau; A) = R(-\tau; A). \tag{2.2}$$

The first expectation is due to the random character of $A(t)$ and the second follows from the assumed behaviour under time reversal: A fluctuation should not be qualitatively different when viewed with the proper and the inverted direction of time.

Figure 4 shows some mean regression curves for different initial values of A. For simplicity these curves are all monotone for $\tau > 0$, but this is not a necessary property of $R(\tau; A)$.

Another function characteristic for the fluctuation is the *auto-correlation* function denoted by $\langle A(0)A(\tau)\rangle$. Basically this function results from averaging over mean regressions or, so as to guarantee a positive initial value, over $R(0; A_\alpha)R(\tau; A_\alpha)$ for all possible A_α's. Specifically we have

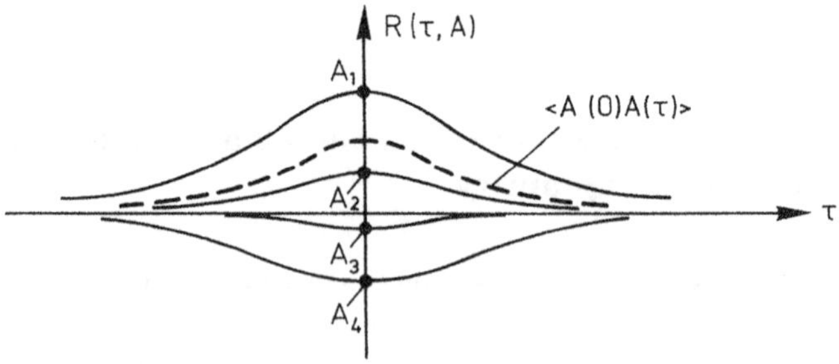

Figure 4. Mean regression for different initial values and the auto-correlation function.

$$\begin{aligned}
\langle A(0)A(\tau)\rangle &= \frac{1}{M}\sum_{\alpha=1}^{M} R(0;A_\alpha)R(\tau;A_\alpha) && \text{or by (2.1)}\\
&= \frac{1}{MN}\sum_{\alpha=1}^{M}\sum_{i=1}^{N} A(t_i^\alpha)A(t_i^\alpha+\tau) && \text{since } A(t_i^\alpha)=A(t_j^\alpha)\\
&= \frac{1}{MN\Delta t}\sum_{\alpha,i=1}^{M,N} A(t_i^\alpha)A(t_i^\alpha+\tau)\Delta t,\\
\langle A(0)A(\tau)\rangle &= \frac{1}{T}\int_0^T A(t)A(t+\tau)\,dt. && (2.3)
\end{aligned}$$

The sequence of steps leading to (1.18) is easily understood; it leads to a form of $\langle A(0)A(\tau)\rangle$ that is the time average of $A(t)A(t+\tau)$. A typical form of the auto-correlation is given by the dashed curve in Figure 4.

2.2 Several Complex Fluctuating Quantities

If $\overset{R}{A}(t)$ is one of J interacting fluctuating quantities, the value of a fluctuation is characterized by the set $\overset{S}{A}$ $(S=1,2,\ldots,J)$ and the mean regressions must be written as

$$R^{\overset{R}{A}}(\tau;\overset{1}{A},\overset{2}{A},\ldots,\overset{J}{A}) = \frac{1}{N}\sum_{i=1}^{N}\overset{R}{A}(t_i+\tau).$$

Accordingly the auto-correlation functions are defined as

$$\begin{aligned}
\langle\overset{R}{A}(0)\,\overset{R}{A}(\tau)\rangle &= \frac{1}{M}\sum_{\alpha=1}^{M} R^{\overset{R}{A}}\left(0;\overset{1}{A}_\alpha,\ldots,\overset{J}{A}_\alpha\right)R^{\overset{R}{A}}\left(\tau;\overset{1}{A}_\alpha,\ldots,\overset{J}{A}_\alpha\right),\\
\langle\overset{R}{A}(0)\,\overset{R}{A}(\tau)\rangle &= \frac{1}{T}\int_0^T \overset{R}{A}(t)\,\overset{R}{A}(t+\tau)\,dt, && (2.4)
\end{aligned}$$

that is, the auto-correlation averages mean regressions over all sets $\overset{1}{A}_\alpha,\ldots,\overset{J}{A}_\alpha$ of initial values.

We shall be particularly interested in fluctuations of Fourier components and these are complex. Therefore, so as to have a positive and real initial value, we define the auto-correlation as

$$\begin{aligned}
\langle\overset{R}{A}(0)\,\overset{R}{A}(\tau)\rangle &= \frac{1}{M}\sum_{\alpha=1}^{M} R^{*\overset{R}{A}}(0;\overset{1}{A}_\alpha,\ldots,\overset{J}{A}_\alpha)R^{\overset{R}{A}}(\tau;\overset{1}{A}_\alpha,\ldots,\overset{J}{A}_\alpha),\\
\langle\overset{R}{A}(0)\,\overset{R}{A}(\tau)\rangle &= \frac{1}{T}\int_0^T \overset{R}{A}{}^*(t)\,\overset{R}{A}(t+\tau)\,dt. && (2.5)
\end{aligned}$$

3 The Spectral Density: A Measurable Quantity

3.1 Auto-Correlation Function of the Scattered Field

We consider the scattered field, given by the real part of (1.14),

$$E^s(\mathbf{R},t) = \operatorname{Re}\hat{E}^s(\mathbf{R},t) = \frac{1}{2}\left(\hat{a}(t)e^{i\omega^i t} + \hat{a}^*(t)e^{-i\omega^i t}\right) \tag{3.1}$$

and calculate the auto-correlation function

$$\langle E^s(R,0)E^s(R,\tau)\rangle = \frac{1}{4T}\int_0^T \left(\hat{a}(t)e^{i\omega^i t} + \hat{a}^*(t)e^{-i\omega^i t}\right)\left(\hat{a}(t+\tau)e^{i\omega^i(t+\tau)} + \hat{a}^*(t+\tau)e^{-i\omega^i(t+\tau)}\right) dt$$
$$= \frac{1}{2T}\int_0^T \mathrm{Re}\left[\hat{a}(t)a(t+\tau)e^{i\omega^i(2t+\tau)}\right] dt + \frac{1}{2T}\int_0^T \mathrm{Re}\left[\hat{a}(t)\hat{a}^*(t+\tau)e^{-i\omega^i\tau}\right] dt.$$

The first term becomes zero for large T, and therefore we have

$$\langle E^s(R,0)E^s(R,\tau)\rangle = \frac{1}{2}\mathrm{Re}\left[\frac{1}{T}\int_0^T \hat{a}^*(t)\hat{a}(t+\tau)dte^{i\omega^i\tau}\right] \quad \text{or by (1.14)}$$
$$= \frac{|\hat{E}_0^i|^2}{(4\pi\varepsilon_0 c^2)^2}\frac{1}{R^2}\omega^{i4}(\mathbf{n}\cdot\mathbf{n}^i)^2\frac{1}{2}\mathrm{Re}\left[\langle\delta\varepsilon^*(\mathbf{q},0)\delta\varepsilon(\mathbf{q},\tau)\rangle e^{i\omega^i\tau}\right]. \tag{3.2}$$

It is not clear a priori that $\langle\delta\varepsilon^*(\mathbf{q},0)\delta\varepsilon(\mathbf{q},\tau)\rangle$ is real, but in a homogeneous gas it can be proved as follows. We write

$$\langle\delta\varepsilon^*(\mathbf{q},0)\delta\varepsilon(\mathbf{q},\tau)\rangle = \frac{1}{T}\int_0^T\int_V\int_V d\mathbf{r}d\mathbf{r}'\delta\varepsilon(\mathbf{r},t)\delta\varepsilon(\mathbf{r}',t+\tau)e^{-i\mathbf{q}(\mathbf{r}-\mathbf{r}')}\,dt$$
$$= \int_V\int_V\left(\frac{1}{T}\int_0^T \delta\varepsilon(\mathbf{r},t)\delta\varepsilon(\mathbf{r}',t+\tau)e^{-i\mathbf{q}(\mathbf{r}-\mathbf{r}')}dt\right)d\mathbf{r}\,d\mathbf{r}'$$
$$= \int_V\int_V \langle\delta\varepsilon(\mathbf{r},0)\delta\varepsilon(\mathbf{r}',\tau)\rangle e^{-i\mathbf{q}(\mathbf{r}-\mathbf{r}')}d\mathbf{r}\,d\mathbf{r}'. \tag{3.3}$$

In a homogeneous body we expect that $\langle\delta\varepsilon(\mathbf{r},0)\delta\varepsilon(\mathbf{r}',\tau)\rangle$ is unchanged when we interchange $\mathbf{r}$ and $\mathbf{r}'$. If this is the case, we conclude from (3.3) that $\langle\delta\varepsilon^*(\mathbf{q},0)\delta\varepsilon(\mathbf{q},\tau)\rangle^* = \langle\delta\varepsilon^*(\mathbf{q},0)\delta\varepsilon(\mathbf{q},\tau)\rangle$, i.e., the auto-correlation of the Fourier components is real.

With this knowledge we simplify (3.2) to read

$$\langle E^s(R,0)E^s(R,\tau)\rangle = \frac{|\hat{E}_0^i|^2}{(4\pi\varepsilon_0 c^2)^2}\frac{1}{R^2}\omega^{i4}(\mathbf{n}\cdot\mathbf{n}^i)^2\frac{1}{2}\langle\delta\varepsilon^*(\mathbf{q},0)\delta\varepsilon(\mathbf{q},\tau)\rangle\cos\omega^i\tau. \tag{3.4}$$

This again is a high frequency oscillation with $\omega^i \approx 10^{15}$ Hz, whose amplitude is modulated by the auto-correlation function of $\delta\varepsilon(\mathbf{q},t)$. Figure 5 shows the time–dependent part of (3.4).

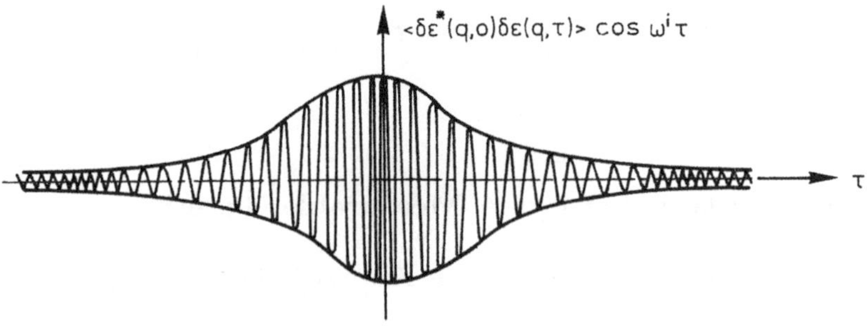

Figure 5. The time-dependent part of the auto-correlation $\langle E^s(R,0)E^s(R,\tau)\rangle$.

3.2 Spectral Density and Dynamic Form Factor

The temporal Fourier transform of the auto-correlation function of the scattered field is called the spectral density. It is denoted by $I(\mathbf{q},\omega)$, and we have

$$I(\mathbf{q},\omega) = \frac{1}{2\pi}\int_{-\infty}^{\infty} \langle E^s(R,0)E^s(R,\tau)\rangle e^{i\omega\tau}\, d\tau \quad \text{or by (3.4)}$$

$$I(\mathbf{q},\omega) = \frac{1}{2}\frac{|\hat{E}_0^i|^2}{(4\pi\varepsilon_0 c^2)^2}\frac{1}{R^2}\omega^{i4}(\mathbf{n}\cdot\mathbf{n}^i)^2 \underbrace{\frac{1}{2\pi}\int_{-\infty}^{\infty} \langle \delta\varepsilon^*(\mathbf{q},0)\delta\varepsilon(\mathbf{q},\tau)\rangle cos\omega^i\tau e^{i\omega\tau}\, d\tau}\,. \tag{3.5}$$

It is this function of ω, the spectral density, which is measured in an interferometer. This fact will not be proved here, because it involves detailed knowledge of the passage of light inside the optical instrument. A clear and simple account is given by Simonsohn [99].

If we wish to *calculate* the spectral density, it suffices to calculate the *dynamic form factor* $S_{\varepsilon\varepsilon}(\mathbf{q},\omega)$, which is defined as

$$S_{\varepsilon\varepsilon}(\mathbf{q},\omega) = \frac{1}{2\pi}\int_{-\infty}^{\infty} \langle \delta\varepsilon^*(\mathbf{q},0)\delta\varepsilon(\mathbf{q},\tau)\rangle e^{i\omega\tau}\, d\tau. \tag{3.6}$$

Indeed, the braced expression in (3.5), which is the important part of the spectral density, is obviously related to $S_{\varepsilon\varepsilon}(\mathbf{q},w)$ by the equation

$$\frac{1}{2\pi}\int_{-\infty}^{\infty} \langle \delta\varepsilon^*(\mathbf{q},0)\delta\varepsilon(\mathbf{q},\tau)\rangle \cos\omega^i\tau e^{i\omega\tau}\, d\tau = \frac{1}{2}\left(S_{\varepsilon\varepsilon}(\mathbf{q},\omega+\omega^i) + S_{\varepsilon\varepsilon}(\mathbf{q},\omega-\omega^i)\right). \tag{3.7}$$

Since $\langle \delta\varepsilon^*(\mathbf{q},0)\delta\varepsilon(\mathbf{q},\tau)\rangle$ is an even function of τ, the dynamic form factor may be rewritten as

$$S_{\varepsilon\varepsilon}(\mathbf{q},\omega) = \frac{1}{2\pi}\int_{-\infty}^{\infty} \langle \delta\varepsilon^*(\mathbf{q},0)\delta\varepsilon(\mathbf{q},\tau)\rangle \cos\omega\tau\, d\tau,$$

$$S_{\varepsilon\varepsilon}(\mathbf{q},\omega) = \frac{1}{\pi}\mathrm{Re}\left[\int_0^{\infty} \langle \delta\varepsilon^*(\mathbf{q},0)\delta\varepsilon(\mathbf{q},\tau)\rangle e^{i\omega\tau} d\tau\right],$$

$$S_{\varepsilon\varepsilon}(\mathbf{q},\omega) = \frac{1}{\pi}\mathrm{Re}\langle \delta\varepsilon^*(\mathbf{q},0)\delta\varepsilon(\mathbf{q},s=i\omega)\rangle. \tag{3.8}$$

Here $\langle \delta\varepsilon^*(\mathbf{q},0)\delta\varepsilon(\mathbf{q},s)\rangle$ is the Laplace transform of $\langle \delta\varepsilon^*(\mathbf{q},0)\delta\varepsilon(\mathbf{q},\tau)\rangle$.

The dielectric constant ε is a function of ϱ and T, but the dependence on T is negligible; therefore, we may write

$$S_{\varepsilon\varepsilon}(\mathbf{q},\omega) = \left(\frac{\partial\varepsilon}{\partial\varrho}\right)^2 \frac{1}{\pi}\mathrm{Re}\left[\int_0^{\infty} \langle \delta\varrho^*(\mathbf{q},0)\delta\varrho(\mathbf{q},\tau)\rangle e^{-i\omega\tau}\, d\tau\right],$$

$$S_{\varepsilon\varepsilon}(\mathbf{q},\omega) = \left(\frac{\partial\varepsilon}{\partial\varrho}\right)^2 \frac{1}{\pi}\mathrm{Re}\langle \delta\varrho^*(\mathbf{q},0)\delta\varrho(\mathbf{q},s=i\omega)\rangle \tag{3.9}$$

where $\langle \delta\varrho^*(q,0)\delta\varrho(q,s)\rangle$ is the Laplace transform of $\langle \delta\varrho^*(\mathbf{q},0)\delta\varrho(\mathbf{q},\tau)\rangle$.

4 Onsager Hypothesis

4.1 Statement

Roughly speaking the Onsager hypothesis states that the mean regression of a fluctuation is equal to the approach to equilibrium of a macroscopic disturbance.

In order to be more specific we concentrate on a Navier–Stokes–Fourier fluid in which we have five fluctuating fields, viz. $\varrho(\mathbf{x},t)$, $\mathbf{v}(\mathbf{x},t)$, $T(\mathbf{x},t)$. These fields are governed by partial differential equations. The Onsager hypothesis applies to the Fourier components. Denoting the five mean regression functions by R^{ϱ_q}, $R^{\mathbf{v}_q}$, and R^{T_q} we may express that hypothesis in the form

$$\begin{aligned} R^{\varrho_q}(\tau;\varrho(\mathbf{q},0),\mathbf{v}(\mathbf{q},0)T(\mathbf{q},0)) &= \varrho(\mathbf{q},\tau;\varrho(\mathbf{q},0),\mathbf{v}(\mathbf{q},0),T(\mathbf{q},0)),\\ R^{\mathbf{v}_q}(\underline{\qquad\qquad\qquad\qquad}) &= \mathbf{v}(\underline{\qquad\qquad\qquad\qquad}),\\ R^{T_q}(\underline{\qquad\qquad\qquad\qquad}) &= T(\underline{\qquad\qquad\qquad\qquad}). \end{aligned} \tag{4.1}$$

These equations state that a fluctuation $\varrho(\mathbf{q},0)$, $\mathbf{v}(\mathbf{q},0)$, $T(\mathbf{q},0)$ evolves, in the mean, in the same manner as the initial state $\varrho(\mathbf{q},0)$, $\mathbf{v}(\mathbf{q},0)$, $T(\mathbf{q},0)$ develops according to the macroscopic field equations.

This appears to be a steep assumption and it has been much criticized, but it has proved its worth in the evaluation of scattering experiments. We proceed to show how the Onsager hypothesis is applied in scattering experiments, always using the Navier–Stokes–Fourier theory as an illustration. Later, of course, we shall employ the hypothesis in extended thermodynamics, see Section 7.

4.2 Field Equations of a Navier–Stokes–Fourier Fluid

The field equations of thermodynamics of a Navier–Stokes–Fourier fluid are based upon the equations of balance of mass, momentum, and energy and on the constitutive equations

$$\begin{aligned} \varepsilon &= \varepsilon(\varrho,T),\\ t_{ij} &= -p(\varrho,T)\delta_{ij} + \lambda\frac{\partial v_n}{\partial x_n}\delta_{ij} + 2\mu\frac{\partial v_{\langle i}}{\partial x_{j\rangle}},\\ q_i &= -\kappa\frac{\partial T}{\partial x_i}. \end{aligned} \tag{4.2}$$

We linearize the field equations about the constant and uniform fields $\tilde{\varrho}, \tilde{v}_i = 0, \tilde{T}$ and obtain

$$\begin{aligned} &\frac{\partial \varrho}{\partial t} + \tilde{\varrho}\frac{\partial v_i}{\partial x_i} = 0,\\ &\tilde{\varrho}\frac{\partial v_i}{\partial t} + \frac{\partial p}{\partial x_i} - \tilde{\mu}\frac{\partial^2 v_i}{\partial x_n \partial x_n} - \left(\tilde{\lambda} + \frac{1}{3}\tilde{\mu}\right)\frac{\partial}{\partial x_i}\left(\frac{\partial v_n}{\partial x_n}\right) = 0,\\ &\tilde{\varrho}\frac{\partial \varepsilon}{\partial t} - \tilde{\kappa}\frac{\partial^2 T}{\partial x_n \partial x_n} = -\tilde{p}\frac{\partial v_n}{\partial x_n}. \end{aligned} \tag{4.3}$$

Equivalent to the momentum balance $(4.3)_2$ is its div and curl, namely,

$$\begin{aligned} &\tilde{\varrho}\frac{\partial}{\partial t}\left(\frac{\partial v_i}{\partial x_i}\right) + \Delta p - \left(\tilde{\lambda} + \frac{4}{3}\tilde{\mu}\right)\Delta\left(\frac{\partial v_i}{\partial x_i}\right) = 0,\\ &\tilde{\varrho}\frac{\partial \operatorname{curl}\mathbf{v}}{\partial t} = -\tilde{\mu}\Delta \operatorname{curl}\mathbf{v}. \end{aligned} \tag{4.4}$$

We note that the equation for curl $\mathbf{v}$ is uncoupled to the other equations; hence, curl $\mathbf{v}$ cannot affect the density fluctuations which determine the scattering. We may therefore ignore $(4.4)_2$ and replace the momentum balance by $(4.4)_1$.

Inserting $p = p(\varrho,T)$, $\varepsilon = \varepsilon(\varrho,T)$, and $\psi \equiv \partial v_n/\partial x_n$ we thus obtain as the relevant system of field equations

$$
\begin{aligned}
&\frac{\partial \varrho}{\partial t} + \tilde{\varrho}\psi = 0, \\
\tilde{\varrho}\ &\frac{\partial \psi}{\partial t} + \tilde{p}_\varrho \Delta \varrho + \tilde{p}_T \Delta T - \tilde{\varrho}\tilde{D}_v \Delta \psi = 0, \\
\tilde{\varrho}\tilde{c}_v &\frac{\partial T}{\partial t} - \frac{\tilde{T}}{\tilde{\varrho}}\tilde{p}_T \frac{\partial \varrho}{\partial t} - \tilde{\varrho}\tilde{c}_p \tilde{D}_T \Delta T = 0.
\end{aligned} \tag{4.5}
$$

Here c_p and c_v are the specific heats whose ratio is γ, and D_v, D_T represent viscosity and heat conductivity. These quantities are defined as

$$
D_v = \frac{1}{\varrho}\left(\lambda + \frac{4}{3}\mu\right) \quad \text{and} \quad D_T = \frac{\kappa}{\varrho c_p}. \tag{4.6}
$$

p_ϱ and p_T denote derivatives of p with respect to ϱ and T.

4.3 Solution of the Field Equations

Since the Onsager hypothesis refers to the spatial Fourier transforms of the fields, it is indicated to rewrite the three equations (4.5) in the Fourier space. Thus we obtain

$$
\begin{aligned}
&\frac{\partial \varrho(\mathbf{q},t)}{\partial t} + \tilde{\varrho}\psi(\mathbf{q},t) = 0, \\
&\frac{\partial \psi(\mathbf{q},t)}{\partial t} - \frac{\tilde{p}_\varrho}{\tilde{\varrho}} q^2 \varrho(\mathbf{q},t) - \frac{\tilde{p}_T}{\tilde{\varrho}} q^2 T(\mathbf{q},t) + \tilde{D}_v q^2 \psi(\mathbf{q},t) = 0, \\
&\frac{\partial T(\mathbf{q},t)}{\partial t} + \tilde{\varrho}(\tilde{\gamma}-1)\frac{\tilde{p}_\varrho}{\tilde{p}_T}\psi(\mathbf{q},t) + \tilde{\gamma}\tilde{D}_T q^2 T(\mathbf{q},t) = 0.
\end{aligned} \tag{4.7}
$$

The solution of (4.7) proceeds by Laplace transformation. Thus the equations (4.7) reduce to a linear algebraic system of equations for the Laplace transforms $\tilde{\varrho}(\mathbf{q},s)$, $\tilde{\psi}(\mathbf{q},s)$, and $\tilde{T}(\mathbf{q},s)$, namely,

$$
\begin{bmatrix}
s & \tilde{\varrho} & 0 \\
-\frac{\tilde{p}_\varrho}{\tilde{\varrho}} q^2 & s + \tilde{D}_v q^2 & -\frac{\tilde{p}_T}{\tilde{\varrho}} q^2 \\
0 & \tilde{\varrho}(\tilde{\gamma}-1)\frac{\tilde{p}_\varrho}{\tilde{p}_T} & s + \tilde{\gamma}\tilde{D}_T q^2
\end{bmatrix}
\begin{bmatrix}
\tilde{\varrho}(\mathbf{q},s) \\ \tilde{\psi}(\mathbf{q},s) \\ \tilde{T}(\mathbf{q},s)
\end{bmatrix}
=
\begin{bmatrix}
\varrho(\mathbf{q},0) \\ \psi(\mathbf{q},0) \\ T(\mathbf{q},0)
\end{bmatrix} \tag{4.8}
$$

with

$$
D(s) = s^3 + \left(\tilde{\gamma}\tilde{D}_T q^2 + \tilde{D}_v q^2\right)s^2 + \left(\tilde{\gamma}\tilde{p}_\varrho q^2 + \tilde{\gamma}\tilde{D}_T q^2 \tilde{D}_v q^2\right)s + \tilde{\gamma}\tilde{D}_T q^2 \tilde{p}_\varrho q^2 \tag{4.9}
$$

as the determinant of this system, omitting the wiggles, we obtain

$$
\begin{aligned}
\tilde{\varrho}(\mathbf{q},s) &= \frac{D_{11}(s)}{D(s)}\ \varrho(\mathbf{q},0) - \frac{\varrho(s+\gamma D_T q^2)}{D(s)}\ \psi(\mathbf{q},0) + \frac{-p_T q^2}{D(s)}\ T(\mathbf{q},0), \\
\tilde{\psi}(\mathbf{q},s) &= \frac{-\frac{1}{\varrho}p_\varrho q^2 (s + D_T q^2)}{D(s)}\ \varrho(\mathbf{q},0) + \frac{s^2 + \gamma D_T q^2 s}{D(s)}\ \psi(\mathbf{q},0) + \frac{-\frac{1}{\varrho}p_T q^2 s}{D(s)}\ T(\mathbf{q},0), \\
\tilde{T}(\mathbf{q},s) &= \frac{s^2 + D_v q^2 s + p_\varrho q^2}{D(s)}\ \varrho(\mathbf{q},0) + \frac{\varrho\frac{p_\varrho}{p_T}(\gamma-1)s}{D(s)}\ \psi(\mathbf{q},0) + \frac{s^2 + D_v q^2 s + p_\varrho q^2}{D(s)}\ T(\mathbf{q},0)
\end{aligned} \tag{4.10}
$$

where $D_{11}(s)$ is the (11)-subdeterminant of the matrix in (4.8). It is given by the expression $D_{11}(s) = s^2 + (\gamma D_T q^2 + D_v q^2)s + (\gamma D_T q^2 D_v q^2 + (\gamma - 1)p_\varrho q^2)$.

According to the Onsager hypothesis these expressions provide the Laplace transforms of the mean regression functions R^{ϱ_q}, R^{ψ_q}, and R^{T_q}.

5 Auto-Correlation Function and Its Properties

5.1 The Dynamic Form Factor

We recall that the auto-correlation function averages over mean regression functions corresponding to all possible initial states, here $\varrho(\mathbf{q},0)$, $\psi(\mathbf{q},0)$, $T(\mathbf{q},0)$. Before averaging the regression function is multiplied by the complex conjugate of the appropriate initial value, here $\varrho^*(\mathbf{q},0)$, $\psi^*(\mathbf{q},0)$, $T^*(\mathbf{q},0)$, respectively. Therefore the Laplace transform of the auto-correlation function of density is given by [see $(4.10)_1$]

$$\begin{aligned}\langle\varrho^*(\mathbf{q},0)\tilde{\varrho}(\mathbf{q},s)\rangle &= \frac{s^2 + (\gamma D_T q^2 + D_v q^2)s + \gamma D_T q^2 D_v q^2 + (\gamma-1)p_\varrho q^2}{D(s)}\langle|\varrho(\mathbf{q},0)|^2\rangle \\ &+ \frac{\varrho(s+\gamma D_T q^2)}{D(s)}\langle\varrho^*(\mathbf{q},0)\psi(\mathbf{q},0)\rangle + \frac{-p_T q^2}{D(s)}\langle\varrho^*(\mathbf{q},0)T(\mathbf{q},0)\rangle,\end{aligned} \tag{5.1}$$

where the angular brackets on the right-hand side denote mean values of the products

$$|\varrho(\mathbf{q},0)|^2 = \varrho^*(\mathbf{q},0)\varrho(\mathbf{q},0), \qquad \varrho^*(\mathbf{q},0)\psi(\mathbf{q},0), \qquad \varrho^*(\mathbf{q},0)T(\mathbf{q},0)$$

over all possible initial values. Since the values $\varrho(\mathbf{q},0)$ are uncorrelated to the values of $\psi(\mathbf{q},0)$ and $T(\mathbf{q},0)$, the last two terms vanish and we have

$$\frac{\langle\varrho^*(\mathbf{q},0)\tilde{\varrho}(\mathbf{q},s)\rangle}{\langle|\varrho(\mathbf{q},0)|^2\rangle} = \frac{s^2 + (\gamma D_T q^2 + D_v q^2)s + \gamma D_T q^2 D_v q^2 + (\gamma-1)p_\varrho q^2}{s^3 + (\gamma D_T q^2 + D_v q^2)s^2 + (\gamma p_\varrho q^2 + \gamma D_T q^2 D_v q^2)s + (\gamma D_T q^2 p_\rho q^2)}. \tag{5.2}$$

This is the simplest form of the Laplace transform of the density auto-correlation function. From (3.9) we conclude that this function determines the dynamic form factor $S_{\varepsilon\varepsilon}$. We have

$$\frac{S_{\varepsilon\varepsilon}(\mathbf{q},\omega)}{\left(\frac{\partial\varepsilon}{\partial\varrho}\right)^2\frac{1}{\pi}\langle|\varrho(\mathbf{q},0)|^2\rangle} \tag{5.3}$$

$$= \mathcal{R}e\left[\frac{s^2 + (\gamma D_T q^2 + D_v q^2)s + \gamma D_T q^2 D_v q^2 + (\gamma-1)p_\varrho q^2}{s^3 + (\gamma D_T q^2 + D_v q^2)s^2 + (\gamma p_\varrho q^2 + \gamma D_T q^2 D_v q^2)s + (\gamma D_T q^2 p_\rho q^2)}\right]_{s=i\omega}$$

$$= \mathrm{Re}\left[\frac{\overbrace{\left\{\kappa D_T q^2 D_v q^2 + (\gamma-1)p_\varrho q^2 - \omega^2\right\}}^{C} + i\overbrace{\left\{\gamma D_T q^2 + D_v q^2\right\}\omega}^{D}}{\underbrace{\left\{\gamma D_T q^2 p_\varrho q^2 - (\gamma D_T q^2 + D_v q^2)\omega^2\right\}}_{A} + i\underbrace{\left\{(\gamma p_\varrho q^2 + \gamma D_T q^2 + D_v q^2)\omega - \omega^3\right\}}_{B}}\right] \tag{5.4}$$

$$= \frac{AC+BD}{A^2+B^2}. \tag{5.5}$$

This is really all we need, because we have argued that $S_{\varepsilon\varepsilon}(q,\omega)$ is the essential ingredient of the spectral density that is measured in light scattering. Now we see that $S_{\varepsilon\varepsilon}$ can also be *calculated,* to within a constant factor, provided the material coefficients of the gas, namely, $\gamma, p_\varrho, D_T, D_v$, are given. Alternatively those coefficients may be adjusted so as to make the function on the right-hand side of (5.5) fit to the measured spectral density. In this manner we may say that $\gamma, p_\varrho, D_T, D_v$ are "measured."

While indeed equation (5.5)—with A through D given by (5.4)—is all we need, it is instructive to derive an alternative form of the dynamic form factor, for this will provide an idea of its appearance and, indeed, of the shape and meaning of the auto-correlation function.

5.2 Alternative Form of the Dynamic Form Factor and the Auto-Correlation Function

The expression (5.2) may be split into three terms by decomposition into partial fractions. We use $N(s)$ and $D(s)$ for numerator and denominator of the right-hand side of (5.2) and denote the roots of $D(s)$ by s_i $(i = 0, 1, 2)$. Thus we may write

$$\frac{\langle \varrho^*(\mathbf{q},0)\tilde{\varrho}(\mathbf{q},s)\rangle}{\langle |\varrho(\mathbf{q},0)|^2\rangle} = \sum_{i=0}^{2} \frac{N(s_i)}{\dot{D}(s_i)} \frac{1}{s-s_i} \quad \text{where } \dot{D} = \frac{dD}{ds}. \tag{5.6}$$

We let s_0 be the real root while $s_{1,2}$ are complex conjugate. Hence it follows for the relevant part of (5.6), that is, the part that by (3.9) determines the dynamic form factor $S_{\varepsilon\varepsilon}(\mathbf{q},w)$, that

$$\begin{aligned} \operatorname{Re} \frac{\langle \varrho^*(\mathbf{q},0)\tilde{\varrho}(\mathbf{q},iw)\rangle}{\langle |\varrho(\mathbf{q},0)|^2\rangle} = \frac{N(s_0)}{\dot{D}(s_0)} \frac{-s_0}{s_0^2+\omega^2} &+ \operatorname{Re}\left[\frac{N(s_1)}{\dot{D}(s_1)}\right] \left(\frac{-s_1'}{s_1'^2 + (\omega+s_1'')^2} + \frac{-s_1'}{s_1'^2+(\omega-s_1'')^2}\right) \\ &- \operatorname{Im}\left[\frac{N(s_1)}{\dot{D}(s_1)}\right] \left(\frac{\omega+s_1''}{s_1'^2+(\omega+s_1'')^2} - \frac{\omega-s_1''}{s_1'^2+(\omega-s_1'')^2}\right). \end{aligned} \tag{5.7}$$

Here s_1' and s_1'' are the real and imaginary parts of s_1.

We conclude that the dynamic form factor consists of three *Lorentz* curves with maxima at $\omega = 0$ and $\omega = \pm s_1''$. The values of the maxima are

$$-\frac{N(s_0)}{\dot{D}(s_0)} \frac{1}{s_0} \quad \text{and} \quad -\operatorname{Re}\left[\frac{N(s_1)}{\dot{D}(s_1)}\right] \frac{1}{s_1'} \tag{5.8}$$

and the half-widths are given by $\sqrt{2|s_0|}$ and $\sqrt{2|s_1'|}$. In addition the dynamic form factor has two terms, which vanish at $\omega = \pm s_1''$ respectively. Figure 6 provides a qualitative picture of the Lorentz curves and of the two *corrective* terms in the second line of (5.7).

The precise shape of the dynamic form factor depends, of course, on the values of the roots s_0 and $s_{1,2}$. It is not always so, for instance, that the maxima of the Lorentz curves can be identified in the spectral density; the maxima may be distorted and disguised by the influence of the corrective terms. The only clear-cut case where we do see three distinct maxima is the case of small values of q, which we shall consider in Paragraph 5.3.

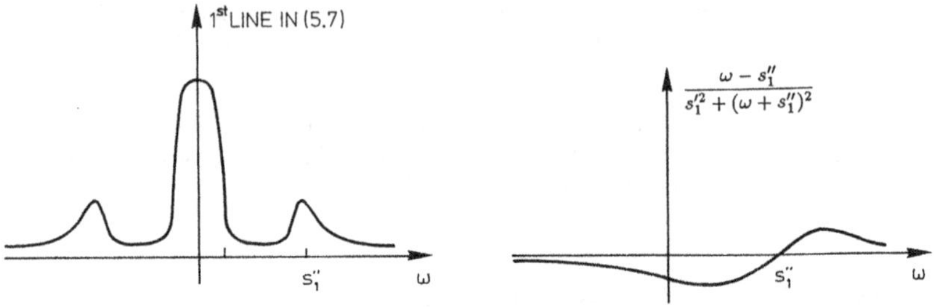

Figure 6. Contributions to the dynamic form factor. Lorentz curves and corrective terms.

We recall that $\langle \varrho^*(\mathbf{q},0)\tilde{\varrho}(\mathbf{q},s)\rangle$ in (5.2) or (5.6) is the Laplace transform of the auto-correlation function. Therefore the function itself is obtained by inverting the Laplace transformation

$$\frac{\langle \varrho^*(\mathbf{q},0)\varrho(\mathbf{q},\tau)\rangle}{\langle |\varrho(\mathbf{q},0)|^2\rangle} = \sum_{i=0}^{2} \frac{N(s_i)}{\dot{D}(s_i)} \frac{1}{2\pi i} \int_{x-i\infty}^{x+i\infty} \frac{1}{s-s_i} e^{|\tau| s}\, ds, \tag{5.9}$$

$$\frac{\langle \varrho^*(\mathbf{q},0)\varrho(\mathbf{q},\tau)\rangle}{\langle |\varrho(\mathbf{q},0)|^2\rangle} = \sum_{i=0}^{2} \frac{N(s_i)}{\dot{D}(s_i)} e^{|\tau| s_i}. \tag{5.10}$$

We make use of the fact that s_0 is real while $s_{1,2}$ are complex conjugates and obtain

$$\frac{\langle \varrho^*(\mathbf{q},0)\varrho(\mathbf{q},\tau)\rangle}{\langle |\varrho(\mathbf{q},0)|^2\rangle} = \left\{ \frac{N(s_0)}{\dot{D}(s_0)} e^{s_0|\tau|} + 2\mathrm{Re}\left[\frac{N(s_1)}{\dot{D}(s_1)}\right] e^{s_1'|\tau|} \cos s_1''\tau - 2\,\mathrm{Im}\left[\frac{N(s_1)}{\dot{D}(s_1)}\right] e^{s_1'|\tau|} \sin s_1''|\tau| \right\}. \tag{5.11}$$

We conclude that the auto-correlation function is the sum of three terms of which two are oscillatory, so that the function itself is certainly non-monotone.

5.3 Form Factor and Auto-Correlation Function in Forward Scattering

In forward scattering the scattering angle ϑ is small, and hence, by Figure 1, q is small. It is instructive to investigate this special case, because the roots s_0 and $s_{1,2}$ can be determined explicitly in terms of γ, p_ϱ, D_T, D_v, and therefore the dynamic form factor and the auto-correlation function become explicit functions of these material coefficients.

It is an easy matter to obtain the roots $s_0, s_{1,2}$ of $D(s)$ to within terms of order q^3. They read

$$s_0 = -D_T q^2, \quad s_{1,2} = -\Gamma q^2 \pm i\sqrt{\gamma p_\varrho}\, q \qquad \text{where } \Gamma \equiv \frac{1}{2}\left[(\gamma - 1)D_T + D_v\right]. \tag{5.12}$$

Insertion into (5.7) provides the essential part of the dynamic form factor as an explicit expression in this case of forward scattering. We obtain

$$\begin{aligned} \mathrm{Re}\,\frac{\langle \varrho^*(\mathbf{q},0)\tilde{\varrho}(q,i\omega)\rangle}{\langle |\varrho(\mathbf{q},0)|^2\rangle} &= \left(1 - \frac{1}{\gamma}\right) \frac{D_T q^2}{(D_T q^2)^2 + \omega^2} \\ &+ \frac{1}{2\gamma} \frac{\Gamma q^2}{(\Gamma q^2)^2 + (\omega - \sqrt{\gamma p_\varrho}\, q)^2} + \frac{1}{2\gamma} \frac{\Gamma q^2}{(\Gamma q^2)^2 + (\omega + \sqrt{\gamma p_\varrho}\, q)^2} \\ &- \frac{1}{2\gamma\sqrt{\gamma p_\varrho}} \frac{(3\Gamma - D_v) q (\omega - \sqrt{\gamma p_\varrho}\, q)}{(\Gamma q^2)^2 + (\omega - \sqrt{\gamma p_\varrho}\, q)^2} + \frac{1}{2\gamma\sqrt{\gamma p_\varrho}} \frac{(3\Gamma - D_v) q (\omega + \sqrt{\gamma p_\varrho}\, q)}{(\Gamma q^2)^2 + (\omega + \sqrt{\gamma p_\varrho}\, q)^2} \end{aligned} \tag{5.13}$$

Again the first line represents the superposition of three Lorentz curves with characteristic features as shown in Table 1.

	Central Maximum	Lateral Maxima
Site of Maxima	$\omega = 0$	$\omega = \pm\sqrt{\gamma p_\varrho} q$
Height of Maxima	$\left(1 - \frac{1}{\gamma}\right) \frac{1}{D_T q^2}$	$\frac{1}{2\gamma} \frac{1}{\Gamma q^2}$
$\frac{1}{2}$-Width of Maxima	$D_T q^2$	Γq^2

Table 1. Characteristic features of the maxima of the dynamic form factor.

The two corrective terms in the third line of (5.13) play no important role in forward scattering. We conclude therefore, that the material coefficients γ, p_ϱ, D_T, and Γ—thus, in particular, heat conductivity κ and the viscosity coefficient $\lambda + \frac{4}{3}\mu$—may be read from the measured form of the dynamic form factor by measuring the sites, heights, and widths of the maxima.

The central maxima is often referred to as the Rayleigh peak, while the lateral maxima are called Brillouin peaks.

The auto-correlation function (5.11) reads explicitly in the case of forward scattering

$$\frac{\langle \varrho^*(\mathbf{q},0)\varrho(\mathbf{q},\tau)\rangle}{\langle |\varrho(\mathbf{q},0)|^2\rangle} = \frac{\gamma-1}{\gamma} e^{-D_T q^2|\tau|} + \frac{1}{\gamma} e^{-\Gamma q^2|\tau|} \cos\sqrt{\gamma p_\varrho} q\tau + \frac{1}{\gamma}(3\Gamma - D_v)\frac{q}{\sqrt{\gamma p_\varrho}} e^{-\Gamma q^2|\tau|} \sin\sqrt{\gamma p_\varrho} q|\tau|. \tag{5.14}$$

This expression is obtained by insertion of the explicit roots (5.12) into (5.11). We point out again that the auto-correlation function is not monotone; its slope at $\tau = 0$ is zero, however, as it must be.

5.4 Heat Modes and Sound Modes

Fourier transforms and Laplace transforms are not easily interpreted—except by an expert—and certainly a combined Fourier and Laplace transform such as $(4.10)_1$ is hard to understand in its physical significance. In this paragraph we attempt to provide an intuitive interpretation of the Rayleigh and Brillouin peaks in the dynamic form factor; this will explain why these peaks are referred to as the "heat mode" and the "sound mode," respectively, in the literature.

We consider the special initial condition $\psi(\mathbf{q},0) = 0$, $T(\mathbf{q},0) = 0$ and obtain from $(4.10)_1$

$$\frac{\tilde{\varrho}(\mathbf{q},s)}{\varrho(\mathbf{q},0)} = \frac{s^2 + (\gamma D_T q^2 + D_v q^2)\, s + \gamma D_T q^2 D_v q^2 + (\gamma - 1) p_\varrho q^2}{s^3 + (\gamma D_T q^2 + D_v q^2)\, s^2 + (\gamma p_\varrho q^2 + \gamma D_T q^2 D_v q^2)\, s + (\gamma D_T q^2 p_\rho q^2)}. \tag{5.15}$$

This is actually the solution, in the Fourier–Laplace space, of macroscopic field equations for a particular initial state. It is identical to the Laplace transform of the auto-correlation function [see (5.2)], which is obtained by averaging over all possible initial states in a fluctuation. Despite this difference we consider the solution (5.15) as typical for the auto-correlation function (5.2); after all, the two are related by the Onsager hypothesis.

Inverting the Laplace transforms we obtain as in the transition from (5.2) or (5.6) to (5.11) and, for small values of q, to (5.14)

$$\frac{\varrho(\mathbf{q},\tau)}{\varrho(\mathbf{q},0)} = \frac{N(s_0)}{\dot{D}(s_0)}e^{s_0|\tau|} + 2\mathrm{Re}\left[\frac{N(s_1)}{\dot{D}(s_1)}\right]e^{s_1'|\tau|}\cos s_1''\tau - 2\mathrm{Im}\left[\frac{N(s_1)}{\dot{D}(s_1)}\right]e^{s_1'|\tau|}\sin s_1''|\tau| \tag{5.16}$$
$$\underset{q\to 0}{\longrightarrow}\left(1-\frac{1}{\gamma}\right)e^{-D_T q^2|\tau|} + \frac{1}{\gamma}e^{-\Gamma q^2|\tau|}\cos\sqrt{\gamma p_\varrho}q\tau + \frac{1}{\gamma}(3\Gamma - D_v)\frac{q}{\sqrt{\gamma p_\varrho}}e^{-\Gamma q^2|\tau|}\sin\sqrt{\gamma p_\varrho}|\tau|.$$

For illustrative purposes we are interested in the initial state

$$\varrho(\mathbf{x},0) = \varrho_0\delta(\mathbf{x}-\mathbf{x}_0) \quad \text{hence} \quad \varrho(\mathbf{q},0) = \varrho_0 e^{i\mathbf{q}\mathbf{x}_0}, \tag{5.17}$$

which represents a singular initial density concentration at $\mathbf{x} = \mathbf{x}_0$.

We insert this into (5.16) and invert the Fourier transform. For simplicity we consider the one-dimensional case and obtain

$$\varrho(x,\tau) = \frac{\varrho_0}{2\pi}\left\{\left(1-\frac{1}{\gamma}\right)\int_{-\infty}^{\infty} e^{-D_T q^2|\tau|}e^{-i(x-x_0)q}\,dq + \frac{1}{\gamma}\int_{-\infty}^{\infty} e^{-\Gamma q^2|\tau|}\cos\sqrt{\gamma p_\varrho}q\tau e^{-i(x-x_0)q}\,dq\right.$$
$$\left. + \frac{1}{\gamma}(3\Gamma - D_v)\int_{-\infty}^{\infty}\frac{q}{\sqrt{\gamma p_\varrho}}e^{-\Gamma q^2|\tau|}\sin\sqrt{\gamma p_\varrho}q|\tau|e^{-i(x-x_0)q}dq\right\},$$

$$\varrho(x,\tau) = \frac{\varrho_0}{2\pi}\left\{\left(1-\frac{1}{\gamma}\right)\sqrt{\frac{\pi}{D_T|\tau|}}e^{-\frac{(x-x_0)^2}{4D_T|\tau|}}\right.$$
$$+ \frac{1}{\gamma}\sqrt{\frac{\pi}{\Gamma|\tau|}}\left(e^{-\frac{(x-x_0-\sqrt{\gamma p_\varrho}|\tau|)^2}{4\Gamma|\tau|}} + e^{-\frac{(x-x_0+\sqrt{\gamma p_\varrho}|\tau|)^2}{4\Gamma|\tau|}}\right)$$
$$+ \frac{1}{\gamma}(3\Gamma - D_v)\sqrt{\frac{\pi}{\Gamma|\tau|}}\left(-\frac{x-x_0-\sqrt{\gamma p_\varrho}|\tau|}{4\Gamma\sqrt{\gamma p_\varrho}|\tau|}e^{-\frac{(x-x_0-\sqrt{\gamma p_\varrho}|\tau|)^2}{4\Gamma|\tau|}}\right.$$
$$\left.\left. + \frac{x-x_0+\sqrt{\gamma p_\varrho}|\tau|}{4\Gamma\sqrt{\gamma p_\varrho}|\tau|}e^{-\frac{(x-x_0+\sqrt{\gamma p_\varrho}|\tau|)^2}{4\Gamma|\tau|}}\right)\right\} \tag{5.18}$$

Inspection shows that the solution has a central peak at the site of the initial concentration. This peak diffuses according to the laws of linear heat conduction; it is the *heat mode.* Also there are diffusing peaks propagating to the right and left with the sound speed $\sqrt{\gamma p_\varrho}$; these are the *sound modes.* Finally there are propagating contributions, which vanish at the sites of the lateral peaks.

Note that the first two lines of (5.18) upon Fourier and Laplace transformation are converted into the three Lorentz curves in the first two lines of (5.13), which represent the Rayleigh and Brillouin peaks. Thus we may say that the heat mode gives rise to the Rayleigh peak, while the sound modes are reflected by the Brillouin peaks in the scattering spectrum.

6 How to Use Light Scattering

6.1 Two Philosophies

There are two ways in which light scattering experiments may be used, namely,

i. for the determination of material coefficients in a given theory of non-equilibrium processes, such as the Navier–Stokes–Fourier theory; and

ii. for the evaluation of the range of validity of a given theory, for example, the Navier–Stokes–Fourier theory, if the material coefficients are known.

For clarification of these points let us always consider a Navier–Stokes–Fourier theory. In forward scattering, that is, for a certain (small) value of q, the dynamic form factor is given, to within a multiplicative factor, by the function (5.13), which is characterized by a central maximum and two lateral ones. At the same time a measurement of the dynamic form factor for the same (small) q in the noble gas Xe provides a function of the type shown in Figure 7.

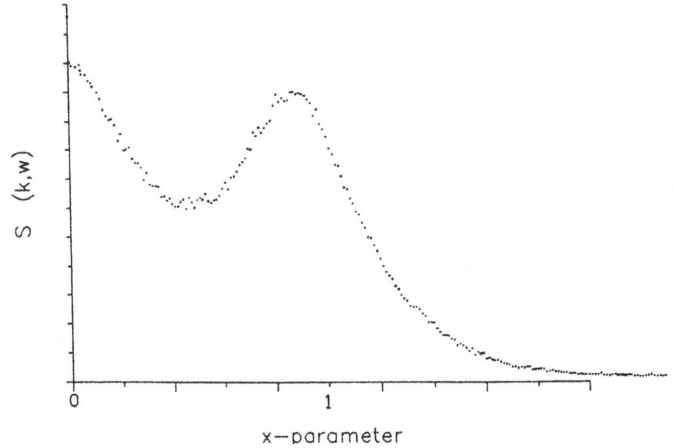

Figure 7. Dynamic form factor in Xe at $p = 1bar, T = 293K$, and $\vartheta = 28^0$, that is, $q = 0.5 \cdot 10^7 \frac{1}{m}$.

As we explained in Paragraph 5.3 a comparison of the sites, the relative heights, and the widths of the maxima of the two curves, calculated and measured, will determine the material coefficients γ, p_ϱ, κ, and $\lambda + \frac{4}{3}\mu$. Once these values have been obtained the two curves will fall on top of each other, so we conclude that the Navier–Stokes–Fourier theory is good—*for the chosen (small) value of q.*

In this way we have satisfactorily "measured" γ, p_ϱ, κ, and $\lambda + \frac{4}{3}\mu$ without ever leaving the equilibrium state. This is a truly remarkable feat, made possible by the Onsager hypothesis, because κ and $\lambda + \frac{4}{3}\mu$ are transport coefficients that determine the heat flux and the stress, both non–equilibrium quantities. Most of the application of light scattering experiments concerns itself with such "measurements" of material coefficients. The results are good, meaning that the value of κ (say), so determined, fits well to values that have been obtained by other, less roundabout methods. Thus confidence in the validity of the Onsager hypothesis has become a certainty among physicists.

Now let us use the measured values of γ, p_ϱ, D_T, D_v and apply them to calculate the dynamic form factor for a large value of q; and let us compare the calculated dynamic form factor with a measured one at the same (large) q. Figure 8 shows both curves, and we find poor agreement, or better, no agreement at all!

The conclusion is clear: The Navier–Stokes–Fourier theory does not describe small scale fluctuations, that is, those that are detected for big values of ϑ and q, (see Paragraph 1.2). The biggest value of q for which the measured and calculated functions $S_{\varepsilon\varepsilon}(q,\omega)$ agree is called the limit of applicability of the Navier–Stokes–Fourier theory.

Thus light scattering may be used to determine the range of validity of a given theory. Of course, now we must ask whether extended thermodynamics has a wider range of applicability than the Navier–Stokes–Fourier theory, and this question will be answered in Section 7.

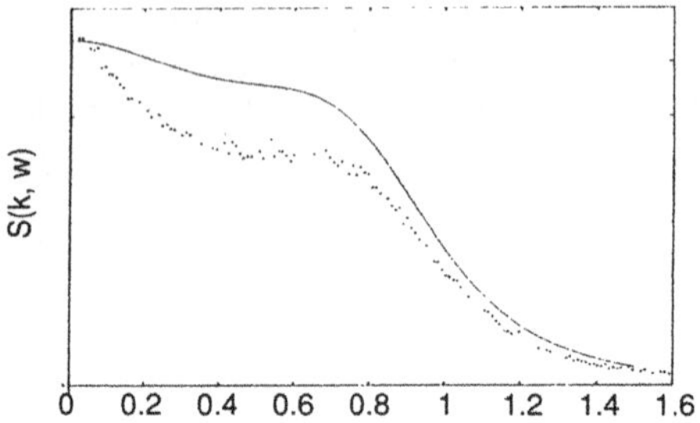

Figure 8. Dynamic form factor in Xe at $p = 1bar$, $T = 293K$, and $\vartheta = 94^0$, that is, $q = 1.5 \cdot 10^7 \frac{1}{m}$.

6.2 Dimensionless Parameters and Limiting Characteristics

The notion of a small or large q need a dimensionless measure in order to be definite. We proceed to introduce this measure, restricting the attention to noble gases, as we shall exclusively be concerned with those.

In a noble gas we have

$$p = \varrho \frac{k}{m} T, \qquad c_p = \frac{5}{2}\frac{k}{m}, \qquad \gamma = \frac{5}{3}, \qquad \lambda = 0, \qquad \frac{\kappa}{\mu} = \frac{15}{4}\frac{k}{m}, \tag{6.1}$$

and we introduce the dimensionless parameters

$$\hat{s} = \frac{s}{\sqrt{2\frac{k}{m}T}q}, \qquad x = \frac{\omega}{\sqrt{2\frac{k}{m}T}q}, \qquad y = \frac{\varrho}{3}\frac{\sqrt{2\frac{k}{m}T}}{\mu}\frac{1}{q} \tag{6.2}$$

so that the dynamic form factor (5.3) reads

$$S_{\varepsilon\varepsilon}(y,x) = \left(\frac{\partial \varepsilon}{\partial \varrho}\right)^2 \frac{1}{\pi}\langle|\varrho(\mathbf{q},0)|^2\rangle \frac{1}{\sqrt{2\frac{k}{m}T}q}\mathrm{Re}\left[\frac{\hat{s}^2 + \frac{23}{18}\frac{1}{y}\hat{s} + \left(\frac{1}{3} + \frac{10}{27}\frac{1}{y^2}\right)}{\hat{s}^3 + \frac{23}{18}\frac{1}{y}\hat{s}^2 + \left(\frac{5}{6} + \frac{10}{27}\frac{1}{y^2}\right)\hat{s} + \frac{5}{12}\frac{1}{y}}\right]_{\hat{s}=ix}, \tag{6.3}$$

$$= \left(\frac{\partial \varepsilon}{\partial \varrho}\right)^2 \frac{1}{\pi}\langle|\varrho(\mathbf{q},0)|^2\rangle \frac{1}{\sqrt{2\frac{k}{m}T}q}\mathrm{Re}\left[\frac{\overbrace{\left\{\left(\frac{1}{3} + \frac{10}{27}\frac{1}{y^2}\right) - x^2\right\}}^{c} + i\overbrace{\left\{\frac{23}{18}\frac{1}{y}x\right\}}^{d}}{\underbrace{\left\{\frac{5}{12}\frac{1}{y} - \frac{23}{18}\frac{1}{y}x^2\right\}}_{a} + i\underbrace{\left\{\left(\frac{5}{6} + \frac{10}{27}\frac{1}{y^2}\right)x - x^3\right\}}_{b}}\right], \tag{6.4}$$

$$= \left(\frac{\partial \varepsilon}{\partial \varrho}\right)^2 \frac{1}{\pi}\langle|\varrho(\mathbf{q},0)|^2\rangle \frac{1}{\sqrt{2\frac{k}{m}T}q}\frac{ac+bd}{a^2+b^2}. \tag{6.5}$$

This formula provides values for arbitrary values of y, however small. On the other hand we have

$$y = \frac{\varrho}{3}\frac{\sqrt{2\frac{k}{m}T}}{\mu}\frac{1}{q} = \frac{\sqrt{2}}{3}\frac{p}{\sqrt{\frac{k}{m}T}}\frac{1}{\mu}\frac{1}{q}, \tag{6.6}$$

and for a fixed pressure p and temperature T the parameter y has a minimal value given by [see (1.13)]

$$q_{\max} = \frac{4\pi}{\lambda^i} \approx 2.1 \cdot 10^7 \frac{1}{m} \quad \text{for } \lambda^i \approx 0.6 \cdot 10^{-6} m$$

corresponding to a scattering angle of $\vartheta = \pi$. For p = 1 bar, T = 293 K Table 2 provides the minimal values of y for different noble gases. The values of the viscosity μ are taken from [100]. Of course, if the pressure in the scattering volume is lowered, the minimal value of y goes down accordingly.

	He	Ne	Ar	Kr	Xe
$\mu\left[10^{-5}\frac{Ns}{m^2}\right]$	1,97	3,16	2,24	2,51	2,26
$\sqrt{\frac{k}{m}T}\ \left[\frac{m}{\text{sec}}\right]$	780	347	247	170	136
$y_{\min}$	0,15	0,20	0,41	0,53	0,73

Table 2. The minimal possible values of y at p=1 bar, T = 293 K.

Long before the minimal value of y is reached the Navier–Stokes–Fourier theory is invalid, but the more refined theories of extended thermodynamics may well be good for small values of $y > y_{\min}$. We proceed to investigate this proposition.

7 Light Scattering and Extended Thermodynamics

7.1 The Case of 13 Moments: An Illustration

We recall the linearized field equations of extended thermodynamics for a mon-atomic gas, which read [see (9.1.41)].

$$\begin{aligned}
&\frac{\partial \varrho}{\partial t} + \tilde{\varrho}\frac{\partial v_i}{\partial x_i} = 0,\\
&\tilde{\varrho}\frac{\partial v_i}{\partial t} + \left(\frac{k}{m}\tilde{T}\right)\frac{\partial \varrho}{\partial x_i} + \tilde{\varrho}\frac{\partial \frac{k}{m}T}{\partial x_i} - \frac{\partial t_{\langle ij\rangle}}{\partial x_j} = 0,\\
&\frac{3}{2}\tilde{\varrho}\frac{\partial \frac{k}{m}T}{\partial t} + \frac{k}{m}\tilde{T}\frac{\partial v_i}{\partial x_i} + \frac{\partial q_i}{\partial x_i} = 0, \qquad (7.1)\\
&\frac{\partial t_{\langle ij\rangle}}{\partial t} - \frac{4}{5}\frac{\partial q_{\langle i}}{\partial x_{j\rangle}} - 2\tilde{\varrho}\frac{k}{m}\tilde{T}\frac{\partial v_{\langle i}}{\partial x_{j\rangle}} = \sigma t_{\langle ij\rangle},\\
&\frac{\partial q_i}{\partial t} - \frac{k}{m}\tilde{T}\frac{\partial t_{\langle ij\rangle}}{\partial x_j} + \frac{5}{2}\tilde{\varrho}\frac{k}{m}\tilde{T}\frac{\partial \frac{k}{m}T}{\partial x_i} = \frac{2}{3}\sigma q_i.
\end{aligned}$$

We reduce this system of 13 equations by an argument similar to the one explained for Navier–Stokes–Fourier fluids in Paragraph 4.2. The system can be decomposed into two uncoupled systems for

$$\varrho, \quad \psi = \frac{\partial v_i}{\partial x_i}, \quad T, \quad \tau = \frac{\partial^2 t_{\langle ij\rangle}}{\partial x_i \partial x_j}, \quad Q = \frac{\partial q_i}{\partial x_i} \quad \text{and} \quad \text{curl } \mathbf{v}, \quad \varepsilon_{ijk} \frac{\partial^2 t_{\langle jn\rangle}}{\partial x_k \partial x_n}, \quad \text{curl } \mathbf{q}. \tag{7.2}$$

The latter quantities do not affect the density field and, therefore, are not of interest for us. Thus for our purposes the system (7.1) is equivalent to the system of five equations, viz.

$$\begin{aligned}
&\frac{\partial \varrho}{\partial t} + \tilde{\varrho}\psi = 0,\\
&\tilde{\varrho}\frac{\partial \psi}{\partial t} + \frac{k}{m}\tilde{T}\Delta\varrho + \tilde{\varrho}\Delta\left(\frac{k}{m}T\right) - \tau = 0,\\
&\tfrac{3}{2}\tilde{\varrho}\frac{\partial \frac{k}{m}T}{\partial t} + \tilde{\varrho}\tfrac{k}{m}\tilde{T}\psi + Q = 0,\\
&\frac{\partial \tau}{\partial t} - \frac{8}{15}\Delta Q - \frac{4}{3}\tilde{\varrho}\frac{k}{m}\tilde{T}\Delta\psi = \sigma\tau,\\
&\frac{\partial Q}{\partial t} - \frac{k}{m}\tilde{T}\tau + \frac{5}{2}\tilde{\varrho}\frac{k}{m}\tilde{T}\Delta\left(\frac{k}{m}T\right) = \frac{2}{3}\sigma Q.
\end{aligned} \tag{7.3}$$

The spatial Fourier and temporal Laplace transform provides a system of algebraic equations, viz.

$$\begin{bmatrix}
s & \tilde{\varrho} & 0 & 0 & 0\\
-\frac{k}{m}\tilde{T}q^2 & \tilde{\varrho}s & -\tilde{\varrho}q^2 & -1 & 0\\
0 & \tilde{\varrho}\frac{k}{m}\tilde{T} & \frac{3}{2}\tilde{\varrho}s & 0 & 1\\
0 & \frac{4}{3}\tilde{\varrho}\frac{k}{m}\tilde{T}q^2 & 0 & s-\sigma & \frac{8}{15}q^2\\
0 & 0 & -\frac{5}{2}\tilde{\varrho}\frac{k}{m}\tilde{T}q^2 & -\frac{k}{m}\tilde{T} & s-\frac{2}{3}\sigma
\end{bmatrix}
\begin{bmatrix}
\tilde{\varrho}(\mathbf{q},s)\\ \tilde{\psi}(\mathbf{q},s)\\ \frac{k}{m}\tilde{T}(\mathbf{q},s)\\ \tilde{\tau}(\mathbf{q},s)\\ \tilde{Q}(\mathbf{q},s)
\end{bmatrix}
=
\begin{bmatrix}
\varrho(\mathbf{q},0)\\ \psi(\mathbf{q},0)\\ \frac{k}{m}T(\mathbf{q},0)\\ \tau(\mathbf{q},0)\\ Q(\mathbf{q},0)
\end{bmatrix}. \tag{7.4}$$

We use analogous arguments as in Sections 4 and 5, including the Onsager hypothesis for the fields ϱ, v_i, T, $t_{\langle ij\rangle}$, q_i and the assumption that the fluctuation $\varrho(\mathbf{q},0)$ is uncorrelated to the fluctuations $\psi(\mathbf{q},0), T(\mathbf{q},0), \tau(\mathbf{q},0), Q(\mathbf{q},0)$, and obtain

$$\frac{\langle \varrho^*(\mathbf{q},0)\varrho(\mathbf{q},s)\rangle}{\langle |\varrho(\mathbf{q},0)|^2\rangle} \tag{7.5}$$

$$= \frac{s^4 - \frac{5}{3}\sigma s^3 + \left(\sigma^2 + \frac{21}{5}\frac{k}{m}Tq^2\right)s^2 - \frac{11}{3}\frac{k}{m}Tq^2\sigma s + \left(\frac{4}{9}\frac{k}{m}Tq^2\sigma^2 + \frac{4}{5}\left(\frac{k}{m}T\right)^2 q^4\right)}{s^5 - \frac{5}{3}\sigma s^4 + \left(\sigma^2 + \frac{26}{5}\frac{k}{m}Tq^2\right)s^3 - \frac{16}{3}\frac{k}{m}Tq^2\sigma s^2 + \left(\frac{10}{9}\frac{k}{m}Tq^2\sigma^2 + 3\left(\frac{k}{m}T\right)^2 q^4\right)s - \frac{5}{3}\left(\frac{k}{m}T\right)^2 q^4\sigma}.$$

This equation is the 13-moment analogue to the Navier–Stokes–Fourier equation (5.2). Therefore the dynamic form factor $S_{\varepsilon\varepsilon}(\mathbf{q},\omega)$ for the 13-moment theory is given by

$$\frac{S_{\varepsilon\varepsilon}(\mathbf{q},\omega)}{\left(\frac{\partial\varepsilon}{\partial\varrho}\right)^2 \frac{1}{\pi}\langle|\varrho(\mathbf{q},0)^2\rangle} = \mathrm{Re}\left[\underline{\qquad\qquad}\right]_{s=i\omega} \tag{7.6}$$

where the bracket denotes the right-hand side of (7.5).

We recall from (2.4.4) that $\sigma = -\varrho(k/m)T/\mu$ holds and introduce dimensionless variables according to [see (6.2)]

$$\hat{s} = \frac{s}{\sqrt{2p_\varrho q}} = \frac{s}{\sqrt{2\frac{k}{m}T}q}, \qquad x = \frac{\omega}{\sqrt{2\frac{k}{m}T}q}, \qquad y = -\frac{\sqrt{2}}{3}\frac{1}{\sqrt{\frac{k}{m}T}q}\sigma.$$

Thus the dynamic form factor (7.6) assumes the form

$$\frac{S_{\varepsilon\varepsilon}(y,x)}{\left(\frac{\partial\varepsilon}{\partial\varrho}\right)^2\frac{1}{\pi}\langle|\varrho(\mathbf{q},0)|^2\rangle} \tag{7.7}$$

$$= \frac{1}{\sqrt{2\frac{k}{m}T}q}\mathrm{Re}\left[\frac{\hat{s}^4 - 5y\hat{s}^3 + \left(9y^2+\frac{21}{10}\right)\hat{s}^2 - \frac{11}{3}y\hat{s} + \left(2y^2+\frac{1}{5}\right)}{\hat{s}^5 - 5y\hat{s}^4 + \left(9y^2+\frac{13}{5}\right)\hat{s}^3 - 8y\hat{s}^2 + \left(5y^2+\frac{3}{4}\right)\hat{s} - \frac{5}{4}y}\right]_{\hat{s}=ix},$$

$$= \frac{1}{\sqrt{2\frac{k}{m}T}q}\mathrm{Re}\left[\frac{\overbrace{\left\{x^4 - \left(9y^2+\frac{21}{20}x^2\right) + \left(2y^2+\frac{1}{5}\right)\right\}}^{c} + i\overbrace{\left\{5yx^3 - \frac{11}{3}yx\right\}}^{d}}{\underbrace{\left\{-5yx^4 + 8yx^2 - \frac{5}{4}y\right\}}_{a} + i\underbrace{\left\{\left(5y^2+\frac{3}{4}\right)s - \left(9y^2+\frac{13}{5}\right)x^3\right\}}_{b}}\right], \tag{7.8}$$

$$= \frac{1}{\sqrt{2\frac{k}{m}T}q}\frac{ac+bd}{a^2+b^2}. \tag{7.9}$$

This is the same equation as (6.5), but, of course, the quantities a through d are different; here they refer to the 13-moment theory, while those previous referred to the Navier–Stokes–Fourier theory.

7.2 Comparison of the Navier–Stokes–Fourier Theory and of Extended Thermodynamics of 13 and 14 Moments

The preceding calculation can be repeated for all the systems listed in Paragraph 9.1.7, which take more and more moments into account. It can also be made for the system (9.3.4) with the cutoff placed anywhere. In all cases the result is formally the same, namely, (6.5) or (7.9), but the quantities a through d become more and more complex. These values are not given here for higher-moment theories, although they have been used in the preparation of the following graphs. The interested reader will be able to calculate a through d for the 35-moment theory, say, if he is diligent, from the equations (9.1.46).

For given values of y the dynamic form factor is plotted in Figure 9 for three theories: The Navier–Stokes–Fourier theory and extended thermodynamics of 13 and 14 moments. Since the function $S_{\varepsilon\varepsilon}(y,x)$ is even, we have plotted only the part for $x > 0$. Also the function $S_{\varepsilon\varepsilon}$ has been divided by the factor $\left(\frac{\partial\varepsilon}{\partial\rho}\right)^2\frac{1}{\pi}\langle|\rho(\mathbf{q},0)|^2\rangle$ before plotting. We observe the following noteworthy features.

First, for the large value $y = 6$ all three theories provide the same curves, which are characterized by well-pronounced Rayleigh and Brillouin peaks.

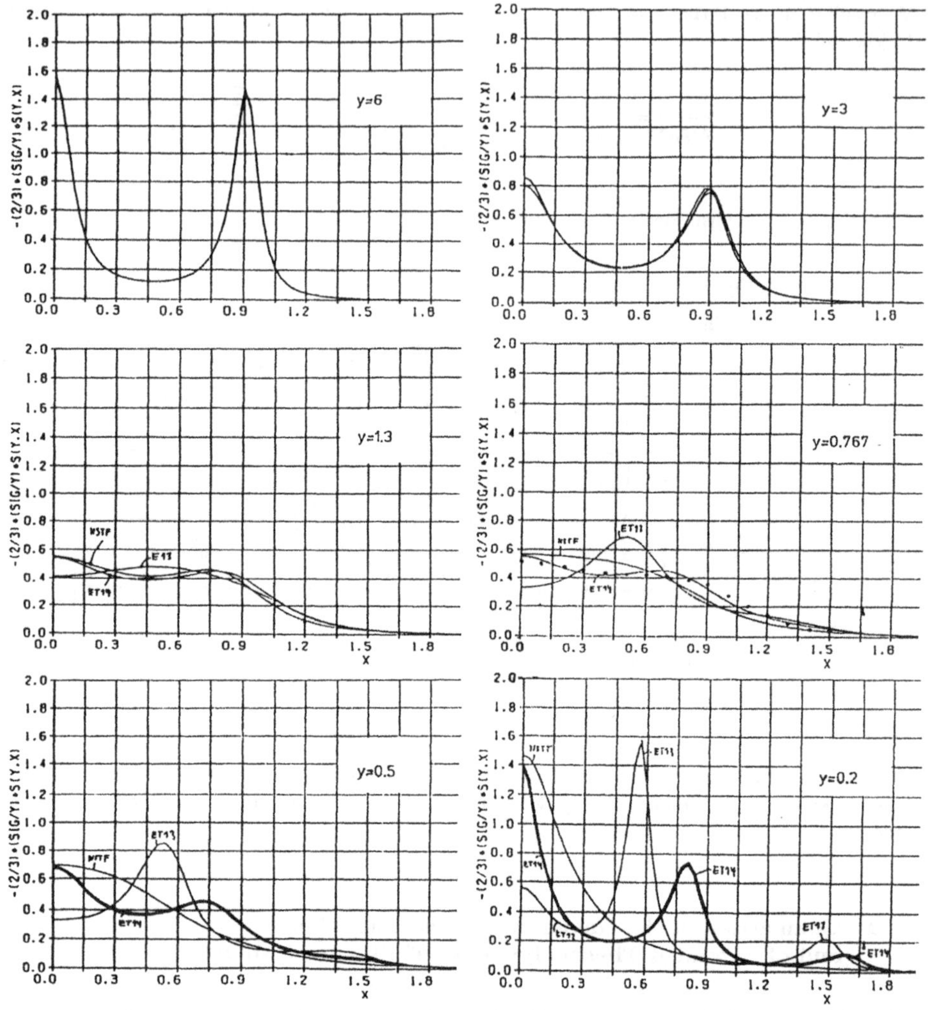

Figure 9. Dynamic form factor for Navier–Stokes–Fourier theory and for extended thermodynamics of 13 and 14 moments. Decreasing values of y as indicated.

Second, for $y = 3$ discrepancies begin to appear between the three theories, and for $y = 1,3$ and $y = 0.767$ these are very noticeable indeed. Of course, there is no way to tell from the theory which graph is correct, that is, the measured one. Therefore we have shown some measured points in Figure 9(d). These points have been taken from the paper [101] by Clark, who has measured $S(y, x)$ for $y = 0.767$. Inspection shows that Clark's results seem to support the idea that the 14-moment theory is good for this value of y; and, indeed, we must then suppose that the 14-moment theory is the only one we could possibly rely on for even smaller values of y.

This is why in the Figures 9(e), (f), which refer to $y = 0.5$ and $y = 0.2$, we have emphasized the 14-moment curve by a thick line. We notice that a second peak appears on the right. This means that in a 14-moment theory—and indeed, even in the 13-moment theory—a second sound mode occurs. We may ask whether that additional peak is ever observed, and the answer is no. The reason for this lack of observability lies in the fact that the 14-moment theory is no longer good at the small values of y where the second peak appears.

Indeed, even for $y = 0.767$ the 14-moment theory, despite its agreement with the experimental points, is not reliable. The agreement it exhibits with the experiments is accidental. In order to justify that judgement, let us consider:

7.3 Extended Thermodynamics with More and More Moments

We may repeat the calculation of $S_{\varepsilon\varepsilon}(y, x)$ for more and more moments and, of course, we work under the assumption that, if the n-moment theory is good down to a given value of y, the $(n + N)$-moment theory is certainly good down to the same value of y. In fact, we expect the $(n + N)$-moment theory to be good down to *smaller* values of y.

Now, since the 14-moment theory seemed to be good at $y = 0.767$—according to Clark's experiments—the theories with 20, 35, 56, and 84 moments should also be good at that value of y; *but they are not* as is demonstrated in Figure 10. Indeed, the functions $S(y, x)$ for these many moment theories disagree widely among each other and with the 14-moment theory and, hence, with experiment. Therefore we must conclude that the good results of the 14-moment theory at $y = 0.767$ were purely fortuitous.

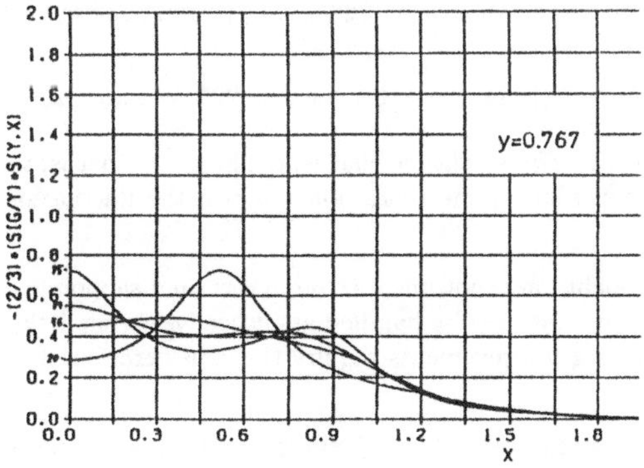

Figure 10. $S(y, x)$ at $y = 0.767$ for 20, 35, 56, and 84 moments.

This conclusion is confirmed, when we take even more moments into account. Figure 11 shows perfect agreement at $y = 0.767$ between the curves for $n = 120, 165, 220$, and 286. Also these curves agree well with the dots in Figure 9(d) as they should.

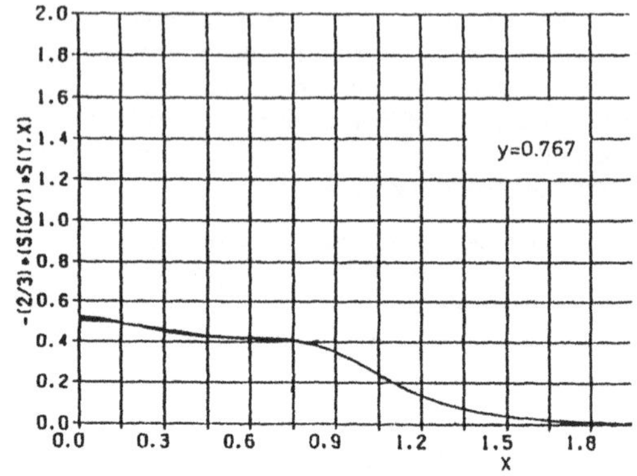

Figure 11. $S(y, x)$ at $y = 0.767$ for 120, 165, 220, and 286 moments.

This does not mean that we can stop with 220 moments, say, or 165 moments. Indeed, for smaller y's even these theories disagree as is shown in Figure 12, which refers to $y = 0.1$ with $n = 210$, 225, 240, and 256.

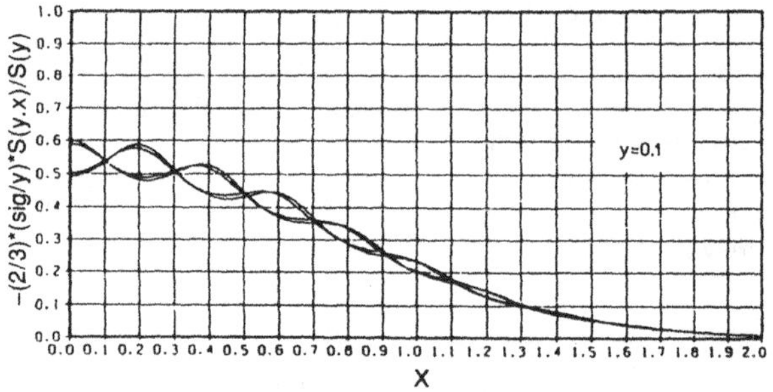

Figure 12. $S(y, x)$ at $y = 0.1$ for 210, 225, 240, and 256 moments.

The final conclusion reads: *The smaller* y, that is, the bigger q, that is, the smaller the wave length of the Fourier harmonic, that is, the more non-uniform the fluctuation, *the more moments are needed.*

Moreover, while this conclusion is not unexpected, a certain disappointment lies in the fact that 13 or 14 moments—numbers that may be handled analytically—bring little advance over the Navier–Stokes–Fourier theory. Big improvements require the consideration of hundreds of moments. Let us make an estimate.

According to Figure 11 we needed 120 moments for an adequate description of the dynamic form factor $S(y, x)$ for $y = 0.767$. From $(6.2)_3$ we have

$$y = \frac{\varrho}{3}\frac{\sqrt{2\frac{k}{m}T}}{\mu}\frac{1}{q} = \frac{\sqrt{2}}{3}\frac{p}{\sqrt{\frac{k}{m}T}}\frac{1}{\mu}\frac{1}{q}. \tag{7.10}$$

With $y = 0.767$ and p = 1 bar, T = 293 K, and $\mu = 2.26 \cdot 10^{-5}\frac{Ns}{m^2}$, $\sqrt{\frac{k}{m}T} = 136\frac{m}{s}$ as in Table 2 for Xe we thus obtain $q = 3.13 \cdot 10^6 \frac{1}{m}$, that is, $\lambda = \frac{2\pi}{q} \approx 2 \cdot 10^{-6}m$. This is in the order of magnitude of the mean free path of the atoms under the given conditions. We conclude that processes with significant density gradients over one mean free path require a theory with at least 120 moments!

Finally we note that, of course, the range of validity of a given theory is not exclusively determined by the wave length but rather by the value of y, which depends on p. It follows that a decrease in pressure has the same effect as a decrease in wave length. Thus a proper description of a dilute gas requires more moments sooner than the description of a dense gas.

Chapter 11

Testing Extended Thermodynamics by Sound

Experimentalists measure phase speeds and attenuations in acoustic resonators. On the other hand, any thermodynamic theory will provide a mathematical expression for the phase speeds and attenuations. If the theory contains unknown parameters, a comparison of the empirical data with the theoretical prediction delivers values for these parameters; but, if a theory is specific—as is the case with molecular extended thermodynamics—the experiments provide an estimate of the range of validity of the theory.

The measurements of dispersion of sound and of scattering of light complement each other. They show that a given theory of n moments is satisfactory up to a frequency ω_n and down to a wave length λ_n respectively. Thus we obtain information about the number of moments that we need for dealing with a given initial-boundary value problem: If ω_n and λ_n are significantly represented in the Fourier spectra of the initial and boundary values, we shall have to use the n-moment theory.

The chapter begins with a short description of how an acoustic resonator works. Our misgivings about the interpretation of the acquired data are registered and explained.

1 Basic Acoustics

1.1 How the Acoustic Resonator Measures Phase Speeds in Principle

The acoustic resonator basically consists of a gas-filled tube with a harmonic sound source at one end and an adjustable piston at the other end (see Figure 1). The resonator is stimulated by an alternating voltage of amplitude U_0 and adjustable frequency ω. On the left end at $x = 0$ this creates a velocity $v(0,t) = V_0 \cos \omega t$, while on the right end at $x = L$ the velocity is zero.

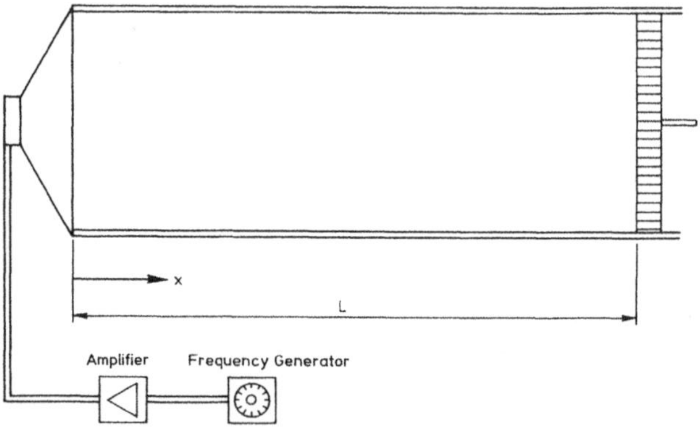

Figure 1. Acoustic resonator (schematic).

So as to explain the principle of measuring phase speeds we first concentrate on the idealized case of a Eulerian fluid in which the viscosity and the heat conductivity may be neglected. The linearized equations of balance of mass, momentum, and energy read

$$\begin{aligned} \frac{\partial \varrho}{\partial t} + \tilde{\varrho}\frac{\partial v_i}{\partial x_i} &= 0, \\ \tilde{\varrho}\frac{\partial v_i}{\partial t} + \tilde{p}_T\frac{\partial T}{\partial x_i} + \tilde{p}_\varrho \frac{\partial \varrho}{\partial x_i} &= 0, \\ \tilde{\varrho}\tilde{\varepsilon}_T \frac{\partial T}{\partial t} + \tilde{T}\tilde{p}_T \frac{\partial v_i}{\partial x_i} &= 0. \end{aligned} \tag{1.1}$$

We set $\varrho = \varrho(x,t)$, $\mathbf{v} = (v(x,t),0,0)$, $T = T(x,t)$ and eliminate ϱ and T to obtain

$$\frac{\partial^2 v}{\partial t^2} - c_0^2 \frac{\partial^2 v}{\partial x^2} = 0 \quad \text{where } c_0^2 = p_\varrho + \frac{T p_T^2}{\varrho^2 \varepsilon_T}. \tag{1.2}$$

The solution appropriate to the boundary conditions

$$v(0,t) = V_0 \cos \omega t, \qquad v(L,t) = 0 \tag{1.3}$$

is a superposition of two harmonic waves of wave length $\lambda = 2\pi \frac{c_0}{\omega}$ running to the right and left, viz.

$$v(x,t) = -\frac{1}{2}\frac{V_0}{\sin(\omega L/c_0)}\left[\sin\left\{(x + c_0 t)\frac{\omega}{c_0} - \frac{\omega L}{c_0}\right\} + \sin\left\{(x - c_0 t)\frac{\omega}{c_0} - \frac{\omega L}{c_0}\right\}\right], \tag{1.4}$$

$$v(x,t) = \frac{V_0}{\sin(\omega L/c_0)} \cos \omega t \sin\left\{(L - x)\frac{\omega}{c_0}\right\}. \tag{1.5}$$

This represents a standing wave with knots at the points

$$x_n = L - n\pi\frac{c_0}{\omega} = L - n\frac{\lambda}{2} \tag{1.6}$$

and singular bulges, if the length L is adjusted to the value

$$L = m\pi\frac{c_0}{\omega} = m\frac{\lambda}{2}. \tag{1.7}$$

The bulges indicate that even a tiny amplitude V_0 will create strong velocity oscillations between the knots, provided that L has the values (1.7). In that case we say that the gas is in resonance or that the resonator is well-tuned.

An old-fashioned method of measuring the sound speed in a gas uses sand on the bottom of the tube. The strong oscillations at the bulges displace the sand and sweep it to the motionless regions at the knots. Thus inspection determines λ and hence $c_0 = \lambda\frac{\omega}{2\pi}$, since ω is known.

1.2 Piezoelectric Transducer and the Mechanical Impedance

In modern days the source of sound is a piezoelectric crystal with a thin layer of metal on opposite sides. An alternating voltage of frequency ω is applied to those sides. It creates a current I and induces oscillations in the crystal, which are transmitted to the gas. Thus we have what is called a piezoelectric transducer. We proceed to describe the properties of this instrument.

The metal layers provide a capacity C so that the current I is split as

$$I = I_C + I_P \quad \text{where } I_C = C\frac{dU}{dt}. \tag{1.8}$$

Here I_C is the current through the capacitor and I_P is the current through the piezocrystal. The power UI_P equals the power of the piezo force F_P on the crystal, which is $F_P v(0,t)$. Thus we conclude that

$$F_P = \alpha U \quad \text{and} \quad I_P = \alpha v(0,t) \tag{1.9}$$

hold, where α is a property of the transducer.

The crystal may be considered to be a linearly viscoelastic body. To fix the ideas we represent it by a mass m, which is connected to a support by an elastic spring and a dashpot (see Figure 2). In addition to the piezo force the crystal is subject to the force

$$F_G = At_{xx}(0,t) \tag{1.10}$$

exerted by the gas. A is the cross-sectional area and t_{xx} is the normal stress in x-direction. The equation of motion reads

$$m\ddot{l} + D\dot{l} + cl = F_P + At_{xx}(0,t). \tag{1.11}$$

We note that $\dot{l}$ equals $v(0,t)$ and differentiate equation (1.1). Thus we obtain by use of (1.9)

$$m\ddot{I}_P + D\dot{I}_P + cI_P - \alpha A\dot{t}_{xx}(0,t) = \alpha^2\dot{U}. \tag{1.12}$$

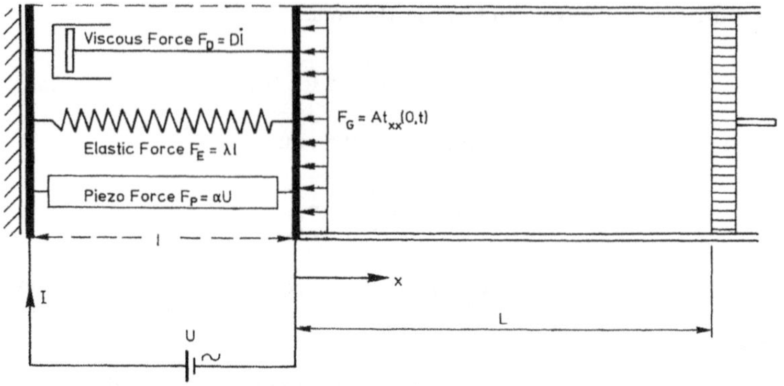

Figure 2. Acoustic resonator and schematic representation of the piezo crystal.

A sinusoidal input $U = U_0 \mathrm{Re}\, e^{i\omega t}$ will create sinusoidal functions

$$I_C = \mathrm{Re}\left(\bar{I}_C e^{i\omega t}\right), \quad I_P = \mathrm{Re}\left(\bar{I}_P e^{i\omega t}\right), \quad t_{xx}(0,t) = \mathrm{Re}\left(\bar{t}_{xx} e^{i\omega t}\right), \quad v(0,t) = \mathrm{Re}\left(\bar{v} e^{i\omega t}\right) \tag{1.13}$$

where $\bar{I}_C$, $\bar{I}_P$, $\bar{t}_{xx}$, and $\bar{v}$ are complex amplitudes. Insertion into (1.8) and (1.12) provides

$$\bar{I} = \left(i\omega C + \frac{\alpha^2}{i\omega m + D - i\frac{c}{\omega} - \alpha A(\bar{t}_{xx}/\bar{I}_P)} \right) U_0 \quad \text{or by} \quad (1.9)_2$$

$$\bar{I} = \underbrace{\left(i\omega C + \frac{\alpha^2}{i\omega m + D - i\frac{c}{\omega} - A(\bar{t}_{xx}/\bar{v})} \right)}_{1/Z} U_0. \tag{1.14}$$

We define the *impedance* Z of the transducer as indicated in (1.14). The expression

$$Z_m = A\frac{\bar{t}_{xx}}{\bar{v}} \tag{1.15}$$

is called the *external mechanical impedance.*

For a given piezoelectric transducer the quantities C, α, m, D, c, and A are known. Thus measuring $\hat{I}$ for a given U_0 we may calculate Z_m. Now arises the question: how to use that knowledge to calculate the wave-length λ in the gas. We proceed to answer that question, first once again for a Eulerian fluid.

1.3 External Mechanical Impedance and Wave-Length

We recall from (1.5) the form of $v(x,t)$ for a Eulerian fluid, viz.

$$v(x,t) = \frac{V_0}{\sin 2\pi\frac{L}{\lambda}} \sin\left\{(L-x)\frac{2\pi}{\lambda}\right\} \cos\omega t = \frac{V_0}{\sin 2\pi\frac{L}{\lambda}} \mathrm{Re}\left[\sin\left\{(L-x)\frac{2\pi}{\lambda}\right\} e^{i\omega t}\right]. \tag{1.16}$$

Also for a Eulerian fluid $t_{xx}(x,t) = -p(x,t)$, and it follows from $(1.1)_2$ that we have

$$-p(x,t) = \tilde{\varrho}c_0 \frac{V_0}{\sin 2\pi\frac{L}{\lambda}} \cos\left\{(L-x)\frac{2\pi}{\lambda}\right\} \sin\omega t = \tilde{\varrho}c_0 \frac{V_0}{\sin 2\pi\frac{L}{\lambda}} \mathrm{Re}\left[-i\cos\left\{(L-x)\frac{2\pi}{\lambda}\right\} e^{i\omega t}\right] \tag{1.17}$$

Therefore we obtain

$$Z_m = A\frac{\bar{t}_{xx}(0,t)}{\bar{v}(0,t)} = -i\tilde{\varrho}c_0 ctg\frac{2\pi L}{\lambda}. \tag{1.18}$$

Inspection shows that Z_m is singular for

$$L = n\frac{\lambda}{2}, \tag{1.19}$$

that is, when the resonator is well-tuned. In that case the impedance Z has a maximum and the current I is minimal. Also Z_m vanishes for

$$L = \left(n+\frac{1}{2}\right)\frac{\lambda}{2} \tag{1.20}$$

and I assumes a maximum. Since I may be measured as a function of L, the wave-length—hence the speed c_0—may be read from the distance of the maxima or minima of I on the L-axis.

All this is perfectly simple and convincing—for a Eulerian fluid.

1.4 Difficulties with Many Modes and Damping

Things are less clearcut when damping occurs, as it does in a Navier–Stokes–Fourier fluid or in the extended theories of various numbers of moments. The problems that occur have been discussed by Weiss [96] in some detail. They arise from the following observations:

i. In a Navier–Stokes–Fourier fluid we have two modes of propagation, that is, waves with different phase speeds and damping and each one runs in both directions in the resonator tube. The fields in the tube are superpositions of those four waves, and the solutions cannot be determined from the boundary values $v(0,t)$ and $v(L,t)$ alone. Rather we need additional boundary conditions on $T(x,t)$ (say), for example, conditions of adiabaticity or fixed temperatures at the ends.

   In extended thermodynamics of many moments this problem is compounded: There are n modes of propagation (say) and the field in the tube is therefore a superposition of $2n$ waves. Thus we need $2n$ boundary conditions and nobody knows how to fix them.

ii. Even if this problem was solved, we immediately face another one. Each one of the n modes has its own phase speed and its own attenuation, and all of these influence the external mechanical impedance.

   From what we were able to determine, however, the experimentalists extract only a single phase speed and a single attenuation from the measured impedance. How can that be? Is it the case that only *one* mode dominates the impedance, because the others are damped too strongly? And which mode is that?

iii. Let us suppose that we know the answer to these questions and that indeed one wave prevails. Even this one wave will be damped and, as soon as there is damping, there are no standing waves any more and the maxima and minima of the impedance have different distances along the L-axis. So, where do we measure that distance in order to obtain the wave-length of that privileged wave?

iv. Calculations have shown that at low frequencies indeed all modes but one are strongly damped and this one is the *sound mode*, that is, the mode whose phase speed tends to $\sqrt{\frac{5}{3}\frac{k}{m}T}$ for $\omega \to 0$.

At higher frequencies, however, even this mode picks up considerable damping and eventually, according to extended thermodynamics, it looses its singular role to the *mode* $V_{\max}$, so-called by Weiss [96], because it has the maximum phase speed. It stands to reason that, where the two modes interchange their role as the least damped one, both have equal influence on the impedance. It therefore must be hard to interpret this function.

We are bothered and confused by these uncertainties and we are unable to remove them. This being so, we stick to a simple strategy: We calculate the phase speeds of only two modes, the sound mode and the mode $V_{\max}$, the latter in the range where its attenuation is smaller than that of the sound mode.

2 Dispersion Relations

2.1 Navier–Stokes–Fourier Theory

According to the Navier–Stokes–Fourier theory the constitutive relations for a mon–atomic ideal gas have the forms

$$t_{ij} = -\varrho\frac{\mathrm{k}}{m}T\delta_{ij} + 2\mu\frac{\partial v_{\langle i}}{\partial x_{j\rangle}}, \qquad q_i = -\kappa\frac{\partial T}{\partial x_i}, \qquad u = \frac{3}{2}\frac{\mathrm{k}}{m}T. \tag{2.1}$$

We introduce these into the equations of balance of mass, momentum, and energy and linearize about a uniform and constant state $\tilde{\varrho}$, $\tilde{\mathbf{v}} = 0, \tilde{T}$. Furthermore we assume that $\mathbf{v}$ points in the x-direction and that all fields depend only on x and t. Thus we obtain

$$\begin{bmatrix} \frac{\partial}{\partial t} & \tilde{\varrho}\frac{\partial}{\partial x} & 0 \\ \frac{\mathrm{k}}{m}\tilde{T}\frac{\partial}{\partial x} & \tilde{\varrho}\frac{\partial}{\partial t} - \frac{4}{3}\mu\frac{\partial^2}{\partial x^2} & \tilde{\varrho}\frac{\partial}{\partial x} \\ 0 & \frac{2}{3}\tilde{\varrho}\frac{\mathrm{k}}{m}\tilde{T}\frac{\partial}{\partial x} & \tilde{\varrho}\frac{\partial}{\partial t} - \frac{5}{2}\mu\frac{\partial^2}{\partial x^2} \end{bmatrix} \begin{bmatrix} \varrho \\ v \\ \frac{\mathrm{k}}{m}T \end{bmatrix} = 0 \tag{2.2}$$

where we have assumed that $\kappa = \frac{15}{4}\frac{\mathrm{k}}{m}\mu$ holds, which is exact for Maxwellian molecules.

We consider harmonic waves of frequency ω, complex wave number k, and complex amplitudes $\bar{\varrho}$, $\bar{v}$, and $\bar{T}$. These amplitudes may be calculated from the algebraic equations

$$\begin{bmatrix} \omega & -\tilde{\varrho}k & 0 \\ -\frac{\mathrm{k}}{m}\tilde{T}k & \tilde{\varrho}\omega - i\frac{4}{3}\mu k^2 & -\tilde{\varrho}k \\ 0 & -\frac{2}{3}\tilde{\varrho}\frac{\mathrm{k}}{m}\tilde{T}k & \tilde{\varrho}\omega - i\frac{5}{2}\mu k^2 \end{bmatrix} \begin{bmatrix} \bar{\varrho} \\ \bar{v} \\ \frac{\mathrm{k}}{m}\bar{T} \end{bmatrix} = 0. \tag{2.3}$$

The dispersion relation results from the requirement that the system (2.3) have a nontrivial solution. We introduce the dimensionless inverse frequency

$$z = \frac{\tilde{\varrho}c_0^2}{\omega\mu} \quad \text{where } c_0 = \sqrt{\frac{5}{3}\frac{\mathrm{k}}{m}\tilde{T}} \tag{2.4}$$

and obtain

$$\left(\frac{10}{3}\frac{1}{z^2}+i\frac{3}{2}\frac{1}{z}\right)\left(\frac{k}{\omega/c_0}\right)^4+\left(1-i\frac{23}{6}\frac{1}{z}\right)\left(\frac{k}{\omega/c_0}\right)^2-1=0. \tag{2.5}$$

This is a biquadratic equation, which has two relevant solutions, called modes 1 and 2. Both $\operatorname{Re} k$ and $\operatorname{Im} k$ are functions of ω and the phase speeds and attenuations are given by

$$v_{Ph}=\frac{\omega}{\operatorname{Re} k} \quad \text{and} \quad \alpha=-\operatorname{Im} k. \tag{2.6}$$

Figure 3 shows the phase speeds and attenuations of both modes as functions of z. The figure also shows experimental points measured by Meyer and Sessler [102]. We see that for small ω, that is, large z, these agree with the calculated values of v_{Ph} and α of mode 1. There is no agreement for large frequencies and, in particular, the phase speeds tend to ∞ for $\omega \to \infty$. We have previously, in Chapter 1, commented on this fact, which is known as the paradox of heat conduction; it has motivated the development of extended thermodynamics.

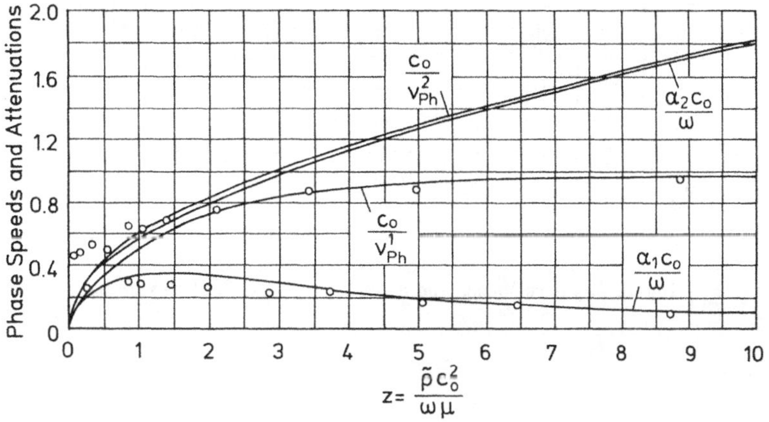

Figure 3. Phase speeds and attenuations in the NSF theory. Experimental points by Meyer and Sessler [102].

2.2 Extended Thermodynamics of 13 Fields

Extended thermodynamics in its simplest form is the 13-field theory whose equation we have written down on several occasions in this book [e.g., see (9.1.41)]. For a longitudinal harmonic wave we obtain the dispersion relation

$$\left(i\frac{3}{2}-\frac{9}{2}\frac{1}{z}\right)\left(\frac{k}{\omega/c_0}\right)^4-\left(z+8i-13\frac{1}{z}\right)\left(\frac{k}{\omega/c_0}\right)^2+\left(z+i\frac{25}{2}-\frac{25}{6}\frac{1}{z}\right)=0. \tag{2.7}$$

This is a biquadratic equation so that there are again two essential modes, that is, two phase speeds and attenuations. These are shown in Figure 4. We conclude that in extended thermodynamics the phase speeds v_{Ph} are finite for all values of ω and, in particular, for $\omega \to \infty$. Therefore, as explained in Chapter 1, the paradox seems to be resolved by the transition to extended thermodynamics.

Although the phase speed v^1_{Ph} is finite for $\omega \to \infty$, however, it does not agree with the measured values for high frequencies. This fact motivates us to use further extensions of thermodynamics.

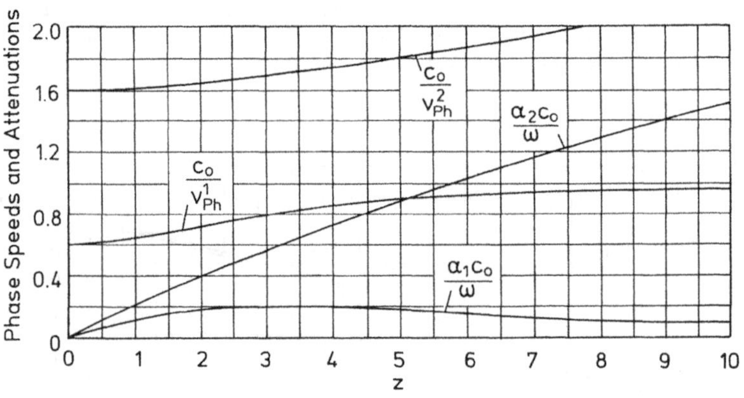

Figure 4. Phase speeds and attenuations of E.T. of 13 variables.

2.3 Extended Thermodynamics with Many Variables

In Chapter 9, when we explained molecular extended thermodynamics, we wrote the systems of field equations for up to 35 variables and gave the recipe for an arbitrary number of variables. All these systems imply dispersion relations that allow us to calculate the phase speed v^1_{Ph} of the sound mode as a function of ω. Figure 5 shows the corresponding graphs. Inspection shows that the range of agreement between theories of differently many moments extends to higher values of ω, if we increase the number of moments. The natural interpretation says that the range of ω in which the theories with the highest moments agree contains reliable results; and, indeed, the coincident curves of different theories lie close to the experimental values as we may read from the Figures 5(c) and 5(d).

We have to go up to close to 5000 moments, if we wish to obtain reliable speeds down to $z = 1$ [see Figure 5(c)], in which, once again we have plotted the result of the Navier–Stokes–Fourier theory so as to demonstrate how much better the moment theories are.

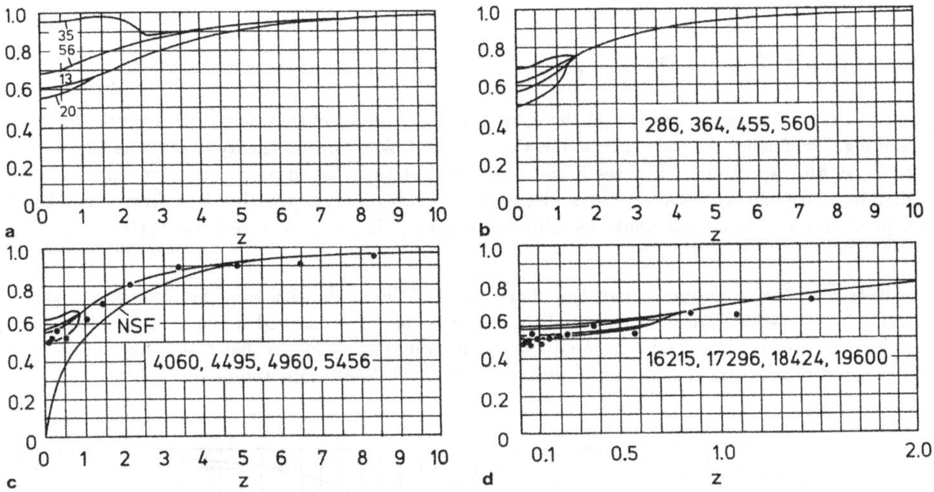

Figure 5. Phase speed of the sound mode in theories with more and more moments. [Note the change of scale in Figure 5(d).]

Figure 5(d) shows that there is no point in going beyond 20,000 moments, because between 16,000 and 19,000 the theories disagree only as much as the experimental points scatter.

2.4 Interpretation and Criticism

The conclusion from these findings reads: The smaller z, that is, the bigger the frequency ω, the more moments are needed. This conclusion complements the conclusion of Chapter 10, which said that many moments are needed for the proper description of the scattering from fluctuations of small wave-length. Here as there, 13 or 14 moments bring little improvement over the Navier–Stokes–Fourier theory. Big improvements require the consideration of hundreds of moments. We make an estimate.

According to Figure 5(c) we need 300 moments for an adequate description of the phase speed v_{Ph}^1 for $z = 3$. From (2.4) we have

$$z = \frac{\tilde{\varrho} c_0^2}{\omega \mu} = \frac{5}{3} \tilde{p} \frac{1}{\omega \mu}.$$

With $z = 3$ and $\tilde{p} = 1 bar$, and $\mu = 2.26 \cdot 10^{-5} \frac{Ns}{m^2}$ as in Table 10.2 for Xe we thus obtain $\omega \approx 2.4\ 10^9 Hz$. This is of the order of magnitude of the collision frequency of the atoms under the given circumstances.

Once again z depends on p so that lowering p has the same effect as increasing ω. A dilute gas requires many moments for its description sooner than a dense gas.

3 Maximum Speed

3.1 Modes of Least Damping

The reason for the concentration on the sound mode is the belief that this mode experiences least damping. This belief is surely satisfied in the Navier–Stokes–Fourier theory and in the 13-moment theory as an inspection of Figures 3 and 4 shows; but then, when more and more moments are taken into account, we obtain more modes and the question arises whether it is always and for all frequencies the sound mode that has minimal damping. In fact, Weiss [96] found that this is not so.

He instructed the computer to plot the phase speed of the mode with least damping and obtained graphs like those in Figure 6: For small frequencies the least damped mode is always the sound mode, but starting with the 57-moment theory the sound mode looses that privileged role for high frequencies to the mode of maximum phase speed, which Weiss calls the mode $V_{\max}$.

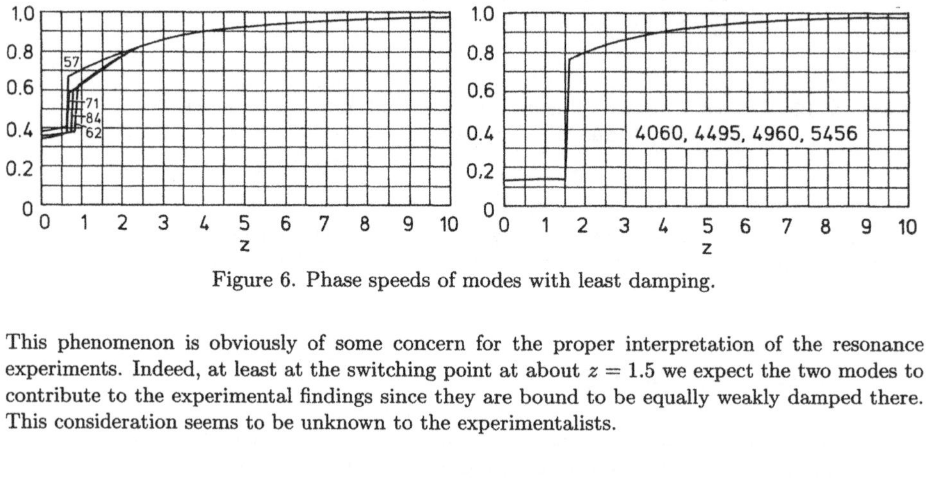

Figure 6. Phase speeds of modes with least damping.

This phenomenon is obviously of some concern for the proper interpretation of the resonance experiments. Indeed, at least at the switching point at about $z = 1.5$ we expect the two modes to contribute to the experimental findings since they are bound to be equally weakly damped there. This consideration seems to be unknown to the experimentalists.

3.2 The Maximum Speed

If we only consider the sound mode, inspection of Figure 5 shows that the phase speed remains finite for $\omega \to \infty$. Thus the paradox of infinite pulse speeds would not seem to arise and, indeed, such was the original motivation for formulating extended thermodynamics.

There are, however, the other modes, and in particular, there is the mode $V_{\max}$. We see from Figure 6 that $V_{\max}$ becomes bigger as the theory is improved by taking more and more moments into account. In fact, Weiss calculated $V_{\max}$ and came up with Table 1 in which $V_{\max}$ shows a steady growth as the number of moments is increased to 15,180. There seems little doubt that $V_{\max}$ keeps on increasing to infinity as the number of moments is raised. Nor is the mode $V_{\max}$ irrelevant; after all it is least damped.

This is somewhat of an anticlimax because, after all, extended thermodynamics started with the desire to have finite pulse speeds, but then, it has already been indicated by Cercignani and Majorana [103] that the non-relativistic kinetic theory—and hence molecular extended thermodynamics of infinitely many moments—permits infinite speeds, for example, in the Maxwell distribution. The present result, therefore, does not come as a surprise.

No. of Moments	v_{max}/c_0
0004	0.77459667
0010	1.34164079
0020	1.80822948
0035	2.21299946
0056	2.57495874
0084	2.90507811
0120	3.21035245
0165	3.49555791
0220	3.76412372
0286	4.01860847
0364	4.26098014
0455	4.49279023
0560	4.71528716
0680	4.92949284
0816	5.13625617
0969	5.33629130
1140	5.53020569
1330	5.71852112
1540	5.90168962
1771	6.08010585
2024	6.25411673
2300	6.42402919
2600	6.59011627
2925	6.75262213
3276	6.91176615
3654	7.06774631
4060	7.22074198
4495	7.37091629
4960	7.51841807
5456	7.66338362
5984	7.80593804
6545	7.94619654
7140	8.08426549
7770	8.22024331
8436	8.35422129
9139	8.48628432
9880	8.61651144
10660	8.74407644
11480	8.87174833
12341	8.99689171
13244	9.12046722
14190	9.24253184
15180	9.36313918

Table 1. Pulse speeds for an increasing number of moments.

Anyway extended thermodynamics has outgrown its original motif of determining a finite speed of heat propagation. Indeed, we may now say that it provides a hierarchy of symmetric hyperbolic systems, which are relevant for the solutions of initial boundary value problems, if their data contain frequencies and wave numbers that do not exceed certain limits.

Bibliography

[1] **C. Eckart** *The thermodynamics of irreversible processes I: the simple fluid.* Phys. Rev. **58** (1940).

[2] **C. Eckart** *The thermodynamics of irreversible processes II: fluid mixtures.* Phys. Rev. **58** (1940).

[3] **C. Eckart** *The thermodynamics of irreversible processes III: relativistic theory of the simple fluid.* Phys. Rev. **58** (1940).

[4] **C. Cattaneo** *Sulla conduzione del calore.* Atti Sem. Mat. Fis. Univ. Modena **3** (1948).

[5] **I. Müller** *Zur Ausbreitungsgeschwindigkeit von Störungen in kontinuierlichen Medien.* Dissertation TH, Aachen (1966).

[6] **I. Müller** *Zum paradox der Wärmeleitungstheorie.* Zeitschrift für Physik **198** (1967).

[7] **H. Grad** *On the kinetic theory of rarefied gases.* Comm. Pure Appl. Math., vol. **2**, Wiley, New York (1949).

[8] **D. Jou, J. Casas-Vázquez, G. Lebon** *Extended irreversible thermodynamics.* Rep. Prog. Phys. **51** (1988).

[9] **M.E. Gurtin, A.C. Pipkin** *A general theory of heat conduction with finite wave speeds.* Arch. Rational Mech. Anal. **31** (1969).

[10] **A.E. Green, K. A. Lindsay** *Thermoelasticity.* J. Elasticity **2** (1972).

[11] **G. Grioli** *Sulla propagazione di onde termomeccaniche nei continui.* Atti Acc. Naz. Lincei **67** (1979).

[12] **D.D. Joseph, L. Preziosi** *Heat waves.* Rev. Mod. Phys. **61** (1989). *Addendum to the paper "Heat Waves."* Rev. Mod. Phys. **62** (1990).

[13] **S.C. Chapman, T.G. Cowling** *The Mathematical Theory of Nonuniform Gases.* Cambridge University Press, London and New York (1961).

[14] **C. Truesdell, R.G. Muncaster** *Fundamentals of Maxwell's Kinetic Theory of a Simple Monatomic Gas.* Academic Press, New York (1980).

[15] **I. Müller** *Thermodynamics.* Pitman, New York (1985).

[16] **B.D. Coleman, V.J. Mizel** *Existence of caloric equations of state in thermodynamics.* J. Chem. Physics **40** (1964).

[17] **I. Müller** *On the entropy inequality.* Arch. Rational Mech. Anal. **26** (1967).

[18] **I.S. Liu** *Method of Lagrange multipliers for exploitation of the entropy principle.* Arch. Rational Mech. Anal. **46** (1972).

[19] **T. Ruggeri** *Symmetric hyperbolic systems of conservative equations for a viscous heat conducting fluid.* Acta Mechanica **47** (1983).

[20] **I.S. Liu, I. Müller** *Extended thermodynamics of classical and degenerate gases.* Arch. Rational Mech. Anal. **83** (1983).

[21] **T. Ruggeri** *Galilean invariance and entropy principle for systems of balance laws.* Cont. Mech. Thermodyn. **1** (1989).

[22] **G.F. Smith** *On isotropic integrity bases.* Arch. Rational Mech. Anal. **18** (1965).

[23] **E. Ikenberry, C. Truesdell** *On the pressures and the flux of energy in a gas according to Maxwell's kinetic theory.* J. Rational Mech. Anal. **5** (1956).

[24] **I. Müller** *On the frame dependence of stress and heat flux.* Arch. Rational Mech. Anal. **45** (1972).

[25] **I. Müller** *On the frame dependence of electric current and heat flux in a metal.* Acta Mechanica **24** (1976).

[26] **M. Abramowitz, I.A. Stegun,** eds. *Handbook of Mathematical Functions.* Dover, New York (1965).

[27] **C. Truesdell** *Sulle basi della termodinamica.* Acc. Naz. dei Lincei (8) **22** (1957).

[28] **C. Truesdell** *Mechanical basis for diffusion.* J. Chem. Phys. **37** (1962).

[29] **I. Müller** *A new approach to thermodynamics of simple mixtures.* Zeitschrift für Naturforschung **28a** (1973).

[30] **S.R. de Groot, P. Mazur** *Non-Equilibrium Thermodynamics.* North-Holland, Amsterdam (1963).

[31] **C.C. Wang** *On the concept of frame indifference in continuum mechanics and in the kinetic theory of gases.* Arch. Rational Mech. Anal. **58** (1975).

[32] **C. Truesdell** *Connection of two errors in the kinetic theory that have been used to cust unfounded upon the principle of material frame indifference.* Meccanica **11** (1976).

[33] **W. Israel** *Nonstationary irreversible thermodynamics: a causal relativistic theory.* Annals of Physics **100** (1976).

[34] **I.S. Liu, I. Müller, T. Ruggeri** *Relativistic thermodynamics of gases.* Annals of Physics **169** (1986).

[35] **T. Ruggeri** *Relativistic extended thermodynamics. General assumptions and mathematical procedures.* (A. Anile and Y. Choquet-Bruhat, eds.). Lecture Notes in Mathematics, vol. **1358**, Springer-Verlag, New York (1989).

[36] **I. Müller** *Relativistic extended thermodynamics II.* (A. Anile and Y. Choquet-Bruhat, eds.). Lecture Notes in Mathematics, vol. **1358**, Springer-Verlag, New York (1989).

[37] **S. Chandrasekhar** *An Introduction to the Study of Stellar Structure.* Dover Publ., New York (1958).

[38] **L. Seccia, A. Strumia** *Wave propagation in relativistic extended thermodynamics.* Cont. Mech. Thermodynamics **2** (1990).

[39] **S. Pennisi** *Some representation theorems in a 4-dimensional product space.* Fisica Matematica Suppl. B.U.M.I. **5** (1986).

[40] **T. Ruggeri** *Convexity and symmetrization in relativistic theories. Privileged time-like congruence and entropy.* Cont. Mech. Thermodyn. **2** (1990), 163–177.

[41] **F. Jüttner** *Das Maxwell'sche Gesetz der geschwindigkeitsverteilung in der relativitätstheorie.* Annalen der Physik **34** (1911).

[42] **F. Jüttner** *Die relativistische quantentheorie des idealen gases.* Zeitschrift für Physik **47** (1928).

[43] **G. Boillat** *Wave velocities in relativistic extended thermodynamics.* Proceedings of the Symposium on Kinetic theory and Extended Thermodynamics, Bologna, May 18–20th, 1987 (I. Müller and T. Ruggeri, eds.). Pitagora Tecnoprint, Bologna (1987).

[44] **M. Kranys** *Phase and signal velocities of waves in dissipative media. Special relativistic theory.* Arch. Rational Mech. Anal **48** (1972).

[45] **J.M. Stewart** *On transient relativistic thermodynamics and kinetic theory.* Proc. Roy. Soc. London Ser. A **357** (1977).

[46] **W. Israel, J.M. Stewart** *On transient relativistic thermodynamics and kinetic theory. II.* Proc. Roy. Soc. London Ser. A **365** (1979).

[47] **N.A. Chernikov** *The relativistic gas in the gravitational field.* Acta Phys. Polonica **23** (1963).

[48] **N.A. Chernikov** *Equilibrium distribution of the relativistic gas.* Acta Phys. Polonica **26** (1964).

[49] **N.A. Chernikov** *Microscopic foundation of relativistic hydrodynamics.* Acta Phys. Polonica **27** (1964).

[50] **W. Dreyer, W. Weiss** *The classical limit of relativistic extended thermodynamics.* Annales de l'Institut Henri Poincaré **45** (1986).

[51] **G.M. Kremer** *Extended thermodynamics of ideal gases with 14 fields.* Annales de l'Institut Henri Poincaré **45** (1986).

[52] **C.D. Levermore** *Relating Eddington factors to flux limiters.* J. Quant. Spectroscopy Rad. Transfer. **31** (1984).

[53] **M. Anile, S. Pennisi, M. Sammartino** *A thermodynamical approach to Eddington factors.* J. Math. Physics **32** (2) (1991).

[54] **G. M. Kremer, I. Müller** *Radiation thermodynamics.* J. Math. Physics **33** (1992).

[55] **W. Dreyer, S. Seelecke** *Entropy and causality as criteria for the existence of shock wave in low temperature ideal conduction.* Cont. Mech. Thermodyn. **4** (1992).

[56] **R. Courant, K.O. Friedrichs** *Supersonic Flow and Shock Waves.* Interscience Publ., New York (1948).

[57] **W. Larecki** *Symmetric conservative form of low-temperature phonon gas hydrodynamics II. Equations of heat trasport and thermal waves in the case of linear isotropic approximation of phonon frequency spectrum.* Il Nuovo Cimento **1868 D** (1991).

[58] **M. Berger, M. Berger** *Perspectives in Nonlinearity.* W. A. Benjamin, Inc., New York (1968), 137.

[59] **K.O. Friedrichs** *On the laws of relativistic electro-magneto-fluid dynamics.* Comm. Pure Appl. Math., vol. bf 27, Wiley, New York (1974).

[60] **A. Fisher, D.P. Marsden** *The Einstein evolution equations as a first order quasi-linear symmetric hyperbolic system.* Comm. Math. Phys. **28** (1972).

[61] **S. K. Godunov** *An interesting class of quasilinear systems.* Sov. Math. **2** (1961).

[62] **G. Boillat** *Sur l' éxistence et la recherche d'équations de conservation supplémentaires pour les systémes hyperboliques.* C. R. Acad. Sci. Paris, Sér A **278** (1974).

[63] **T. Ruggeri, A. Strumia** *Main field and convex covariant density for quasi-linear hyperbolic systems. Relativistic fluid dynamics.* Annales de l'Institut Henri Poincaré **34** (1981).

[64] **K.O. Friedrichs, P.D. Lax** *Systems of conservation equations with a convex extension.* Proc. Nat. Acad. Sci. U.S.A. **68** (1971).

[65] **A. Muracchini, T. Ruggeri, L. Seccia** *Dispersion relation in high frequency limit and non linear wave stability for hyperbolic dissipative system.* Wave Motion **15**, (2) (1992).

[66] **G. Boillat** *La propagation des ondes.* Gauthier-Villars, Parigi (1965).

[67] **G. Boillat, T. Ruggeri** *On the evolution law of the weak discontinuities for hyperbolic quasi-linear systems.* Wave Motion **1** (1979), 149.

[68] **T. Ruggeri** Stability and discontinuity waves for symmetric hyperbolic systems, *Nonlinear Wave Motion* (A. Jeffrey, ed.). Longman, Harlow (1989).

[69] **A. Muracchini, L. Seccia** *Thermo-acceleration waves and shock formation in Extended Thermodynamics of gravitational gases.* Cont. Mech. Thermodyn. **1** (1989), 227.

[70] **P. Chen** Growth and decay of waves in solids, *Mechanics of Solids III*, Handbuch der Physik, vol. **6A/3**. Springer-Verlag, Berlin, (1973), 303.

[71] **T. Ruggeri** *Shock waves in hyperbolic dissipative systems: non equilibrium gases.* Pitman Research Notes in Mathematics, vol. **227**. (D. Fusco, A. Jeffrey, eds.). Longman, Harlow (1991).

[72] **L.D. Landau, E. Lifsits** *Mècanique des Fluides*, MIR, Moscow, (1971), 418.

[73] **P.D. Lax** Shock Waves and Entropy, *Contribution to Non Linear Functional Analysis.* (Zarantonello, ed.). Academic Press, New York (1971).

[74] **G. Boillat** *Sur une fonction croissant comme l'entropie et generatrice des chocs dans les systemes hyperboliques.* C. R. Acad. Sci. Paris Sér. A **283** (1976), 409.

[75] **P.D. Lax** *Hyperbolic systems of conservation laws.* Comm. Pure Appl. Math., vol. **10**. Wiley, New York, (1957), 537.

[76] **H. Grad** *The profile of a steady plane shock wave.* Comm. Pure Appl. Math., vol. **5**. Wiley, New York, (1952), 257.

[77] **A.M. Anile, A. Majorana** *Shock structure for heat conducting and viscid fluids.* Meccanica **16** (3) (1981), 149.

[78] **L. Tisza** *Transport phenomena in He II.* Nature **141** (1938).

[79] **L.D. Landau** *The theory of superfluidity of Helium II.* J. Phys. USSR **5** (1941), 71.

[80] **V. Peshkov** J. Phys. USSR **8** (1944), 381.

[81] **W. Dreyer** *Zur thermodynamik von He II—superfluides Helium mit und ohne wirbellinien als binäre mischung.* Dissertation TU Berlin (1983).

[82] **V. Peshkov** *Report on an International Conference on Fundamental Particles and Low Temperature Physics, vol. II.* The Physical Society of London (1947), 19.

[83] **C.C. Ackerman, B. Bertman, H.A. Fairbank, R.A. Guyer**, *Second sound in solid Helium.* Phys. Rev. Lett. **16** (18) (1966), 789.

[84] **C.C. Ackerman, W.C. Overton Jr.** *Second sound in solid Helium-3.* Phys. Rev. Lett. **22** (15) (1969), 764.

[85] **H.E. Jackson, C.T. Walker, T.F. McNelly**, *Second sound in NaF.* Phys. Rev. Lett. **25** (1970), 26.

[86] **V. Narayanamurti, R.C. Dynes** *Observation of second sound in Bismuth.* Phys. Rev. Lett. **28** (1972), 1461.

[87] **R.A. Guyer, J.A. Krumhansl** *Solution of the linearized phonon Boltzmann equation.* Phys. Rev. Lett. **148** (2) (1966), 766; *Thermal conductivity, second sound, and phonon hydrodynamic phenomena in nonmetallic crystals.* Phys. Rev. Lett. **148** (2) (1966), 778.

[88] **T. Ruggeri, A. Muracchini, L. Seccia** *Shock waves and second sound in a rigid heat conductor: a critical temperature for NaF and Bi.* Phys. Rev. Lett. **64** (22) (1990), 2640.

[89] **T. Ruggeri, A. Muracchini, L. Seccia** *Continuum approach to phonon gas and shape of second sound via shock waves theory.* Phys. Rev. B, submitted.

[90] **T. Ruggeri** *Struttura dei sistemi alle derivate parziali compatibili con un principio di entropia e termodinamica estesa.* Suppl. BUMI - Fisica Matematica **4** (1985), 261; *Thermodynamics and symmetric hyperbolic systems.* Rend. Sem. Mat. Univ. Torino. Fasc. speciale *Hyperbolic Equations* (1987), 167.

[91] **A. Morro, T. Ruggeri** *Second sound and internal energy in solids.* Int J. Non-Linear Mech. **22** (1987), 27.

[92] **A. Morro, T. Ruggeri** *Non-equilibrium properties of solids obtained from second-sound measurements.* J. Phys. C: Solid State Phys. **21** (1988), 1743.

[93] **B.D. Coleman, D.C. Newman** *Implications of a nonlinearity in the theory of second sound in solids.* Phys. Rev. B **37** (1988), 1492.

[94] **C.C. Ackerman, R.A. Guyer** *Temperature pulses in dielectric solids.* Ann. of Phys. **50** (1968), 128.

[95] **W. Dreyer** *Maximisation of the entropy in non-equilibrium.* J.Phys.A: Math.Gen. **20** (1987).

[96] **W. Weiss** *Zur hierarchie der erweiterten thermodynamik.* Dissertation. TU Berlin (1990).

[97] **L. Waldmann** *Transporterscheinungen in gasen von mittlerem druck.* Handbuch der Physik, vol. XII. Springer-Verlag, Berlin (1958).

[98] **B.J. Berne, R. Pecora** *Light Scattering.* Wiley, New York (1976).

[99] **G. Simonsohn** *The role of the first order auto-correlation function in conventional grating spectroscopy.* Optics Communications **5** (1972).

[100] **J. Kestin, K. Knierim, E.A. Mason, B. Najafi, S.T. Ro, M. Waldmann** *Equilibrium and transport properties of the noble gases and their mixtures at low density.* J. Phys. Chem. Ref. Data **13** (1984).

[101] **N.A. Clark** *Inelastic light scattering from density fluctuations in dilute gases. The kinetic-hydrodynamic transition in a mon-atomic gas.* Phys. Rev. A **12** (1975).

[102] **E. Meyer, G. Sessler** *Schallausbreitung in gasen bei hohen frequenzen und niedrigen drücken.* Zeitschrift für Physik **149** (1957).

[103] **C. Cercignani, A. Majorana** *Analysis of thermal and shear waves according the B.G.K. kinetic model.* ZAMP **36** (1985).

Springer Tracts in Natural Philosophy

Vol. 18 Martin: Introduction to the Theory of Bases
X–I, 149 pages. 1969.

Vol. 19 Knops/Payne

Vol. 20 Edelen/Wilson: Relativity and the Question of Discretization in Astonomy
With 34 figures. XII, 186 pages. 1970.

Vol. 21 McBride: Obtaining Generating Functions
XIII, 100 pages. 1971.

Vol. 22 Day: The Thermodynamics of Simple Materials with Fading Memory
With 8 figures. X, 134 pages. 1972.

Vol. 23 Stetter: Analysis of Discretization Methods for Ordinary Differential Equations
With 12 figures. XVI, 388 pages. 1973.

Vol. 24 Strieder/Aris: Variational Methods Applied to Problems of Diffusion and Reaction
With 12 figures. IX, 109 pages. 1973.

Vol. 25 Bohl: Monotonie: Lösbarkeit und Numerik bei Operatorgleichungen
With 9 figures. IX, 255 pages. 1974.

Vol. 26 Romanov: Integral Geometry and Inverse Problems for Hyperbolic Equations
With 21 figures. VI, 152 pages. 1974.

Vol. 27 Joseph: Stability of Fluid Motions I
With 57 figures. XIII, 282 pages. 1976.

Vol. 28 Joseph: Stability of Fluid Motions II
With 39 figures. XIV, 274 pages. 1976.

Vol. 29 Bressan: Relativistic Theories of Materials
XIV, 290 pages. 1978.

Vol. 30 Day: Heat Conduction Within Linear Thermoelasticity
VII, 82 pages. 1985.

Vol. 31 Valent: Boundary Value Problems of Finite Elasticity
XIII, 191 pages. 1988.

Vol. 32 Day: A Commentary on Thermodynamics
IX, 96 pages. 1988.

Vol. 33 Cohen/Muncaster: The Theory of Pseudo-Rigid Bodies
X, 180 pages. 1988.

Vol. 34 Angeles: Rational Kinematics
XII, 137 pages. 1989.

Vol. 35 Capriz: Continua with Microstructure
X, 92 pages. 1989.

Vol. 36 Tiersten: A Development of the Equations of Electromagnetism
With 40 figures. XI, 156 pages. 1990.

Vol. 37 Müller/Ruggeri: Extended Thermodynamics
With 43 figures. XI, 230 pages. 1993.

Made in the USA
Las Vegas, NV
28 October 2024

10600436R00136